Springer Series on

Wave Phenomena 11

Edited by L.M. Brekhovskikh

Springer Series on
Wave Phenomena

Editors: L.M. Brekhovskikh   L.B. Felsen   H.A. Haus
Managing Editor: H.K.V. Lotsch

N.D. Veksler

Resonance
Acoustic Spectroscopy

With 153 Figures

Springer-Verlag

Berlin Heidelberg New York
London Paris Tokyo
Hong Kong Barcelona
Budapest

Naum D. Veksler, DSc in Physics
Institute of Cybernectics, Estonian Academy of Sciences,
Akadeemia tee 21, EE0108 Tallinn, Estonia

Translation edited by
Professor Herbert Überall, PhD
Department of Physics, Catholic University of America,
Hannan Hall, Washington, DC 20064, USA

Series Editors:
Professor Leonid M. Brekhovskikh, Academician
P.P. Shirsov Institute of Oceanology, Russian Academy of Sciences, Krasikowa Street 23,
117218 Moscow, Russia

Professor Leopold B. Felsen, Ph.D.
Department of Electrical Engineering, Weber Research Institute, Polytechnic University,
Farmingdale, NY 11735, USA

Professor Hermann A. Haus
Department of Electrical Engineering & Computer Science, MIT,
Cambridge, MA 02139, USA

Managing Editor: Dr.-Ing. Helmut K.V. Lotsch
Springer-Verlag, Tiergartenstrasse 17,
W-6900 Heidelberg, Fed. Rep. of Germany

ISBN-13:978-3-642-84797-4   e-ISBN-13:978-3-642-84795-0
DOI: 10.1007/978-3-642-84795-0

Library of Congress Cataloging-in-Publication Data. Veksler, Naum Davidovich. [Akusticheskaia spektro-skopiia. English] Resonance acoustic spectroscopy / N.D. Veksler; [translation edited by Herbert Überall]. p. cm. — (Springer series on wave phenomena; v. 11). Includes bibliographical references and index. ISBN-13:978-3-642-84797-4 1. Acoustic surface waves. 2. Sound-waves — Scattering. 3. Solids — Acoustic properties. I. Title. II. Series: Springer series on wave phe-nomena; 11. QC176.8.A3V4313 1993. 530.4'12 — dc20 92-30272

This work is subject to copyright. All rights are reserved, whether the whole or part of the material is concerned, specifically the rights of translation, reprinting, reuse of illustrations, recitation, broadcasting, reproduction on microfilm or in any other way and storage in data banks. Duplication of this publication or parts thereof is permitted only under the provisions of the German Copyright Law of September 9, 1965, in its current version, and permission for use must always be obtained from Springer-Verlag. Violations are liable for prosecution under the German Copyright Law.

© Springer-Verlag Berlin Heidelberg 1993
Softcover reprint of the hardcover 1st edition 1993

The use of general descriptive names, registered names, trademarks, etc. in this publication does not imply, even in the absence of a specific statement, that such names are exempt from the relevant protective laws and regulations and therefore free for general use.

Typeset by Macmillan India Ltd, Bangalore-25

54/3140/SPS-543210 – Printed on acid-free paper

Preface

This monograph is devoted to the analysis of waves generated in an elastic body by a plane harmonic acoustic wave. It concentrates on the "peripheral" (or "circumferential") elastic waves: Rayleigh and whispering gallery waves, which are generated on solid and thick-walled elastic bodies, and the Lamb waves generated in thin-walled bodies. Franz and Stoneley waves are considered to a lesser extent only. Franz waves have been treated in detail for several two-dimensional scattering problems, and therefore I decided to touch only lightly upon this subject. Franz waves propagating on helical paths are considered in the case of scattering of an obliquely incident plane acoustic wave by a solid elastic cylinder of infinite extent.

The physical phenomena of excitation, propagation, and re-radiation of elastic waves during the scattering of an incident plane wave are investigated in this book. Known methods are applied for solving the traditional problems of scattering by elastic spheres and cylinders. Special emphasis is laid on the interpretation of the solution. I tried to fill the gaps existing between the papers in which new methods in scattering theory are applied to model (test) problems.

The material is presented systematically, including the formulation of the problem, method of solution, algorithm, computation, and analysis. A large number of new computational results concerning the solution of the scattering problem are given as form functions, modal resonances, dispersion curves, and acoustic spectrograms. Each numerical example is carefully constructed to elucidate one or the other aspect of the scattering process. Although each chapter can be read independently, they are all closely connected and are mutually complementary.

The limited number of geometrical shapes of scatterers considered here is occasioned by the aim to analyze the solution in a rather broad frequency band. Analytical methods (including the asymptotic ones) permit one to obtain the solution of very difficult problems, for example, three-dimensional scattering problems by elastic bodies of smooth shape. Sometimes the asymptotics trace back the physics of the scattering process. However, as a rule, they break down in the resonance frequency range, to which the main attention is paid in this book. In spite of the fact that new numerical methods and fast computers now allow the three-dimensional scattering problem for elastic bodies to be solved in principle, only the solution of scattering by a spheroid (at an arbitrary angle of incidence) is actually obtained in the low-frequency range.

In the case of scattering by elastic spheres and cylinders, the availability of the exact solution in series form allows one to obtain the solution in a very broad frequency range, and the specially elaborated procedures permit it to be analyzed qualitatively. The problem is considered as a steady-state one, assuming the loading to be in the form of a plane harmonic wave. Such an approach is common in acoustics. Once the solution of the steady-state problem has been found, one can obtain the solution for a loading that arbitrarily changes in time by using the convolution theorem.

The solutions of two classical problems are very useful for the analysis of the elastic peripheral waves generated in spheres and cylinders, namely, first, the Rayleigh wave on an elastic half-space, and second, the Lamb waves in a plane "dry" layer (without any ambient liquid).

The elastic scatterers considered in this book vary from solid to thick- or thin-walled bodies. The equations of linear elasticity theory are generally used to describe the motion of the elastic body. In one case only, are additionally the equations of Timoshenko-type and of the membrane theory of thin shells used for a thin-walled elastic body. The models of acoustically rigid and soft bodies commonly employed in acoustics are very useful and sometimes indispensable. Such a variety of models is in no degree connected with the wish to enlarge the proportion of numerical results. On the contrary, the models arise naturally and are used only for qualitative understanding of the scattering process; a new model is considered only when it helps the main thought further.

The treatment in this book is descriptive rather than based on rigorous proof. We assume that some of the phenomena considered here will later be rigorously justified on the level of a theorem. We use the physical level of rigor typical of the original papers.

From the point of view of content, the issues presented in the book are closely connected with the well-studied topics of acoustics and mechanics of a solid deformable body: the free vibrations of "dry" elastic spheres and cylinders – solid, thick- and thin-walled; the vibrations of plates and shells; and the dispersion relations. The contact with the liquid surrounding the scatterer and the presence of an incident plane wave are the main differences of the scattering problem considered here from that corresponding to eigenvibrations of the body.

Certainly, realistic problems are more complicated than those which are treated in the book. This is connected both with the geometry of the scatterer (variable thickness, totality of different geometrical shapes bounding a common volume, presence of bulkheads, framing, and reinforcements) and with non-uniformity, for example, the lamination of the liquid surrounding the scatterer, presence of the bottom and the surface, and other circumstances. Therefore the problems considered can be understood as model ones. Without comprehending the nature of the scattering process in the model problem it is difficult to formulate, and to solve, real problems.

The material presented has appeared as original papers during the last ten years and is treated here in a English-language monograph for the first time.

There are some difficulties with terminology in this field because the material is essentially new and not all terms are settled yet. However, I have endeavored to use a consistent terminology throughout the whole book.

Technical applications and problems of identification are not treated here. I consider this to be a book about the physical phenomena which can be discovered in the course of a mathematical experiment. The restricted size of the book forced me to select the material carefully and to omit all extreme ramifications. Recently published monographs and surveys have made the problem of selection substantially easier. Preference was given to the subjects of broad interest that are reasonably well studied and comprehended, and which are important in terms of new methods or new concepts. Much of the material was first published in Russian in my book *Acoustic Spectroscopy*.

Tallinn, September 1991   *N.D. Veksler*

Acknowledgements

The author is thankful to Dr. Victor Korsunskii for help with computer codes, to Mrs. Valentina Furik and Mrs. Monika Perkmann for speedily and accurately typing the text, and to Mr. Ants Kivilo, who prepared all the drawings with great care.

Prof. Herbert Überall, translation editor of this book, has improved the text very carefully and has given it a readable form. The author is very grateful to him for this.

Contents

1 Scattering of an Obliquely Incident Plane Acoustic Wave by a Circular–Cylindrical Shell of Infinite Extent

The formal solution of the problem of an obliquely incident plane acoustic wave scattered by a circular–cylindrical shell is presented here. In the limiting cases the solutions of the following scattering problems can be obtained from it: (i) a wave obliquely incident on a solid elastic cylinder, (ii) a wave normally incident on a solid elastic cylinder, and (iii) a wave normally incident on an elastic shell. The analysis of the form function at oblique incidence is presented in Chap. 11 for scattering from a solid elastic cylinder and in Chap. 12 for that from a shell. The results of computation and their analysis for scattering at normal incidence from a solid elastic cylinder are given in Chap. 7, and from a shell in Chaps. 2, 8 and 9.

1.1 Formulation of the Problem and its Solution in Series Form

Let us consider the steady-state scattering of a plane acoustic wave by a circular–cylindrical shell. In Fig. 1.1, the orientation of the cylindrical coordinate system and the direction of propagation of the incident wave is shown. The z' axis of the coordinate system (r, θ, z') is chosen as the longitudinal axis of the shell. The propagation vector $\mathbf{k}$ of the incident wave is in the x'–z' plane and forms an angle α' with the x' axis. The following notations will be used below: a is the outer radius of the shell, b is the inner radius of the shell; ρ_1, c_1 and c_t are the density, longitudinal and transverse wave velocities of the shell material; ρ and c are the density and the sound velocity of the liquid surrounding the shell. The shell is of infinite extent in the z' direction.

The acoustic pressure of the incident wave is defined by

$$p_i = p_* \exp[i(\mathbf{k}r - \omega t)] , \tag{1.1}$$

where ω is the angular frequency, t is time and p_* is a constant with the dimensions of pressure.

Using the angle α_0 (Fig. 1.1) where

$$\alpha_0 = \pi - \alpha' \tag{1.2}$$

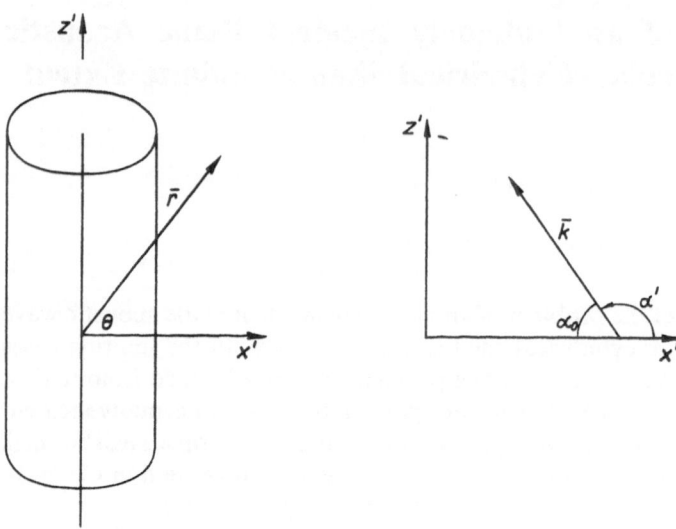

Fig. 1.1. Choice of the coordinate axes for scattering by a cylindrical shell

and introducing the notations

$$k_{x'} = k \cos \alpha' = -k \cos \alpha_0 = \delta \,,$$

$$k_{z'} = k \sin \alpha' = k \sin \alpha_0 = \gamma \,, \tag{1.3}$$

we obtain

$$p_i = p_0 \exp(i\delta x') \,, \qquad p_0 = p_* \exp[i(\gamma z' - \omega t)] \,. \tag{1.4}$$

Using the relation

$$x' = r \cos \theta \tag{1.5}$$

and the expansion

$$\exp(-i\delta r \cos \theta) = \sum_{n=0}^{\infty} \varepsilon_n (-i)^n J_n(\delta r) \cos n\theta \,, \quad \varepsilon_0 = 1, \; \varepsilon_n = 2 \; (n \geqslant 1) \,, \tag{1.6}$$

we find from (1.1)

$$p_i = p_0 \sum_{n=0}^{\infty} \varepsilon_n (-i)^n J_n(\delta r) \cos n\theta \,. \tag{1.7}$$

The acoustic pressure of the scattered field p_s satisfies the wave equation

$$\left(\nabla^2 - \frac{1}{c^2} \frac{\partial^2}{\partial t^2} \right) p_s = 0 \,, \tag{1.8}$$

where ∇^2 is the Laplace operator.

In cylindrical coordinates

$$\nabla^2 p_s = \left(\frac{\partial}{\partial r^2} + \frac{1}{r} \frac{\partial}{\partial r} + \frac{1}{r^2} \frac{\partial^2}{\partial \theta^2} + \frac{\partial^2}{\partial z'^2} \right) p_s \tag{1.9}$$

and (1.8) assumes the form

$$\left(\frac{\partial}{\partial r^2} + \frac{1}{r} \frac{\partial}{\partial r} + \frac{1}{r^2} \frac{\partial^2}{\partial \theta^2} + \frac{\partial^2}{\partial z'^2} - \frac{1}{c^2} \frac{\partial^2}{\partial t^2} \right) p_s = 0 . \tag{1.10}$$

The vector of displacement **u** in linear theory of elasticity satisfies the equation

$$(\lambda + 2\mu) \operatorname{grad} \operatorname{div} \mathbf{u} - \mu \operatorname{curl} \operatorname{curl} \mathbf{u} - \rho_1 \frac{\partial^2 \mathbf{u}}{\partial t^2} = 0 , \tag{1.11}$$

where λ and μ are the Lamé constants.

Introducing scalar (ϕ) and vector (ψ) potentials of displacement, we represent **u** in the form

$$\mathbf{u} = \operatorname{grad} \phi + \operatorname{curl} \psi . \tag{1.12}$$

Using the relation (1.12) in (1.11), we obtain

$$\operatorname{grad}\left((\lambda + 2\mu)\nabla^2 \phi - \rho_1 \frac{\partial^2 \phi}{\partial t^2} \right) + \operatorname{curl}\left(\mu\nabla^2 \psi - \rho_1 \frac{\partial^2 \psi}{\partial t^2} \right) = 0 . \tag{1.13}$$

This equation will be satisfied if

$$(\lambda + 2\mu)\nabla^2 \phi - \rho_1 \frac{\partial^2 \phi}{\partial t^2} = 0 , \tag{1.14}$$

$$\mu\nabla^2 \psi - \rho_1 \frac{\partial^2 \psi}{\partial t^2} = 0 . \tag{1.15}$$

Introducing standard notations

$$c_1 = [(\lambda + 2\mu)/\rho_1]^{1/2}, \qquad c_t = (\mu/\rho_1)^{1/2} , \tag{1.16}$$

we can write (1.14) and (1.15) in the form

$$\left(\nabla^2 - \frac{1}{c_1^2} \frac{\partial^2}{\partial t^2} \right) \phi = 0 , \tag{1.17}$$

$$\left(\nabla^2 - \frac{1}{c_t^2} \frac{\partial^2}{\partial t^2} \right) \bar{\psi} = 0 . \tag{1.18}$$

Using

$$\psi = \psi_r \mathbf{i}_r + \psi_\theta \mathbf{i}_\theta + \psi_z \mathbf{i}_{z'} \tag{1.19}$$

and the expression of the Laplace operator for a vector quantity, we obtain from

(1.18) three scalar equations

$$\nabla^2\psi_r - \frac{1}{r^2}\psi_r - \frac{2}{r}\frac{\partial\psi_\theta}{\partial\theta} - \frac{1}{c_t^2}\frac{\partial^2\psi_r}{\partial t^2} = 0 , \tag{1.20}$$

$$\nabla^2\psi_\theta - \frac{1}{r^2}\psi_\theta + \frac{2}{r}\frac{\partial\psi_r}{\partial\theta} - \frac{1}{c_t^2}\frac{\partial^2\psi_\theta}{\partial t^2} = 0 , \tag{1.21}$$

$$\nabla^2\psi_{z'} - \frac{1}{c_t^2}\frac{\partial^2\psi_{z'}}{\partial t^2} = 0 . \tag{1.22}$$

Equation (1.22) is separated, but (1.20) and (1.21) are coupled.
The solution of (1.10) can be presented in the form

$$p_s = p_0 \sum_{n=0}^{\infty} A_n H_n^{(1)}(\delta r)\cos n\theta , \tag{1.23}$$

where A_n is a factor as yet undefined.
We shall find solutions of (1.17, 1.20–22) in the form

$$\phi = g_0 \sum_{n=0}^{\infty} \phi(r)\cos n\theta , \tag{1.24}$$

$$\psi_r = g_0 \sum_{n=0}^{\infty} h_r(r)\sin n\theta , \tag{1.25}$$

$$\psi_\theta = g_0 \sum_{n=0}^{\infty} h_\theta(r)\cos n\theta , \tag{1.26}$$

$$\psi_{z'} = g_0 \sum_{n=0}^{\infty} h_z(r)\sin n\theta , \tag{1.27}$$

where

$$g_0 = a^2 \exp[i(\gamma z' - \omega t)] .$$

Introducing the expansion (1.24) into (1.17), we obtain

$$\left[\frac{d^2}{dr^2} + \frac{1}{r}\frac{d}{dr} + \left(\alpha^2 - \frac{n^2}{r^2}\right)\right]\phi(r) = 0 , \tag{1.28}$$

where

$$\alpha^2 = \left(\frac{\omega}{c_1}\right)^2 - \gamma^2 . \tag{1.29}$$

The solution of (1.28) can be written in the form

$$\phi(r) = B_n J_n(\alpha r) + C_n N_n(\alpha r) , \tag{1.30}$$

where B_n and C_n are free parameters.

Introducing the expansions (1.25) and (1.26) into (1.20) and (1.21), we obtain

$$\frac{d^2h_r}{dr^2} + \frac{1}{r}\frac{dh_r}{dr} + \left(\beta^2 - \frac{n^2+1}{r^2}\right)h_r + \frac{2n}{r^2}h_\theta \neq 0 ,\tag{1.31}$$

$$\frac{d^2h_\theta}{dr^2} + \frac{1}{r}\frac{dh_\theta}{dr} + \left(\beta^2 - \frac{n^2+1}{r^2}\right)h_\theta + \frac{2n}{r^2}h_r = 0 ,\tag{1.32}$$

where

$$\beta^2 = (\omega/c_t)^2 - \gamma^2 .\tag{1.33}$$

According to [1.1] we shall subtract (1.32) from (1.31), which leads to

$$\left[\frac{d^2}{dr^2} + \frac{1}{r}\frac{d}{dr} + \left(\beta^2 - \frac{(n+1)^2}{r^2}\right)\right](h_r - h_\theta) = 0 .\tag{1.34}$$

The solution of (1.34) can be written in the form

$$h_r - h_\theta = 2D_n J_{n+1}(\beta r) + 2E_n N_{n+1}(\beta r) ,\tag{1.35}$$

where D_n and F_n are free parameters.

Now we sum (1.31) and (1.32)

$$\left[\frac{d^2}{dr^2} + \frac{1}{r}\frac{d}{dr} + \left(\beta^2 - \frac{(n-1)^2}{r^2}\right)\right](h_r + h_\theta) = 0 .\tag{1.36}$$

The solution of this equation can be written in the form

$$h_r + h_\theta = 2L_n J_{n-1}(\beta r) + 2M_n N_{n-1}(\beta r) ,\tag{1.37}$$

where L_n and M_n are free parameters.

Using (1.35) and (1.37), we obtain

$$h_r = D_n J_{n+1}(\beta r) + E_n N_{n+1}(\beta r) + L_n J_{n-1}(\beta r) + M_n N_{n-1}(\beta r) ,\tag{1.38}$$

$$h_\theta = - D_n J_{n+1}(\beta r) - E_n N_{n+1}(\beta r) + L_n J_{n-1}(\beta r) + M_n N_{n-1}(\beta r) .\tag{1.39}$$

Taking

$$L_n = M_n = 0 ,\tag{1.40}$$

we obtain

$$h_r = D_n J_{n+1}(\beta r) + E_n N_{n+1}(\beta r) ,\tag{1.41}$$

$$h_\theta = - h_r .$$

Introducing the expansion (1.27) into (1.22), we obtain

$$\left[\frac{d^2}{dr^2} + \frac{1}{r}\frac{d}{dr} + \left(\beta^2 - \frac{n^2}{r^2}\right)\right]h_z = 0 ,\tag{1.42}$$

which has the solution

$$h_z = F_n J_n(\beta r) + G_n N_n(\beta r) , \tag{1.43}$$

where F_n and G_n are free parameters.

For each summation index n seven constants are to be found from the contact conditions:
on the outer surface of the shell

$$\frac{\partial v_r}{\partial t} = \frac{\partial^2 u_r}{\partial t^2}, \qquad \sigma_{rr} = -(p_i + p_s), \qquad \sigma_{r\theta} = 0, \qquad \sigma_{rz'} = 0 \quad \text{at } r = a , \tag{1.44}$$

and on the inner surface of the shell

$$\sigma_{rr} = 0, \qquad \sigma_{r\theta} = 0, \qquad \sigma_{rz'} = 0 \quad \text{at } r = b . \tag{1.45}$$

In (1.44) $\partial v_r/\partial t$ is the radial acceleration of the liquid and $\partial^2 u_r/\partial t^2$ is the radial acceleration of the elastic medium. The minus sign in the second boundary condition results from the sign rule accepted in elasticity theory, according to which an expanding stress is considered positive.

Using (1.12), we shall write the displacement vector **u** in the form

$$\mathbf{u} = u_r \mathbf{i}_r + u_\theta \mathbf{i}_\theta + u_{z'} \mathbf{i}_{z'} ,$$

$$u_r = \frac{\partial \phi}{\partial r} + \frac{1}{r}\frac{\partial \psi_{z'}}{\partial \theta} - \frac{\partial \psi_\theta}{\partial z'}, \qquad u_\theta = \frac{1}{r}\frac{\partial \phi}{\partial \theta} + \frac{\partial \psi_r}{\partial z'} - \frac{\partial \psi_{z'}}{\partial r} , \tag{1.46}$$

$$u_{z'} = \frac{\partial \phi}{\partial z'} + \frac{\partial \psi_\theta}{\partial r} + \frac{1}{r}\psi_\theta - \frac{1}{r}\frac{\partial \psi_r}{\partial \theta} .$$

According to the Euler equation

$$\frac{\partial \mathbf{v}}{\partial t} = -\frac{1}{\rho} \operatorname{grad} p , \tag{1.47}$$

thus, we shall write the radial acceleration of the liquid in the form

$$\frac{\partial v_r}{\partial t} = -\frac{1}{\rho}\frac{\partial p}{\partial r} . \tag{1.48}$$

The sum of the incident and scattered pressure should be used in it as the acoustic pressure; thus,

$$\frac{\partial v_r}{\partial t} = -\frac{1}{\rho}\frac{\partial}{\partial r}(p_i + p_s) . \tag{1.49}$$

We shall now express the components of the stress tensor by the projections of the scalar and vector potentials of the displacement

$$\sigma_{rr} = \lambda \nabla^2 \phi + 2\mu \left(\frac{\partial^2 \phi}{\partial r^2} + \frac{1}{r}\frac{\partial^2 \psi_{z'}}{\partial r \partial \theta} - \frac{1}{r^2}\frac{\partial^2 \psi_{z'}}{\partial \theta^2} - \frac{\partial^2 \psi_\theta}{\partial r \partial z'} \right),$$

$$\sigma_{r\theta} = \mu\left(\frac{2}{r}\frac{\partial^2\phi}{\partial r\partial\theta} - \frac{2}{r^2}\frac{\partial\phi}{\partial\theta} + \frac{\partial^2\psi_r}{\partial r\partial z} - \frac{1}{r}\frac{\partial\psi_r}{\partial z} - \frac{\partial^2\psi_{z'}}{\partial r^2} + \frac{1}{r}\frac{\partial\psi_{z'}}{\partial r} - \frac{1}{r}\frac{\partial^2\psi_\theta}{\partial\theta\partial z}\right),$$

$$\sigma_{rz'} = \mu\left(2\frac{\partial^2\phi}{\partial r\partial z'} - \frac{1}{r}\frac{\partial^2\psi_r}{\partial r\partial\theta} + \frac{1}{r^2}\frac{\partial\psi_r}{\partial\theta} + \frac{1}{r}\frac{\partial\psi_\theta}{\partial r} - \frac{1}{r^2}\psi_\theta\right.$$

$$\left. + \frac{\partial^2\psi_\theta}{\partial r^2} - \frac{\partial^2\psi_\theta}{\partial z'^2} + \frac{1}{r}\frac{\partial^2\psi_{z'}}{\partial\theta\partial z'}\right). \tag{1.50}$$

Inserting (1.23–27) [while taking account of (1.30, 38, 39 and 43)] into (1.46, 49 and 50), and introducing the latter into (1.44) and (1.45), we obtain, for each n value, a system of algebraic equations written in the order corresponding to that used in the boundary conditions

$$a_{11}A_n + a_{12}B_n + a_{13}C_n + a_{14}D_n + a_{15}E_n + a_{16}F_n + a_{17}G_n = a_{10},$$

$$a_{21}A_n + a_{22}B_n + a_{23}C_n + a_{24}D_n + a_{25}E_n + a_{26}F_n + a_{27}G_n = a_{20},$$

$$a_{32}B_n + a_{33}C_n + a_{34}D_n + a_{35}E_n + a_{36}F_n + a_{37}G_n = 0,$$

$$a_{42}B_n + a_{43}C_n + a_{44}D_n + a_{45}E_n + a_{46}F_n + a_{47}G_n = 0, \tag{1.51}$$

$$a_{52}B_n + a_{53}C_n + a_{54}D_n + a_{55}E_n + a_{56}F_n + a_{57}G_n = 0,$$

$$a_{62}B_n + a_{63}C_n + a_{64}D_n + a_{65}E_n + a_{66}F_n + a_{67}G_n = 0,$$

$$a_{72}B_n + a_{73}C_n + a_{74}D_n + a_{75}E_n + a_{76}F_n + a_{77}G_n = 0.$$

The nonvanishing coefficients in this system of equations $a_{ij}(i, j = 1, 2, \ldots, 7)$ can be written in the form

$$a_{11} = -\frac{p_*}{\rho(\omega a)^2}(\delta a)H_n^{(1)'}(\delta a),$$

$$a_{12} = (\alpha a)J'_n(\alpha a), \qquad a_{13} = (\alpha a)N'_n(\alpha a),$$

$$a_{14} = (\gamma a)J_{n+1}(\beta a), \qquad a_{15} = (\gamma a)N_{n+1}(\beta a),$$

$$a_{16} = nJ_n(\beta a), \qquad a_{17} = nN_n(\beta a),$$

$$a_{10} = \frac{p_*}{\rho(\omega a)^2}\varepsilon_n(-i)^n(\delta a)J'_n(\delta a), \qquad a_{21} = -\frac{p_*}{2\mu}H_n^{(1)}(\delta a),$$

$$a_{22} = \frac{\lambda}{2\mu}\left(\frac{\omega a}{c_1}\right)^2 J_n(\alpha a) - (\alpha a)^2 J''_n(\alpha a),$$

$$a_{23} = \frac{\lambda}{2\mu}\left(\frac{\omega a}{c_1}\right)^2 N_n(\alpha a) - (\alpha a)^2 N''_n(\alpha a),$$

$$a_{24} = -(\beta a)(\gamma a)J'_{n+1}(\beta a), \qquad a_{25} = -(\beta a)(\gamma a)N'_{n+1}(\beta a),$$

$$a_{26} = n[J_n(\beta a) - (\beta a)J'_n(\beta a)], \qquad a_{27} = n[N_n(\beta a) - (\beta a)N'_n(\beta a)],$$

$$a_{20} = \frac{p_*}{2\mu}\varepsilon_n(-i)^n J_n(\delta a),$$

$$a_{32} = 2n[J_n(\alpha a) - (\alpha a)J'_n(\alpha a)], \qquad a_{33} = 2n[N_n(\alpha a) - (\alpha a)N'_n(\alpha a)],$$

$$a_{34} = -(\gamma a)[(n+1)J_{n+1}(\beta a) - (\beta a)J'_{n+1}(\beta a)],$$

$$a_{35} = -(\gamma a)[(n+1)N_{n+1}(\beta a) - (\beta a)N'_{n+1}(\beta a)],$$

$$a_{36} = 2[(\tfrac{1}{2}\beta^2 a^2 - n^2)J_n(\beta a) + (\beta a)J'_n(\beta a)],$$

$$a_{37} = 2[(\tfrac{1}{2}\beta^2 a^2 - n^2)N_n(\beta a) + (\beta a)N'_n(\beta a)],$$

$$a_{42} = 2(\alpha a)(\gamma a)J'_n(\alpha a), \qquad a_{43} = 2(\alpha a)(\gamma a)N'_n(\alpha a),$$

$$a_{44} = [-(\beta a)^2 + (\gamma a)^2 + n^2 + n]J_{n+1}(\beta a) + n(\beta a)J'_{n+1}(\beta a),$$

$$a_{45} = [-(\beta a)^2 + (\gamma a)^2 + n^2 + n]N_{n+1}(\beta a) + n(\beta a)N'_{n+1}(\beta a),$$

$$a_{46} = n(\gamma a)J_n(\beta a), \qquad a_{47} = n(\gamma a)N_n(\beta a),$$

$$a_{52} = \frac{\lambda}{2\mu}\left(\frac{\omega b}{c_1}\right)^2 J_n(\alpha b) - (\alpha b)^2 J''_n(\alpha b),$$

$$a_{53} = \frac{\lambda}{2\mu}\left(\frac{\omega b}{c_1}\right)^2 N_n(\alpha b) - (\alpha b)^2 N''_n(\alpha b),$$

$$a_{54} = -(\beta b)(\gamma b)J'_{n+1}(\beta b), \qquad a_{55} = -(\beta b)(\gamma b)N'_{n+1}(\beta b),$$

$$a_{56} = n[J_n(\beta b) - (\beta b)J'_n(\beta b)], \qquad a_{57} = n[N_n(\beta b) - (\beta b)N'_n(\beta b)],$$

$$a_{62} = 2n[J_n(\alpha b) - (\alpha b)J'_n(\alpha b)], \qquad a_{63} = 2n[N_n(\alpha b) - (\alpha b)N'_n(\alpha b)],$$

$$a_{64} = -(\gamma b)[(n+1)J_{n+1}(\beta b) - (\beta b)J'_{n+1}(\beta b)],$$

$$a_{65} = -(\gamma b)[(n+1)N_{n+1}(\beta b) - (\beta b)N'_{n+1}(\beta b)],$$

$$a_{66} = 2[(\tfrac{1}{2}\beta^2 b^2 - n^2)J_n(\beta b) + (\beta b)J'_n(\beta b)],$$

$$a_{67} = 2[(\tfrac{1}{2}\beta^2 b^2 - n^2)N_n(\beta b) + (\beta b)N'_n(\beta b)],$$

$$a_{72} = 2(\alpha b)(\gamma b)J'_n(\alpha b), \qquad a_{73} = 2(\alpha b)(\gamma b)N'_n(\alpha b),$$

$$a_{74} = [-(\beta b)^2 + (\gamma b)^2 + n^2 + n]J_{n+1}(\beta b) + n(\beta b)J'_{n+1}(\beta b),$$

$$a_{75} = [-(\beta b)^2 + (\gamma b)^2 + n^2 + n]N_{n+1}(\beta b) + n(\beta b)N'_{n+1}(\beta b),$$

$$a_{76} = n(\gamma b)J_n(\beta b), \qquad a_{77} = n(\gamma b)N_n(\beta b). \qquad\qquad (1.52)$$

For simplicity, the index n is omitted here in the notation of the coefficients. The prime means the derivative with respect to the argument.

By solving the system of the algebraic equations, the free parameters can be obtained. We shall give the resulting expression for only one of them, namely for A_n , which is contained in the series (1.23) for the acoustic pressure

$$A_n = D_1/D_2, \qquad\qquad (1.53)$$

where

$$
D_1 = \begin{vmatrix}
a_{10} & a_{12} & a_{13} & a_{14} & a_{15} & a_{16} & a_{17} \\
a_{20} & a_{22} & a_{23} & a_{24} & a_{25} & a_{26} & a_{27} \\
0 & a_{32} & a_{33} & a_{34} & a_{35} & a_{36} & a_{37} \\
0 & a_{42} & a_{43} & a_{44} & a_{45} & a_{46} & a_{47} \\
0 & a_{52} & a_{53} & a_{54} & a_{55} & a_{56} & a_{57} \\
0 & a_{62} & a_{63} & a_{64} & a_{65} & a_{66} & a_{67} \\
0 & a_{72} & a_{73} & a_{74} & a_{75} & a_{76} & a_{77}
\end{vmatrix} ,
$$

$$
D_2 = \begin{vmatrix}
a_{11} & a_{12} & a_{13} & a_{14} & a_{15} & a_{16} & a_{17} \\
a_{21} & a_{22} & a_{23} & a_{24} & a_{25} & a_{26} & a_{27} \\
0 & a_{32} & a_{33} & a_{34} & a_{35} & a_{36} & a_{37} \\
0 & a_{42} & a_{43} & a_{44} & a_{45} & a_{46} & a_{47} \\
0 & a_{52} & a_{53} & a_{54} & a_{55} & a_{56} & a_{57} \\
0 & a_{62} & a_{63} & a_{64} & a_{65} & a_{66} & a_{67} \\
0 & a_{72} & a_{73} & a_{74} & a_{75} & a_{76} & a_{77}
\end{vmatrix} .
$$

(1.54)

The determinants of seventh rank, D_1 and D_2, differ only in the first column. It is convenient to present the formula for A_n , (1.53), in the form

$$
A_n = B_1 \frac{1 - B_2 B}{1 - B_3 B} \tag{1.55}
$$

in which

$$
B_1 = \frac{a_{10}}{a_{11}}, \qquad B_2 = \frac{a_{20}}{a_{10}}, \qquad B_3 = \frac{a_{21}}{a_{11}}, \qquad B = \frac{A_{20}}{A_{10}},
$$

$$
A_{10} = \begin{vmatrix}
a_{22} & a_{23} & a_{24} & a_{25} & a_{26} & a_{27} \\
a_{32} & a_{33} & a_{34} & a_{35} & a_{36} & a_{37} \\
a_{42} & a_{43} & a_{44} & a_{45} & a_{46} & a_{47} \\
a_{52} & a_{53} & a_{54} & a_{55} & a_{56} & a_{57} \\
a_{62} & a_{63} & a_{64} & a_{65} & a_{66} & a_{67} \\
a_{72} & a_{73} & a_{74} & a_{75} & a_{76} & a_{77}
\end{vmatrix} ,
$$

(1.56)

$$
A_{20} = \begin{vmatrix}
a_{12} & a_{13} & a_{14} & a_{15} & a_{16} & a_{17} \\
a_{32} & a_{33} & a_{34} & a_{35} & a_{36} & a_{37} \\
a_{42} & a_{43} & a_{44} & a_{45} & a_{46} & a_{47} \\
a_{52} & a_{53} & a_{54} & a_{55} & a_{56} & a_{57} \\
a_{62} & a_{63} & a_{64} & a_{65} & a_{66} & a_{67} \\
a_{72} & a_{73} & a_{74} & a_{75} & a_{76} & a_{77}
\end{vmatrix} .
$$

The determinants of sixth rank, A_{10} and A_{20} , differ only in the first row.

It is appropriate to use normalized variables in the computations. We shall introduce the following notations:

$$u_1 = \alpha a, \qquad v_1 = \beta a, \qquad w_1 = \gamma a,$$

$$u_2 = \alpha b, \qquad v_2 = \beta b, \qquad w_2 = \gamma b,$$

$$\lambda_n'(u_1) = \frac{J_n'(u_1)}{J_n(u_1)}, \qquad \lambda_n''(u_1) = \frac{J_n''(u_1)}{J_n(u_1)},$$

$$l_n'(u_1) = \frac{N_n'(u_1)}{N_n(u_1)}, \qquad l_n''(u_1) = \frac{N_n''(u_1)}{N_n(u_1)},$$

$$\lambda_n'(v_1) = \frac{J_n'(v_1)}{J_n(v_1)}, \qquad l_n'(v_1) = \frac{N_n'(v_1)}{N_n(v_1)}, \tag{1.57}$$

$$\lambda_{n+1}'(v_1) = \frac{J_{n+1}'(v_1)}{J_{n+1}(v_1)}, \qquad l_{n+1}'(v_1) = \frac{N_{n+1}'(v_1)}{N_{n+1}(v_1)},$$

$$p_n(\alpha) = \frac{J_n(u_2)}{J_n(u_1)}, \qquad q_n(\alpha) = \frac{N_n(u_2)}{N_n(u_1)},$$

$$p_n(\beta) = \frac{J_n(v_2)}{J_n(v_1)}, \qquad q_n(\beta) = \frac{N_n(v_2)}{N_n(v_1)},$$

$$p_{n+1}(\beta) = \frac{J_{n+1}(v_2)}{J_{n+1}(v_1)}, \qquad q_{n+1}(\beta) = \frac{N_{n+1}(v_2)}{N_{n+1}(v_1)}.$$

The coefficients of the system of (1.51) now take the form

$$a_{12} = u_1 \lambda_n'(u_1), \qquad a_{13} = u_1 l_n'(u_1),$$

$$a_{14} = w_1, \qquad a_{15} = w_1, \qquad a_{16} = n, \qquad a_{17} = n,$$

$$a_{22} = \frac{\lambda}{2\mu} w_1^2 + u_1^2 \left[1 + \frac{\lambda}{2\mu} + \frac{\lambda_n'(u_1)}{u_1} - \frac{n^2}{u_1^2} \right],$$

$$a_{23} = \frac{\lambda}{2\mu} w_1^2 + u_1^2 \left[1 + \frac{\lambda}{2\mu} + \frac{l_n'(u_1)}{u_1} - \frac{n^2}{u_1^2} \right],$$

$$a_{24} = -v_1 w_1^2 \lambda_{n+1}'(v_1), \qquad a_{25} = -v_1 w_1^2 l_{n+1}'(v_1),$$

$$a_{26} = n[1 - v_1 \lambda_n'(v_1)], \qquad a_{27} = n[1 - v_1 l_n'(v_1)],$$

$$a_{32} = n[1 - u_1 \lambda_n'(u_1)], \qquad a_{33} = n[1 - u_1 l_n'(u_1)],$$

$$a_{34} = -\tfrac{1}{2} w_1 [(n+1) - v_1 \lambda_{n+1}'(v_1)],$$

$$a_{35} = -\tfrac{1}{2} w_1 [(n+1) - v_1 l_{n+1}'(v_1)],$$

$$a_{36} = (\tfrac{1}{2} v_1^2 - n^2) + v_1 \lambda_n'(v_1),$$

$$a_{37} = (\tfrac{1}{2} v_1^2 - n^2) + v_1 l_n'(v_1),$$

$$a_{42} = 2u_1 w_1 \lambda_n'(u_1), \qquad a_{43} = 2u_1 w_1 l_n'(u_1),$$

$$a_{44} = (-v_1^2 + w_1^2 + n^2 + n) + nv_1 \lambda'_{n+1}(v_1),$$

$$a_{45} = (-v_1^2 + w_1^2 + n^2 + n) + nv_1 l'_{n+1}(v_1),$$

$$a_{46} = nw_1, \qquad a_{47} = nw_1,$$

$$a_{52} = p_n(\alpha)\left\{\frac{\lambda}{2\mu}w_2^2 + u_2^2\left[1 + \frac{\lambda}{2\mu} + \frac{\lambda'_n(u_2)}{u_2} - \frac{n^2}{u_2^2}\right]\right\},$$

$$a_{53} = q_n(\alpha)\left\{\frac{\lambda}{2\mu}w_2^2 + u_2^2\left[1 + \frac{\lambda}{2\mu} + \frac{l'_n(u_2)}{u_2} - \frac{n^2}{u_2^2}\right]\right\},$$

$$a_{54} = -p_{n+1}(\beta)v_2 w_2 \lambda'_{n+1}(v_2),$$

$$a_{55} = -q_{n+1}(\beta)v_2 w_2 l'_{n+1}(v_2),$$

$$a_{56} = np_n(\beta)[1 - v_2 \lambda'_n(v_2)], \qquad a_{57} = nq_n(\beta)[1 - v_2 l'_n(v_2)],$$

$$a_{62} = np_n(\alpha)[1 - u_2 \lambda'_n(u_2)], \qquad a_{63} = nq_n(\alpha)[1 - u_2 l'_n(u_2)],$$

$$a_{64} = -\tfrac{1}{2}p_{n+1}(\beta)w_2[(n+1) - v_2 \lambda'_{n+1}(v_2)],$$

$$a_{65} = -\tfrac{1}{2}q_{n+1}(\beta)w_2[(n+1) - v_2 l'_{n+1}(v_2)],$$

$$a_{66} = p_n(\beta)[(\tfrac{1}{2}v_2^2 - n^2) + v_2 \lambda'_n(v_2)],$$

$$a_{67} = q_n(\beta)[(\tfrac{1}{2}v_2^2 - n^2) + v_2 l'_n(v_2)],$$

$$a_{72} = 2p_n(\alpha)u_2 w_2 \lambda'_n(u_2), \qquad a_{73} = 2q_n(\alpha)u_2 w_2 l'_n(u_2),$$

$$a_{74} = p_{n+1}(\beta)[(-v_2^2 + w_2^2 + n^2 + n) + nv_2 \lambda'_{n+1}(v_2)],$$

$$a_{75} = q_{n+1}(\beta)[(-v_2^2 + w_2^2 + n^2 + n) + nv_2 l'_{n+1}(v_2)],$$

$$a_{76} = np_n(\beta)w_2, \qquad a_{77} = nq_n(\beta)w_2. \tag{1.58}$$

Using the notation

$$\Delta_n(\delta a) = \frac{J'_n(\delta a)}{N'_n(\delta a)} \equiv \frac{J_n(\delta a)}{N_n(\delta a)} \frac{\lambda'_n(\delta a)}{l'_n(\delta a)}, \tag{1.59}$$

we shall present B_1, B_2 and B_3, see (1.56), in the form

$$B_1 = -\varepsilon_n(-i)^n \frac{\Delta_n(\delta a)}{i + \Delta_n(\delta a)}, \qquad B_2 = \frac{\rho(\omega a)^2}{2\mu(\delta a)} \frac{1}{\lambda'_n(\delta a)},$$

$$B_3 = \frac{\rho(\omega a)^2}{2\mu(\delta a)} \frac{\left[i + \dfrac{J_n(\delta a)}{N_n(\delta a)}\right]}{l'_n(\delta a)[i + \Delta_n(\delta a)]}. \tag{1.60}$$

In the far field (at $\delta r \gg 1$), $H_n^{(1)}(\delta r)$ in (1.23) can be replaced by its asymptotic expansion

$$H_n^{(1)}(\delta r) \sim \sqrt{\frac{2}{\pi \delta r}} \exp(i\delta r) \exp\left[-i\frac{\pi}{2}\left(n + \frac{1}{2}\right)\right][P(n, \delta r) + iQ(n, \delta r)],$$

$$P(n, \delta r) = 1 - \frac{(\mu - 1)(\mu - 9)}{2!(8\delta r)^2} + \frac{(\mu - 1)(\mu - 9)(\mu - 25)(\mu - 49)}{4!(8\delta r)^4} - \cdots ,$$

$$\mu = 4n^2 , \tag{1.61}$$

$$Q(n, \delta r) = \frac{(\mu - 1)}{8\delta r} - \frac{(\mu - 1)(\mu - 9)(\mu - 25)}{3!(8\delta r)^3} + \cdots .$$

Introducing the notation

$$H(\delta r) = \frac{1}{\sqrt{\pi}} \exp\left[-i\frac{\pi}{2}\left(n + \frac{1}{2}\right)\right][P(n, \delta r) + iQ(n, \delta r)] , \tag{1.62}$$

we shall write

$$H_n^{(1)}(\delta r) \sim \sqrt{\frac{2}{\delta r}} \exp(i\delta r) H(\delta r) . \tag{1.63}$$

Inserting this expression in (1.23), we obtain

$$p_s = p_* \exp[i(\delta r + \gamma z' - \omega t)] \sqrt{\frac{2}{\delta r}} \sum_{n=0}^{\infty} A_n H(\delta r) \cos n\theta . \tag{1.64}$$

In the far field it is appropriate to compute the quantity

$$p(r, \theta; x) = \frac{p_s}{p_* \sqrt{\frac{2}{\delta r}} \exp[i(\delta r + \gamma z' - \omega t)]} \equiv \sum_{n=0}^{\infty} A_n H(\delta r) \cos n\theta . \tag{1.65}$$

Usually, the computations are carried out at a fixed observation point $(r = r_*, \theta = \theta_*)$. The quantity $|p(x)| \equiv |p(r_*, \theta_*; x)|$ is termed the form function.

The algorithm of computation of $p(x)$ in the case of oblique incidence of the plane wave is similar to that in the case of normal incidence. The latter is described in detail in [1.2], therefore we shall not discuss this question here. We shall note only that the coefficients (1.58) in the determinants A_{10} and A_{20} [see (1.56)] and in B_1, B_2 and B_3 [see (1.60)] are presented now in a form not containing the derivatives with respect to the argument.

At $\alpha_0 = 0$ [see (1.2)], i.e. when the direction of propagation of the incident wave is normal to the longitudinal axis of the shell, the above-presented solution gets transformed into one known earlier [1.3]. This problem is considered in Chaps. 2, 7 and 9, where the results of computations are presented and analyzed.

At $b = 0$, i.e. for a solid elastic cylinder, from the solution (1.65) it is easy to find the solution obtained in [1.4], for which numerical results are presented in [1.5]. In this case, in order to obtain a bounded solution at $r = 0$ in the formulae (1.30), (1.41) and (1.43) one must set

$$C_n = E_n = G_n = 0 . \tag{1.66}$$

The other free parameters A_n , B_n , D_n and F_n are determined from the four boundary conditions on the outer surface of the cylinder.

From the above-given solution of the scattering problem it is easy to find the solution of the problem concerning the free vibrations of a "dry" shell [1.5]. In this case the wave equation describing the motion of the liquid can be omitted and the boundary conditions on the outer and inner surfaces of the shell can be written in the form

$$\sigma_{rr} = \sigma_{r\theta} = \sigma_{rz'} = 0 \quad \text{at } r = a \quad \text{and} \quad r = b . \tag{1.67}$$

In such a case the dispersion equation can be represented as

$$D = 0 , \tag{1.68}$$

where

$$D = \begin{vmatrix} a_{22} & a_{23} & a_{24} & a_{25} & a_{26} & a_{27} \\ a_{32} & a_{33} & a_{34} & a_{35} & a_{36} & a_{37} \\ a_{42} & a_{43} & a_{44} & a_{45} & a_{46} & a_{47} \\ a_{52} & a_{53} & a_{54} & a_{55} & a_{56} & a_{57} \\ a_{62} & a_{63} & a_{64} & a_{65} & a_{66} & a_{67} \\ a_{72} & a_{73} & a_{74} & a_{75} & a_{76} & a_{77} \end{vmatrix} . \tag{1.69}$$

1.2 Numerical Results

Below, we present the numerical results from [1.4] for a limiting case, namely the scattering by a solid elastic cylinder. The computation was carried out for the case of an aluminum cylinder immersed in water with the following parameters:

aluminum: $\rho_1 = 2.7118 \times 10^3$ kg/m^3 , $c_l = 6370$ m/s , $c_t = 3136$ m/s , (1.70)

water: $\rho = 0.998 \times 10^3$ kg/m^3 , $c = 1482$ m/s .

The computation step size was $l_x = 0.05$, in the domain considered $0 \leqslant x \leqslant 15$. The angle α_0 (Fig. 1.1) plays the role of the main parameter. The form functions were computed at $\alpha_0 = 0°$, $10°$, $20°$, $25°$ and $30.3°$. In Fig. 1.2, plots of the form functions are shown. The analysis of the form function for $\alpha = 0°$ is presented in Chaps. 2, 5 and 7.

At oblique incidence, the form function receives contributions from the specularly reflected wave, waves refracted into the elastic cylinder, and peripheral and creeping waves propagating along helical lines. In the low-frequency range ($x \lesssim 4.5$) the resonances of the Franz-type wave can be clearly observed.

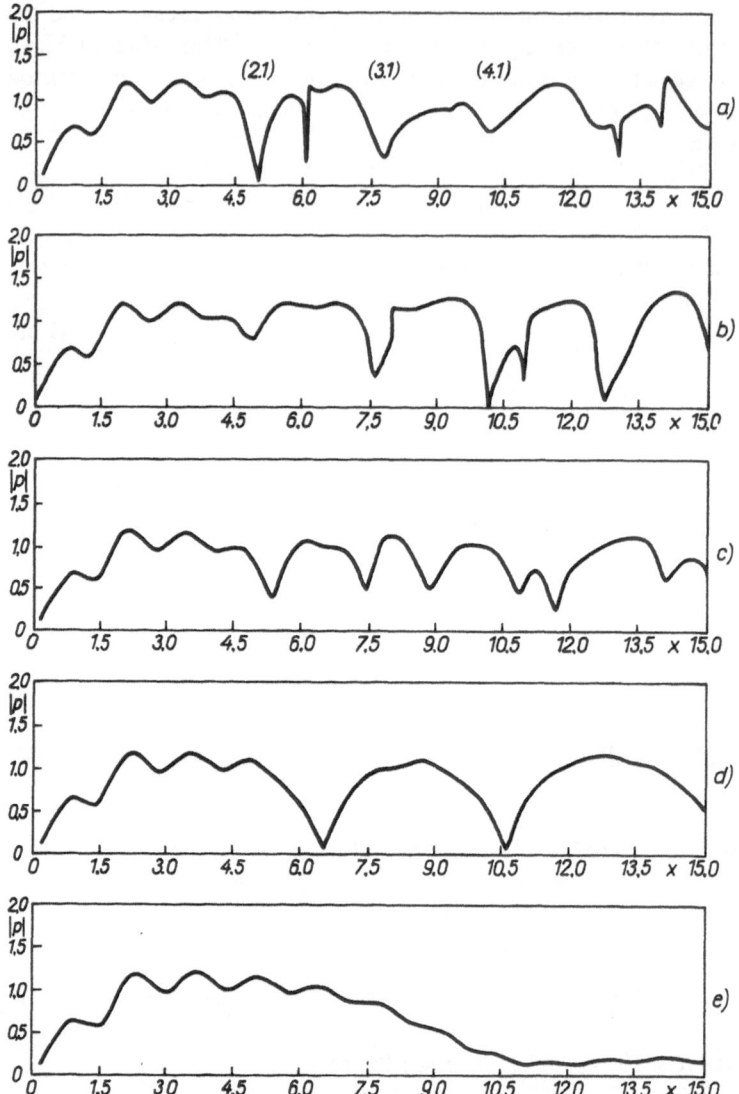

Fig. 1.2. The form function of the acoustic pressure scattered by a solid aluminum cylinder immersed in water at various values of the angle α_0: **a** $\alpha_0 = 0°$, **b** $\alpha_0 = 10°$, **c** $\alpha_0 = 20°$, **d** $\alpha_0 = 25°$, **e** $\alpha_0 = 30.3°$ [1.4, Figs. 2–6]

At normal incidence of the plane wave, as is well known, the Franz-type waves revolving around an elastic cylinder are, to a considerable extent, similar to the waves of this type revolving around an acoustically rigid cylinder.

At oblique incidence the velocities of the Franz-type waves, revolving around an acoustically rigid cylinder along helical lines, were determined in

[1.6] by using the method of the Sommerfeld–Watson integral transformation. The phase and group velocity of the lth Franz-type wave are defined by the formulae

$$c_l^{ph} = \frac{c}{1 + \frac{1}{2}q_l 6^{-1/3}\left(\dfrac{\sin^2 \alpha_0}{x}\right)^{2/3}} \, ,$$

$$c_l^{gr} = \frac{c}{1 + q_l 6^{-4/3}\left(\dfrac{\sin^2 \alpha_0}{x}\right)^{2/3}} \, , \tag{1.71}$$

$$q_1 = 1.469354, \ q_2 = 4.684712, \ q_3 = 6.951786 \, .$$

For $l > 3$ the constants q_l are tabulated in ([1.7], p. 353, Table 1), where they are denoted by $q_\perp^{(s)}$. The formulae (1.71) are obtained under the assumption that $x \gg 1$, therefore at a fixed x the value of the angle α_0 cannot be small.

At oblique incidence, the analysis of the form function is presented in Chap. 11 for scattering from a solid elastic cylinder and in Chap. 12 for that from a shell.

2 Resonance Scattering by Elastic Bodies of Cylindrical Shape

In this chapter the investigation of the resonances of partial-wave modes is described. This method was transferred into acoustics from quantum scattering theory by American acousticians [2.1] at the end of the 1970s. It is often called the *resonance scattering theory* (RST). According to this method, the amplitude of each partial-wave mode is divided into two components – those of the resonances, and the non-resonant background. (In nuclear scattering theory, this background is denoted as "potential scattering".) A qualitative description of the resonance component is given. The isolated resonances of partial waves of the same kind are grouped into families (the Regge poles). Using the positions of the isolated modal resonances, the phase and the group velocities of the peripheral waves revolving around the elastic body are found. A correspondence between the resonance frequencies of partial waves and the eigenfrequencies of the elastic body is established.

The procedure is applied to the scattering problem for a solid elastic cylinder, and then extended to the scattering problem for a shell. The computation is carried out with step size $l_x = 0.05$ for the case of an aluminum cylinder (and shell) in water, with the observation point situated in the far field at backscattering. The dependence of the Regge poles and the phase and group velocities of peripheral waves on the relative thickness of the shell is obtained. It is shown that the calculated acoustic spectrogram is in good agreement with the measured one.

2.1 Scattering by a Solid Elastic Cylinder

Let a plane harmonic acoustic pressure wave

$$p_i = p_* \exp[i(k\xi - \omega t)] \tag{2.1}$$

be incident on a solid elastic cylinder, and be scattered by the latter. At a fixed observation point $P(r, \phi)$, the acoustic pressure scattered by the cylinder can be presented in the form [2.1–4]

$$p_s = p_* \exp(-i\omega t) \sum_{n=0}^{\infty} \varepsilon_n i^n \frac{L_n J_n(x) - x J_n'(x)}{L_n H_n^{(1)}(x) - x H_n^{(1)\prime}(x)} H_n^{(1)}(kr) \cos n\phi . \tag{2.2}$$

Here

$$\varepsilon_0 = 1, \qquad \varepsilon_n = 2 \quad (\text{at } n > 0), \qquad L_n = \frac{\rho D_n^{(1)}}{\rho_1 D_n^{(2)}}, \tag{2.3}$$

where $D_n^{(1)}$ and $D_n^{(2)}$ are determinants of second rank

$$D_n^{(1)} = \begin{vmatrix} b_{21} & b_{23} \\ b_{31} & b_{33} \end{vmatrix}, \qquad D_n^{(2)} = \begin{vmatrix} b_{11} & b_{13} \\ b_{31} & b_{33} \end{vmatrix} \tag{2.4}$$

with the coefficients

$$b_{21} = x_1 J'_n(x_1), \qquad b_{23} = n J_n(x_2),$$
$$b_{31} = -2n[J_n(x_1) - x_1 J'_n(x_1)],$$
$$b_{33} = -2[(\tfrac{1}{2}x_2^2 - n^2)J_n(x_2) + x_2 J'_n(x_2)], \tag{2.5}$$
$$b_{11} = 2\frac{x_1^2}{x_2^2}\left[\frac{\lambda}{2\mu}J_n(x_1) - J''_n(x_1)\right],$$
$$b_{13} = \frac{2n}{x_2^2}\left[J_n(x_2) - x_2 J'_n(x_2)\right].$$

Here the prime denotes the derivative with respect to the argument.
The following notations are used

$$c_l = \left(\frac{\lambda + 2\mu}{\rho_1}\right)^{1/2} = \left(\frac{E(1-\nu)}{\rho_1(1+\nu)(1-2\nu)}\right)^{1/2},$$

$$c_t = \left(\frac{\mu}{\rho_1}\right)^{1/2} = \left(\frac{E}{2\rho_1(1+\nu)}\right)^{1/2}, \tag{2.6}$$

$$x = ka, \quad k = \frac{\omega}{c}, \quad n_1 = \frac{c}{c_l}, \quad n_2 = \frac{c}{c_t}, \quad x_j = n_j x \ (j = 1, 2),$$

where c_l, c_t are the longitudinal and transverse velocities in linear elasticity theory; λ, μ are the Lamé constants; E is the Young modulus, ν is Poisson's ratio; ρ, c are the density and the sound velocity in the liquid, a is the outer radius of the cylinder, p_* is a constant of dimension of the pressure, ω is the frequency; x denotes the wave radius in the liquid.

Some differences between the accepted notations and those used in [2.4] should be noted. The coefficients b_{kl}, (2.5), are related to the coefficients a_{kl} of ([2.4], (1.31) and (1.23)), in the following way:

$$b_{11} = \frac{2}{x_2^2}a_{21}, \qquad b_{13} = \frac{2}{x_2^2}a_{11}, \qquad b_{21} = a_{11}, \qquad b_{23} = a_{13},$$

$$b_{31} = -2a_{31}, \qquad b_{33} = -2a_{33}. \tag{2.7}$$

In (2.2) there is a factor i^n but in x_{7n} of [2.4], (1.31), there is a factor $(-i)^n$. This difference is connected with the direction of propagation of the incident wave.

Here it propagates in the positive ξ direction, but in [2.4] it was assumed that the incident wave propagates in the negative ξ direction. Here the direction of the source is $\phi = \pi$, but in [2.4] it was defined as $\theta = 0$. When the normalizations used in [2.4] are done, these differences in notation do not affect the result.

For the sake of simplicity we shall use the notations

$$R_n = -\frac{L_n J_n(x) - x J'_n(x)}{L_n H_n^{(1)}(x) - x H_n^{(1)\prime}(x)} , \tag{2.8}$$

$$S_n = 2R_n + 1 \equiv \exp(2i\delta_n) . \tag{2.9}$$

The function S_n is called the partial-wave scattering function and δ_n is called the phase shift of scattering.

With regard to (2.8), the acoustic pressure scattered by a cylinder takes the form

$$p_s = \frac{p_*}{2} \exp(-i\omega t) \sum_{n=0}^{\infty} \varepsilon_n i^n (S_n - 1) H_n^{(1)}(kr) \cos n\phi . \tag{2.10}$$

In the far field (at $kr \gg 1$), instead of $H_n^{(1)}(kr)$ its asymptotic representation can be used

$$H_n^{(1)}(kr) \sim \left(\frac{2}{i\pi kr}\right)^{1/2} i^{-n} \exp(ikr) . \tag{2.11}$$

Then in the far field (2.10) takes the form

$$p_s \sim p_* \left(\frac{a}{2r}\right)^{1/2} \exp[i(kr - \omega t)] \sum_{n=0}^{\infty} \varepsilon_n (S_n - 1)(i\pi x)^{-1/2} \cos n\phi . \tag{2.12}$$

If the form function $f(\phi)$, consisting of the partial-wave form functions $f_n(\phi)$, is introduced

$$f(\phi) = \sum_{n=0}^{\infty} f_n(\phi) , \tag{2.13}$$

$$f_n(\phi) = (i\pi x)^{-1/2} \varepsilon_n (S_n - 1) \cos n\phi , \tag{2.14}$$

then (2.12) can be written as

$$p_s \sim p_* \left(\frac{a}{2r}\right)^{1/2} \exp[i(kr - \omega t)] f(\phi) . \tag{2.15}$$

In the first limiting case, when $\rho/\rho_1 \to \infty$ [see (2.3)], (2.8) determines the function $R_n^{(s)}$, corresponding to scattering by an "acoustically soft" cylinder

$$R_n^{(s)} = -\frac{J_n(x)}{H_n^{(1)}(x)} , \tag{2.16}$$

and in the second limiting case, when $\rho/\rho_1 \to 0$, the corresponding function

$R_n^{(r)}$ for scattering by an "acoustically hard" cylinder, is obtained as

$$R_n^{(r)} = -\frac{J_n'(x)}{H_n^{(1)'}(x)} \, . \tag{2.17}$$

The expression for S_n , (2.9), can be presented in two alternative forms [2.1, 2]

$$S_n = S_n^{(s)}\frac{L_n - z_2}{L_n - z_1}, \qquad S_n^{(s)} = -\frac{H_n^{(2)}(x)}{H_n^{(1)}(x)}, \tag{2.18}$$

$$S_n = S_n^{(r)}\frac{L_n^{-1} - z_2^{-1}}{L_n^{-1} - z_1^{-1}}, \qquad S_n^{(r)} = -\frac{H_n^{(2)'}(x)}{H_n^{(1)'}(x)}, \tag{2.19}$$

where

$$z_1 = x\frac{H_n^{(1)'}(x)}{H_n^{(1)}(x)}, \qquad z_2 = x\frac{H_n^{(2)'}(x)}{H_n^{(2)}(x)}, \tag{2.20}$$

$$\tan \zeta_n^{(s)} = \frac{J_n(x)}{Y_n(x)} \, . \tag{2.21}$$

Equation (2.18) can be written as

$$S_n^{(s)} = -\frac{\tan \zeta_n^{(s)} - i}{\tan \zeta_n^{(s)} + i} \, . \tag{2.22}$$

Using

$$S_n^{(s)} = \rho_n^{(s)} \exp(i\phi_n^{(s)}) \, ,$$

we find from (2.22)

$$\rho_n^{(s)} = 1, \qquad \tan \phi_n^{(s)} = \frac{2\tan \zeta_n^{(s)}}{1 - \tan^2 \zeta_n^{(s)}}, \qquad \phi_n^{(s)} = 2\zeta_n^{(s)} \, . \tag{2.23}$$

Then

$$S_n^{(s)} = \exp(2i\zeta_n^{(s)}), \quad |S_n^{(s)}| \equiv 1 \, . \tag{2.24}$$

Analogously,

$$S_n^{(r)} = \exp(2i\zeta_n^{(r)}), \quad |S_n^{(r)}| \equiv 1, \qquad \tan \zeta_n^{(r)} = \frac{J_n'(x)}{Y_n'(x)} \, . \tag{2.25}$$

Here, $S_n^{(s)}$ and $S_n^{(r)}$ are the partial-wave scattering functions for scattering by acoustically soft and rigid cylinders, respectively, and $\zeta_n^{(s)}$ and $\zeta_n^{(r)}$ correspond to the phase shifts.

Each of the quantities z_1, z_2 [see (2.20)] and those inverse to them can be separated into the real and imaginary parts

$$z_1 = \Delta_n^{(s)} + is_n^{(s)}, \qquad z_2 = \Delta_n^{(s)} - is_n^{(s)} \, , \tag{2.26}$$

$$z_1^{-1} = \Delta_n^{(r)} + is_n^{(r)}, \qquad z_2^{-1} = \Delta_n^{(r)} - is_n^{(r)} \, , \tag{2.27}$$

where

$$\Delta_n^{(s)} = x \frac{J_n(x)J_n'(x) + Y_n(x)Y_n'(x)}{J_n^2(x) + Y_n^2(x)},$$

$$S_n^{(s)} = \frac{2}{\pi} \frac{1}{J_n^2(x) + Y_n^2(x)},$$

(2.28)

$$\Delta_n^{(r)} = \frac{1}{x} \frac{J_n(x)J_n'(x) + Y_n(x)Y_n'(x)}{[J_n'(x)]^2 + [Y_n'(x)]^2},$$

$$S_n^{(r)} = \frac{2}{\pi x^2} \frac{1}{[J_n'(x)]^2 + [Y_n'(x)]^2}.$$

(2.29)

The resonance frequency x_n of the partial-wave mode is determined from the equation

$$L_n - \Delta_n^{(s)} = 0, \qquad L_n^{-1} - \Delta_n^{(r)} = 0,$$

(2.30)

i.e. at this frequency

$$L_n(x_n^{(s)}) = \Delta_n^{(s)}, \qquad L_n^{-1}(x_n^{(r)}) = \Delta_n^{(r)}.$$

It is supposed that in the vicinity of the resonance frequency $L_n(x)$ and $L_n^{-1}(x)$ may be expanded in a Taylor series in x

$$L_n(x) \cong \Delta_n^{(s)} + \beta_n^{(s)}(x - x_n^{(s)}),$$

(2.31)

$$L_n^{-1}(x) \cong \Delta_n^{(r)} + \beta_n^{(r)}(x - x_n^{(r)}).$$

If now the "width of the resonance" Γ_n is introduced according to the definition

$$\Gamma_n^{(s)} = -2S_n^{(s)}/\beta_n^{(s)}, \qquad \Gamma_n^{(r)} = -2S_n^{(r)}/\beta_n^{(r)},$$

(2.32)

then the function S_n can be presented in the resonance form [2.1]

$$S_n = S_n^{(s)} \frac{x - x_n^{(s)} - \frac{1}{2}i\Gamma_n^{(s)}}{x - x_n^{(s)} + \frac{1}{2}i\Gamma_n^{(s)}}, \qquad S_n = S_n^{(r)} \frac{x - x_n^{(r)} - \frac{1}{2}i\Gamma_n^{(r)}}{x - x_n^{(r)} + \frac{1}{2}i\Gamma_n^{(r)}}.$$

(2.33)

According to (2.22), (2.24) and (2.25), we obtain

$$|S_n| = 1.$$

(2.34)

This means that during the scattering no energy is being absorbed. For the elastic scatterer this is connected with the fact that both x_1 and x_2 in (2.6) are real.

The function S_n must possess not only a resonance pole at the complex frequency

$$x = x_p^{(n)} = x_n - \frac{1}{2}i\Gamma_n,$$

(2.35)

but also a resonance zero at

$$x = x_z^{(n)} = x_n + \tfrac{1}{2}i\Gamma_n \, . \tag{2.36}$$

This is necessary to satisfy the unitarity condition (2.34).

Since the resonance widths are positive quantities, the pole is situated in the lower complex half-plane, and the zero is situated in the upper half-plane.

The quantity $S_n - 1$ in the expression for $f_n(\phi)$, see (2.14), i.e.

$$
\begin{aligned}
S_n - 1 &= 2i \exp(i\delta_n) \sin \delta_n \\
&= 2i \exp(2i\xi_n^{(s)}) \left(\frac{S_n^{(s)}}{L_n - \Delta_n^{(s)} - iS_n^{(s)}} + \exp(-i\xi_n^{(s)}) \sin \xi_n^{(s)} \right) \\
&= 2i \exp(2i\xi_n^{(r)}) \left(\frac{S_n^{(r)}}{L_n^{-1} - \Delta_n^{(r)} - iS_n^{(r)}} + \exp(-i\xi_n^{(r)}) \sin \xi_n^{(r)} \right)
\end{aligned}
\tag{2.37}
$$

can, by using (2.31) and (2.32), be presented in the form

$$
\begin{aligned}
S_n - 1 &= 2i \exp(2i\xi_n^{(s)}) \left(\frac{\tfrac{1}{2}\Gamma_n^{(s)}}{x_n^{(s)} - x - \tfrac{1}{2}i\Gamma_n^{(s)}} + \exp(-i\xi_n^{(s)}) \sin \xi_n^{(s)} \right) \\
&= 2i \exp(2i\xi_n^{(r)}) \left(\frac{\tfrac{1}{2}\Gamma_n^{(r)}}{x_n^{(r)} - x - \tfrac{1}{2}i\Gamma_n^{(r)}} + \exp(-i\xi_n^{(r)}) \sin \xi_n^{(r)} \right) \, .
\end{aligned}
\tag{2.38}
$$

Then the partial-wave form function $f_n(\phi)$ (2.14) can be written as

$$f_n(\phi) = 2i\varepsilon_n(i\pi x)^{-1/2} \exp(i\delta_n) \sin \delta_n \cos n\phi \, , \tag{2.39}$$

$$f_n(\phi) = g_n^{(s)}(\phi) + f_n^{(s)}(\phi), \qquad f_n(\phi) = g_n^{(r)}(\phi) + f_n^{(r)}(\phi) \, , \tag{2.40}$$

where

$$g_n^{(s)}(\phi) = 2i\varepsilon_n(i\pi x)^{-1/2} \exp(2i\xi_n^{(s)}) \frac{\tfrac{1}{2}\Gamma_n^{(s)}}{x_n^{(s)} - x - \tfrac{1}{2}i\Gamma_n^{(s)}} \cos n\phi \, ,$$

$$f_n^{(s)}(\varphi) = 2i\varepsilon_n(i\pi x)^{-1/2} \exp(i\xi_n^{(s)}) \sin \xi_n^{(s)} \cos n\varphi \, ,$$

$$g_n^{(r)}(\varphi) = 2i\varepsilon_n(i\pi x)^{-1/2} \exp(2i\xi_n^{(r)}) \frac{\tfrac{1}{2}\Gamma_n^{(r)}}{x_n^{(r)} - x - \tfrac{1}{2}i\Gamma_n^{(r)}} \cos n\phi \, ,$$

$$f_n^{(r)}(\varphi) = 2i\varepsilon_n(i\pi x)^{-1/2} \exp(i\xi_n^{(r)}) \sin \xi_n^{(r)} \cos n\varphi \, . \tag{2.41}$$

The partial-wave form function consists of the resonance $g_n^{(s)}(\phi)$ (or $g_n^{(r)}(\phi)$) and the background $f_n^{(s)}(\phi)$ (or $f_n^{(r)}(\phi)$) components.

In the calculation of the form function $f(\phi)$ (2.13) one obtains an interference between the background (potential) and resonance components, which can sometimes lead to striking forms of the resonance plots, as will be shown below.

It has been shown earlier in [2.3, 5, 6] that the calculated and measured values of the resonance frequency (in the x scale), both for $f_n(\pi)$ and $f(\pi)$, coincide with the eigenfrequencies of the elastic vibrations of the scatterer. These eigenfrequencies, in fact, are found from the condition that in (2.18) and (2.19) the denominator is equal to zero:

$$L_n - z_1 = 0 \qquad \text{or} \qquad L_n^{-1} - z_1^{-1} = 0 , \tag{2.42}$$

which in the limit $\rho/\rho_1 \to 0$ leads to the real eigenfrequencies of a "dry" body. When this assumption is not used, then from (2.42) the complex eigenfrequencies of an elastic body immersed in a liquid are obtained. The real part of the latter is very close to the real eigenfrequency, as will be seen below for the problem of scattering by an aluminum cylinder immersed in water. It should be mentioned, however, that the elastic scatterer can have more modes of eigenvibrations than are observed in the form function plot if these modes cannot be generated by the incident wave. Thus, for example, the "breathing" mode $n = 0$ is generated only with a very small magnitude in the scattering by cylinders and spheres [2.5-7].

In addition to the above-considered magnitude of the external scattered acoustic pressure, it is instructive to analyze the solution of the same problem inside the scatterer. Here the displacement vector **u** is usually presented as a sum of irrotational and solenoidal fields

$$\mathbf{u} = \text{grad}\, \psi + \text{curl}\, \mathbf{A} , \tag{2.43}$$

where ψ and $\mathbf{A}$ are the scalar and vector potentials of the displacement which can be presented in the form

$$\psi = \exp(-i\omega t) \sum_{n=0}^{\infty} \varepsilon_n i^n C_n J_n(k_1 r) \cos n\varphi ,$$

$$A_z = \exp(-i\omega t) \sum_{n=0}^{\infty} \varepsilon_n i^n E_n J_n(k_t r) \sin n\varphi . \tag{2.44}$$

Here, A_z is the single non-vanishing component of the vector potential.

In the earlier accepted notation the coefficients C_n and D_n can be presented in the form

$$C_n = \frac{2i}{\pi\rho\omega^2} \frac{(x_2^2 - 2n^2) J_n(x_2) + 2x_2 J_n'(x_2)}{x H_n^{(1)\prime}(x) D_n^{(1)}(z_1^{-1} - L_n^{-1})} ,$$

$$E_n = \frac{2i}{\pi\rho\omega^2} \frac{2n[J_n(x_1) - x_1 J_n'(x_1)]}{x H_n^{(1)\prime}(x) D_n^{(1)}(z_1^{-1} - L_n^{-1})} . \tag{2.45}$$

In (2.45), only one possibility out of those given in (2.18) is used, namely that corresponding to the rigid background.

In the vicinity of each resonance, the expansion (2.31) can be used. Then the resonance expressions

$$\psi = \frac{2\exp(-i\omega t)}{\pi i \rho\omega^2} \sum_{n=0}^{\infty} \frac{\varepsilon_n i^n}{\beta_n^{(r)}} \frac{(x_2^2 - 2n^2) J_n(x_2) + 2x_2 J_n'(x_2)}{x H_n^{(1)\prime}(x) D_n^{(1)}} \frac{J_n(k_1 r) \cos n\varphi}{x - x_n^{(r)} + \frac{1}{2} i \Gamma_n^{(r)}} ,$$

$$A_z = \frac{2\exp(-i\omega t)}{\pi i \rho\omega^2} \sum_{n=0}^{\infty} \frac{\varepsilon_n i^n}{\beta_n^{(r)}} \frac{2n[J_n(x_1) - x_1 J_n'(x_1)]}{x H_n^{(1)\prime}(x) D_n^{(1)}} \frac{J_n(k_t r) \sin n\varphi}{x - x_n^{(r)} + \frac{1}{2} i \Gamma_n^{(r)}} \tag{2.46}$$

are obtained. Only resonance terms are present in these formulae. The absence of the background (soft or rigid) components is natural, because the acoustically soft and rigid bodies are impenetrable and do not allow any fields in the interior domain (i.e. at $r < a$).

The formulae (2.46) together with the formulae for the resonance magnitudes (2.37) indicate the astonishing fact that for scattering at a frequency situated between any two eigenfrequencies of the elastic body, the scatterer "behaves" like an impenetrable body and scatters the wave according to "potential scattering". On the other hand, at and in the vicinity of a resonance frequency the incident wave generates resonance scattering and the sound field penetrates inside the body. The resonance and potential scattering interfere, which becomes apparent in a structure of the amplitude of the total scattering which shows strong dips plotted vs frequency.

The resonance phenomena which appear in the scattering of the acoustic waves by elastic bodies have been discussed ealier for both the steady-state and the transient cases (see, for example, [2.8–10]). It was pointed out in these papers that the resonances will appear when both the mechanical and the acoustical impedances tend to zero. In [2.1] it was shown that the background term does not always correspond to the background component for a rigid body. It can correspond to the background component for an acoustically soft body (for example, in the case of air bubbles) or to an intermediate background (for thin-walled elastic shells). For both the resonance and background components, in [2.1] has given explicit computational formulae. This allows us to interpret the resonances in complicated plots of the calculated (or measured) form functions.

The above-given approximate formulae (2.39), (2.40) are used for a qualitative description of the scattering process. In the computations the exact expressions are used. In the far field the difference between the acoustic pressure scattered by elastic (p_s) and acoustically rigid ($p_s^{(r)}$) cylinders can be written as

$$p_s - p_s^{(r)} = p_*' \left(\frac{a}{2r} \right)^{1/2} \exp|i(kr - \omega t)| \psi^{(r)}(\varphi) , \qquad (2.47)$$

where

$$\psi^{(r)}(\varphi) = \sum_{n=0}^{\infty}{}' \psi_n^{(r)}(\varphi) ,$$

$$\psi_n^{(r)}(\varphi) = -2(i\pi x)^{-1/2} \varepsilon_n \left(\frac{L_n J_n(x) - x J_n'(x)}{L_n H_n^{(1)}(x) - x H_n^{(1)\prime}(x)} - \frac{J_n'(x)}{H_n^{(1)\prime}(x)} \right) \cos n\varphi \qquad (2.48)$$

and the resonance components of the partial-wave modes are obtained according to $\psi_n^{(r)}(\phi)$, computed at fixed $\phi = \phi_0$.

2.2 Numerical Results for Scattering by a Solid Elastic Cylinder

Computational results are given below for the case of scattering by an aluminum cylinder in water. The computation was carried out with the following parameters:

aluminum: $\rho_1 = 2.79 \times 10^3 \text{ kg/m}^3$, $c_l = 6380 \text{ m/s}$, $c_t = 3100 \text{ m/s}$,

water: $\rho = 1 \times 10^3 \text{ kg/m}^3$, $c = 1370 \text{ m/s}$.   (2.49)

In Fig. 2.1, the dependence of the form function on $x = ka$ is presented. The resonance features in the graph are labeled by two numbers (n, l). The first index defines the ordinal number (order) of the resonance, and the second determines the type of the peripheral wave revolving around the elastic cylinder that generates the resonance. The index $l = 1$ corresponds to the "Rayleigh wave", and $l = 2, 3, 4, \ldots$ correspond to "Whispering Gallery waves". Here the longitudinal and transverse waves are not separated yet.

In Fig. 2.2, the dependence of $|f_n(\pi)|$ on x for the first six ($n = 0$–5) partial-wave modes is presented.

In Fig. 2.3, a comparison is presented of the dependence of $|f_n(\pi)|$ on x for two partial-wave modes ($n = 2$ and $n = 3$) for the scattering by an elastic cylinder (marked by a solid line) with the corresponding curves for scattering by acoustically soft (dot-and-dash line) and hard cylinders (dashed line).

In Fig. 2.4, we show the dependence of $|\psi_n^{(r)}(\pi)|$ on x (see (2.48)). The second resonances ($n = 2$) of three peripheral waves can be clearly observed here: the Rayleigh wave ($l = 1$) and the Whispering Gallery ($l = 2$ and $l = 3$) waves. From plots similar to that of Fig. 2.4, the ordinal number of the resonance (n) and its type (l) can be found for each $n = 1, 2, 3, \ldots$. This information is then used in the form function and acoustic spectrogram.

In Fig. 2.5, the dependence of phase angle and magnitude of the partial-wave mode on x is shown. As is known, the phase of the resonance component of the

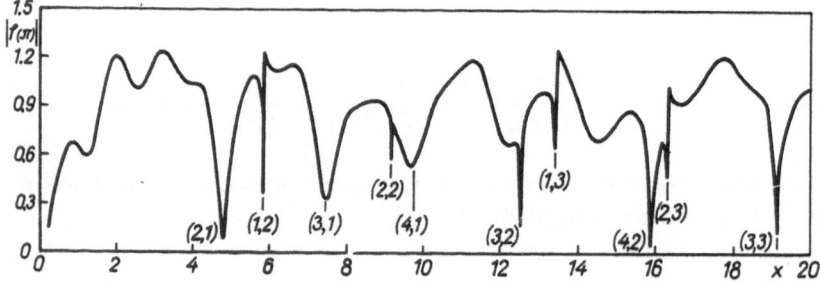

Fig. 2.1. The form function for backscattering by a solid aluminum cylinder in water. The two numbers (n, l) identify the resonance; n determines its ordinal number and l defines its family; $l = 1$ corresponds to the Rayleigh wave and $l = 2, 3, 4, \ldots$ to the Whispering Gallery waves [2.1, Fig. 1a]

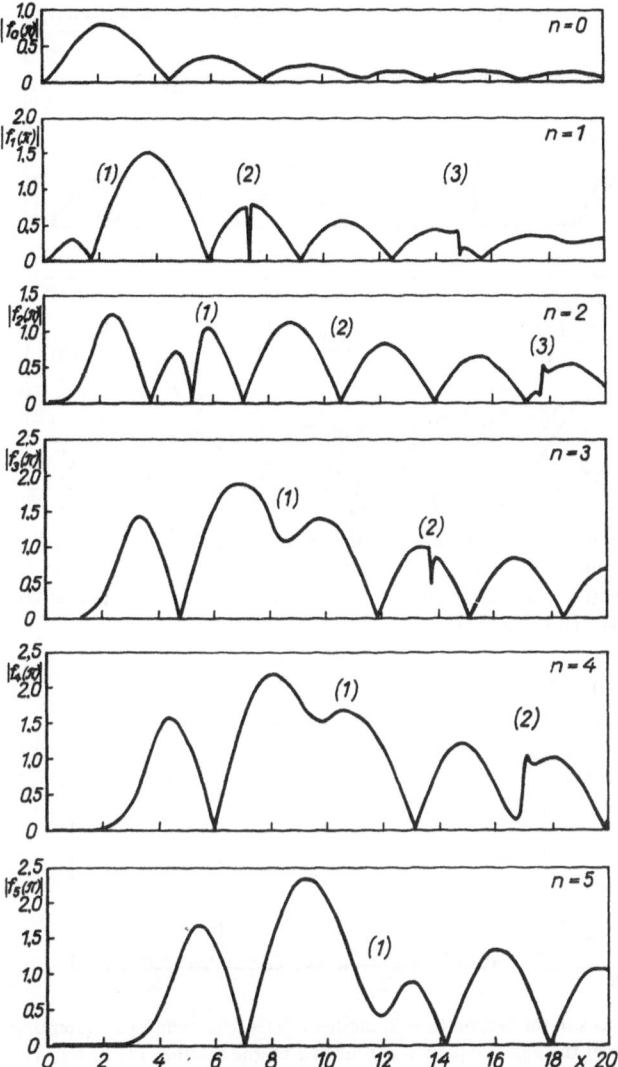

Fig. 2.2. Isolated modal resonances for the first six partial-wave modes ($n = 0$–5) for backscattering by a solid aluminum cylinder in water. The positions of the eigenfrequency of a partial-wave mode are labeled by the number l [2.1, Fig. 2]

partial-wave mode performs a jump equal to π when passing through the resonance. This can easily be checked when the phase of $f_n(\pi)$ (see (2.39)) is considered. This phase is determined by the formula

$$\varphi_n = \delta_n + \tfrac{1}{2}\pi \ . \tag{2.50}$$

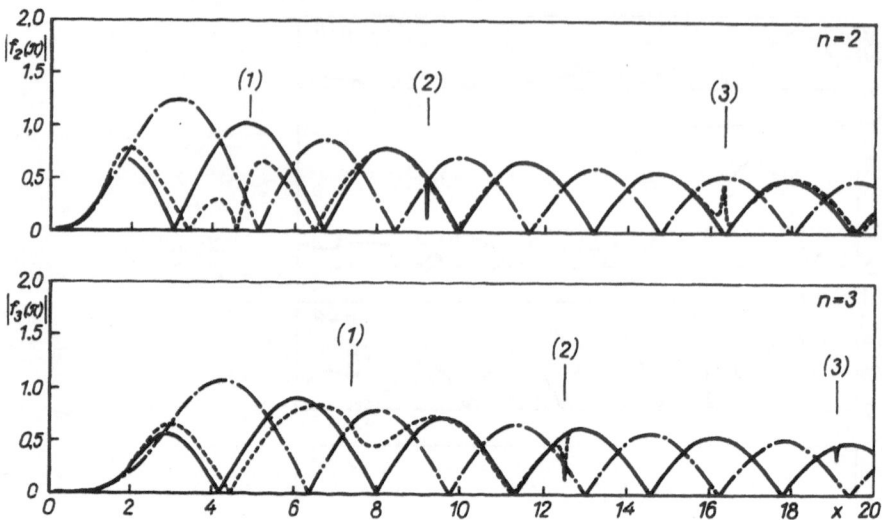

Fig. 2.3. Comparison of the dependence of $|f_n(\pi)|$ on x for the partial-wave modes $n = 2$ and $n = 3$ for scattering by a solid elastic cylinder (solid line), an acoustically soft cylinder (dot-and-dash line) and an acoustically hard cylinder (dashed line). The positions of the eigenfrequency of a partial-wave mode are labeled by the number l [2.1, Fig. 3a]

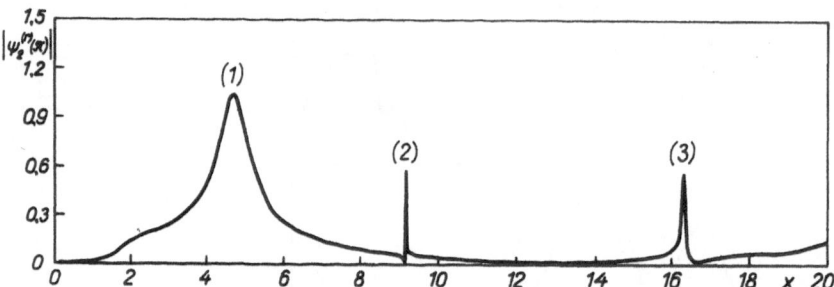

Fig. 2.4. Isolated modal resonances of the second ($n = 2$) mode for backscattering by an aluminum cylinder in water. The positions of the eigenfrequency are labeled by the number l [2.1, Fig. 4]

Using (2.33), we can write

$$\delta_n - \xi_n^{(r)} = \arctan\left(\frac{\Gamma_n^{(r)}}{2(x_n^{(r)} - x)}\right), \qquad (2.51)$$

which under the assumption $\Gamma_n^{(r)} > 0$ (which is usually fulfilled) displays a jump in the difference $\delta_n - \xi_n^{(r)}$ from zero to π, when x changes from $-\infty$ to $+\infty$. At small $\Gamma_n^{(r)}$ this jump happens very sharply in the vicinity of $x = x_n^{(r)}$. The phase jump of the resonance component is superimposed on the slowly changing phase of the background as x varies. This is what can be observed in Fig. 2.5,

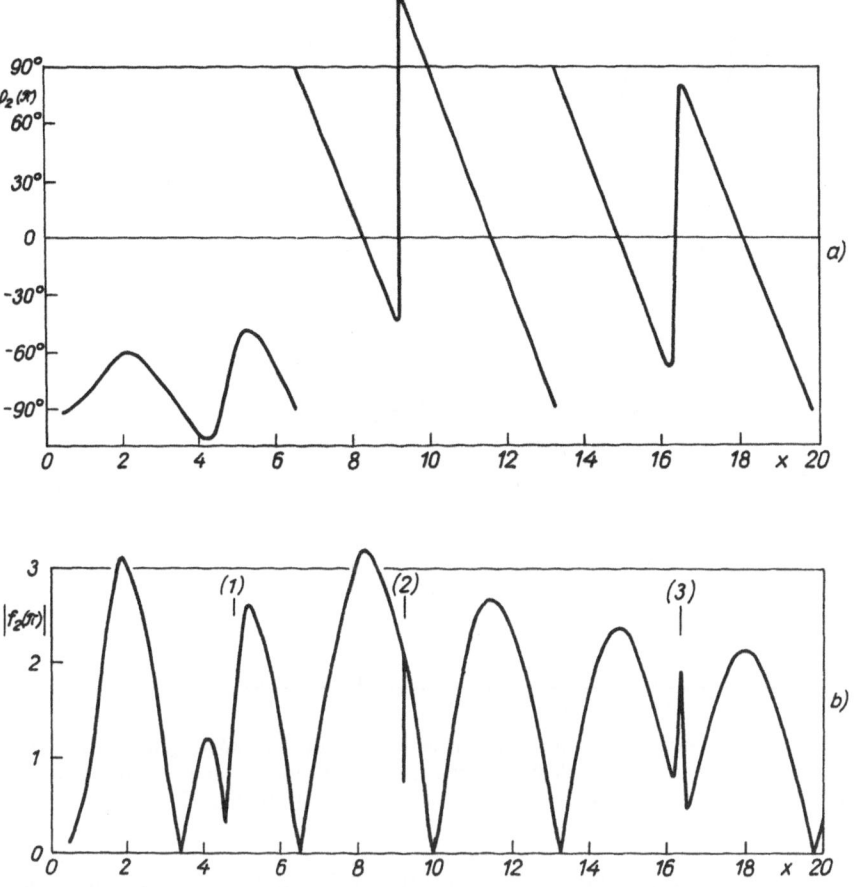

Fig. 2.5. The phase **a** and the modulus **b** of the second ($n = 2$) partial-wave mode for scattering by a solid aluminum cylinder in water. The positions of the eigenfrequency are labeled by the number l [2.1, Fig. 5]

where the eigenfrequencies are marked by the number l. From the practical point of view, the phase jump at the resonance frequency can be useful, especially for the cases when the width of the resonance is small and the computational step size is large. In such a situation, on the curve of a partial-wave form function the resonance can be missed, while the phase jump is more noticeable and can indicate the presence of a resonance.

The phase velocity $c_l^{ph}(x)$ and the group velocity $c_l^{gr}(x)$ of the lth peripheral wave can be found according to the formulae [2.7, 11]

$$c_l^{ph}(x) = c\frac{x}{\mathrm{Re}\{n_l\}}, \qquad c_l^{gr}(x) = \frac{c}{\mathrm{Re}\left\{\dfrac{dn_l}{dx}\right\}}, \tag{2.52}$$

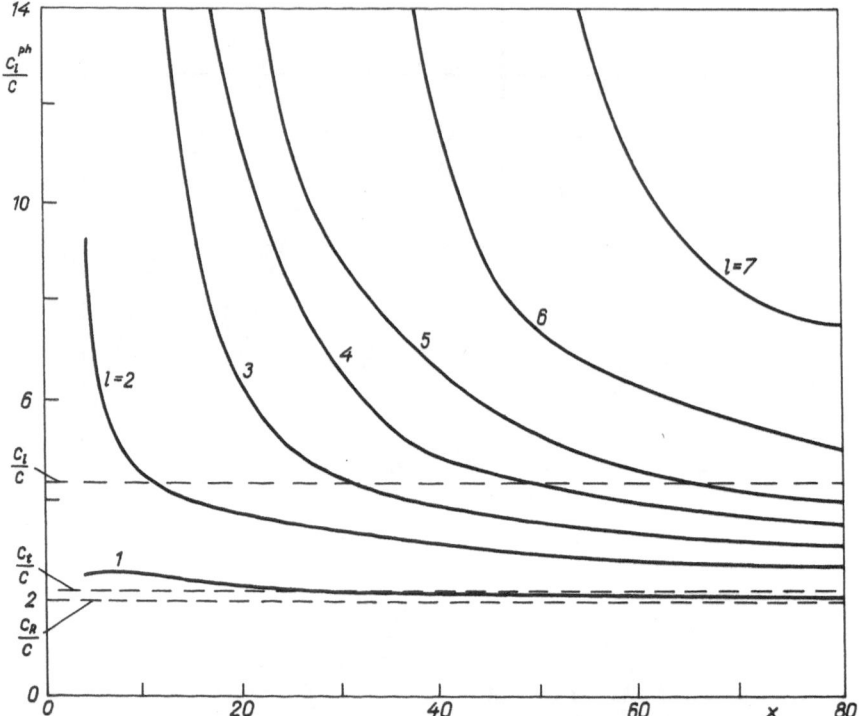

Fig. 2.6. Dispersion curves of relative phase velocities of peripheral waves generated by sound scattering from a solid aluminum cylinder in water. The Rayleigh wave is labeled by the index $l = 1$, and the Whispering Gallery waves by the indices $l = 2, 3, 4, \ldots$ [2.16, Fig. 7]

where n_l is the corresponding value of the mode number n, which is now considered a continuous complex variable. In Fig. 2.6, we show the dispersion curves of the relative phase velocities, and in Fig. 2.7 those relating to the group velocities.

2.3   Scattering by a Circular–Cylindrical Shell

Let the plane acoustic pressure wave (2.1) be incident on a cylindrical shell, and be scattered by the latter. It is assumed that the direction of the initial wave propagation is normal to the longitudinal axis of the shell. The relative thickness of the shell h is determined by

$$h = 1 - \frac{b}{a},\qquad(2.53)$$

where a, b are the outer and the inner radius of the shell.

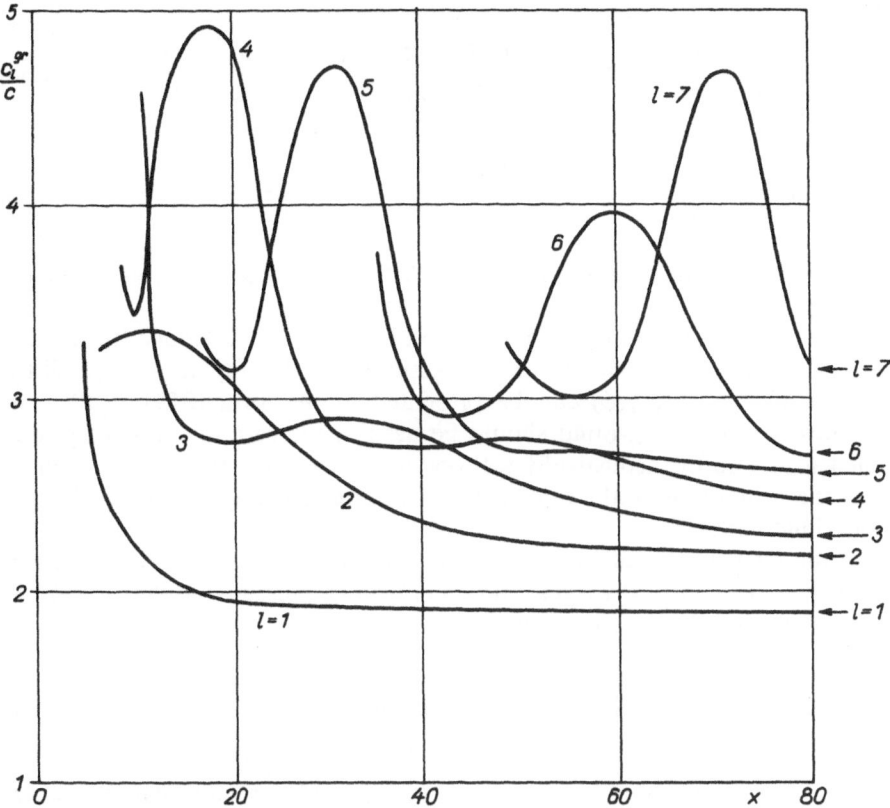

Fig. 2.7. Dispersion curves of relative group velocities of peripheral waves for scattering by an aluminum cylinder in water. The Rayleigh wave is labeled by the index $l = 1$, and the Whispering Gallery waves by the indices $l = 2, 3, 4, \ldots$ [2.16, Fig. 8]

The acoustic pressure scattered by the shell has a form similar to (2.2) (see [2.12, 13])

$$p_s = p_* \exp(-i\omega t) \sum_{n=0}^{\infty} \varepsilon_n i^n R_n H_n^{(1)}(kr) \cos n\varphi , \tag{2.54}$$

where

$$R_n = \frac{b_n}{D_n} , \tag{2.55}$$

and the determinants of the sixth rank b_n and D_n are given in [2.12, 13].

According to the above-described procedure, and following the methodology given in [2.14], the partial scattering function S_n (2.9) can be introduced. Then, using (2.11), the series (2.10) can be presented in the form (2.15). Now, however, S_n has a new meaning. The formula for S_n can be represented in two forms

$$S_n = S_n^{(s)} \frac{F_n - z_2}{F_n - z_1}, \qquad S_n = S_n^{(r)} \frac{F_n^{-1} - z_2^{-1}}{F_n^{-1} - z_1^{-1}}, \qquad (2.56)$$

where

$$F_n = \frac{\rho}{\rho_1} x_2^2 \frac{D_n^{(1)}}{D_n^{(2)}}, \qquad (2.57)$$

and the determinants of fifth rank $D_n^{(1)}$ and $D_n^{(2)}$, and their elements are defined in [2.1, 12].

The resonance frequency x_n of the partial-wave mode is defined from (2.30), where instead of L_n one should use F_n. Assuming such a replacement in all the formulae of Sect. 2.1, $f_n(\phi)$ can be presented in the form (2.39), (2.40). The acoustically rigid background should be used in the case of a not very thin-walled shell, and the acoustically soft one in the case of a thin-walled shell.

In the computations, the exact expressions are generally used instead of the approximate ones.

In the far field, the difference between the pressure scattered by an elastic shell p_s and an acoustically rigid cylinder $p_s^{(r)}$ can be presented in the form (2.47), (2.48) in which $\psi_n^{(r)}(\phi)$ is written as

$$\psi_n^{(r)}(\varphi) = 2(i\pi x)^{-1/2} \varepsilon_n \left(R_n + \frac{J_n'(x)}{H_n^{(1)'}(x)} \right) \cos n\varphi, \qquad (2.58)$$

with R_n given by (2.55) or defined as $R_n \equiv X_{7n}$, and the latter expressions are given in [2.4, Eq. (1.25)]. The elements of X_{7n} are determined by either (1.23) from [2.4]), or (1.30) from [2.4]. In some of the latter expressions, misprints are present. The correct coefficients are

$$a_{33} = \frac{x_2}{2n} [x_2 + 2\alpha_n'(x_2)] - n, \qquad a_{34} = \frac{x_2}{2n} [x_2 + 2\beta_n'(x_2)] - n,$$

$$a_{73} = \frac{\gamma_n(\beta_2)}{n} [\tfrac{1}{2} y_2^2 - n^2 + y_2 \alpha_n'(y_2)], \qquad (2.59)$$

$$a_{74} = \frac{\delta_n(\beta_2)}{n} [\tfrac{1}{2} y_2^2 - n^2 + y_2 \beta_n'(y_2)].$$

Similarly, in the far field the difference between the acoustic pressure scattered by an elastic shell p_s and an acoustically soft cylinder $p_s^{(s)}$ can be presented as

$$p_s - p_s^{(s)} = p_* \left(\frac{a}{2r} \right)^{1/2} \exp[i(kr - \omega t)] \psi^{(s)}(\varphi), \qquad (2.60)$$

where

$$\psi^{(s)}(\varphi) = \sum_{n=0}^{\infty} \psi_n^{(s)}(\varphi) \, , \tag{2.61}$$

$$\psi_n^{(s)}(\varphi) = 2(i\pi x)^{-1/2} \varepsilon_n \left(R_n + \frac{J_n(x)}{H_n^{(1)}(x)} \right) \cos n\varphi \, .$$

2.4 Numerical Results for the Problem of a Plane Acoustic Pressure Wave Scattered by a Circular-Cylindrical Shell

In the following, we present some results obtained in [2.15–17] where the problem of scattering by a cylindrical shell in water was considered. The physical parameters are defined by (2.49). The observation point is situated in the far field at backscattering.

In Fig. 2.8, we show the dependence of $|\psi_n^{(r)}(\pi)|$ on x [see (2.58)] for scattering by a thick-walled shell with $b/a = 0.1$. In order to identify unambiguously the families of resonances (Regge poles), we use, besides the integral n values, also non-integral values ($n = 1.2, 1.5, 2.2, 2.5, 3.5$). The family of rather broad resonances on the extreme left corresponds to the Rayleigh wave, and the subsequent families to the Whispering Gallery waves. There are two kinds of Whispering Gallery waves here: the longitudinal (L) and the transverse (T). The type of the peripheral wave is determined by the short-wave asymptotics (at $x \gg 1$) of the wave's phase velocity.

In Fig. 2.9, for the same tube, we present the result of a systematic investigation of the resonance positions on the x axis. Here $0 \leqslant x \leqslant 200$ and $0 \leqslant n \leqslant 80$. In all, sixteen families of resonances are shown: those corresponding to the Rayleigh wave, to the four longitudinal Whispering Gallery waves and to the eleven transverse Whispering Gallery waves. The low n-number modes ($n = 0$–5) are strongly dispersive. The intersection of lines corresponding to the longitudinal and to the transverse Whispering Gallery waves complicates the identification of resonances, as indicated in [2.17].

Using the Regge trajectories, it is easy to calculate the phase and group velocity of each peripheral wave. In Fig. 2.10, dispersion curves of the group velocities are presented for the Rayleigh wave R and the first three transverse Whispering Gallery waves (T_1, T_2 and T_3). A logarithmic scale is used on the x axis.

As can be seen from the plots, at $x \gg 1$ the R curve tends to the value $c_R/c = 1.916$ which corresponds to the relative velocity of the Rayleigh wave propagating on the "dry" elastic half-space, and the curves T_1, T_2 and T_3 tend to the value $c_t/c = 2.037$, respectively. The tube is very thick walled here and therefore the peripheral waves revolving around it appear as those revolving around a solid elastic cylinder (compare Figs. 2.10 and 2.7). In the latter,

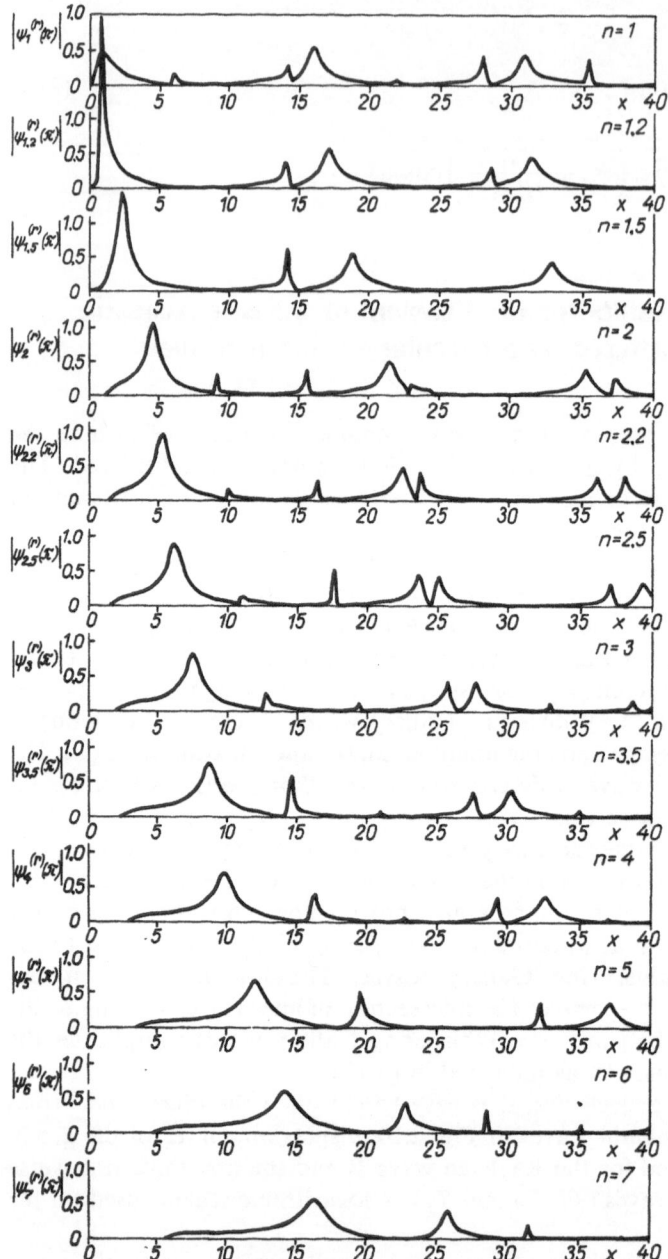

Fig. 2.8. Isolated modal resonances for backscattering by an aluminum tube with $b/a = 0.1$ in water. Both integral and non-integral n values are used [2.17, Fig. 2]

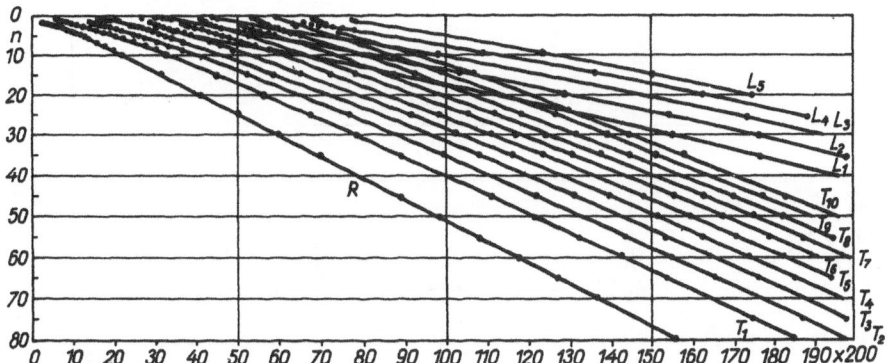

Fig. 2.9. Regge trajectories of the peripheral waves for scattering by an aluminum tube with $b/a = 0.1$ in water. The Rayleigh wave is labeled by R, the longitudinal Whispering Gallery waves by L_j ($j = 1$–5), and the transverse Whispering Gallery waves by T_j ($j = 1$–10) [2.17, Fig. 3]

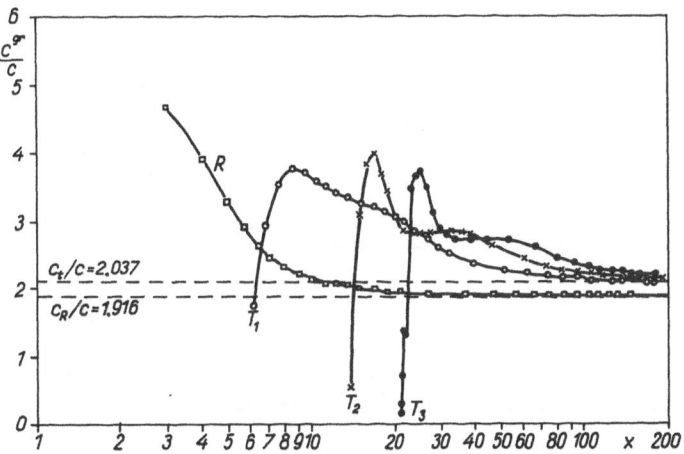

Fig. 2.10. Dispersion curves of relative group velocities of the peripheral waves for scattering by an aluminum tube with $b/a = 0.1$ in water. The Rayleigh wave is labeled by R, and the transverse Whispering Gallery waves by T_j ($j = 1, 2, 3$) [2.17, Fig. 4]

different notations are used. The similarity of the dispersion curves can be easily determined using corresponding notations ($l = 1 \rightarrow R$, $l = 3 \rightarrow T_1$, $l = 4 \rightarrow T_2$, $l = 5 \rightarrow T_3$).

In Fig. 2.11, the dependence of $|\psi_n^{(r)}(\pi)|$ on x is presented for the case of scattering by a thick-walled tube with $b/a = 0.2$. As to be expected, the dependence of $|\psi_n^{(r)}(\pi)|$ on x for two very thick-walled tubes with $b/a = 0.1$ and $b/a = 0.2$ are rather similar (compare Figs. 2.11 and 2.8). The computational step size of the curve in Fig. 2.11 is small ($l_x = 0.05$) and therefore even the narrow resonances can be clearly distinguished.

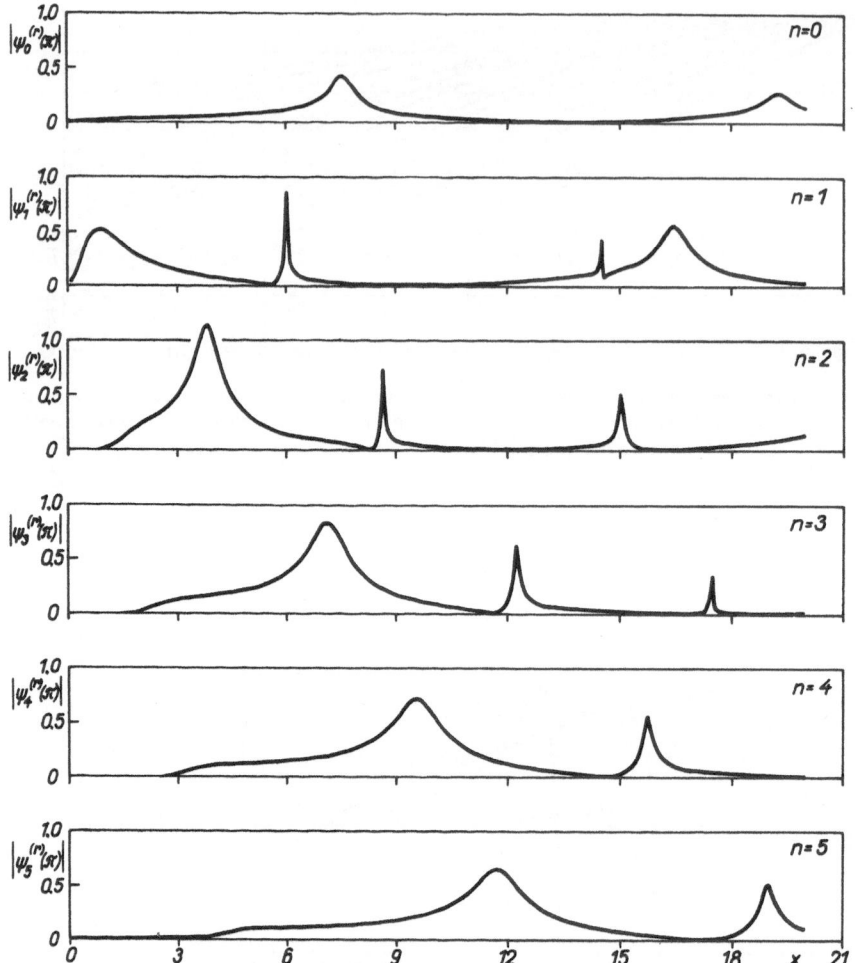

Fig. 2.11. Isolated modal resonances for scattering by an aluminum tube with $b/a = 0.2$ in water [2.14, Fig. 2]

In Fig. 2.12, the dependence of $|\psi_n^{(r)}(\pi)|$ on x is shown for the same thick-walled tube but over a broader range, $0 \leqslant x \leqslant 105$. Here, already at $n = 3$ eighteen families of resonances can be clearly observed.

The results presented in [2.16] concerning the scattering by an aluminum tube with $b/a = 2/3$ in water can serve as a good illustration to the procedure described in Sect. 2.3. In Fig. 2.13, for the modes $n = 1$–5 we show the dependence on x of the modulus of the partial-wave mode $|f_n(\pi)|$ (left column), the modulus of the partial-wave mode for scattering by an acoustically rigid cylinder $|f_n^{(r)}(\pi)|$ (central column), and the isolated modal resonance component $|\psi_n^{(r)}(\pi)|$ (right column). The central column represents the contribution of the background.

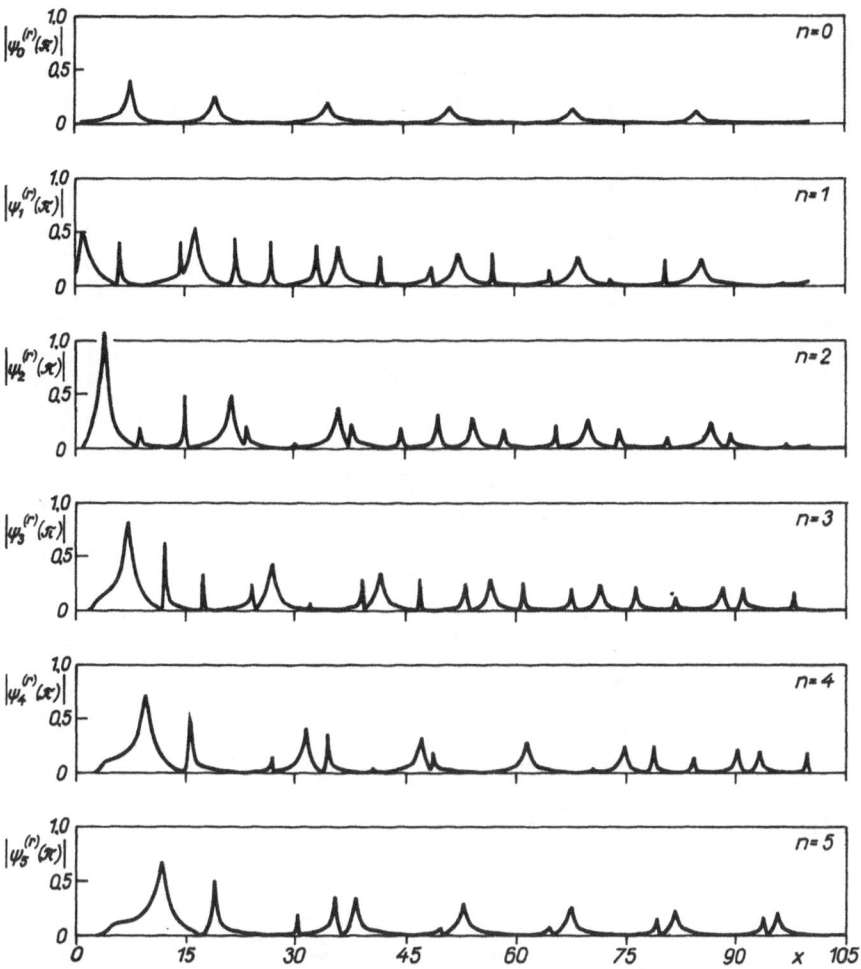

Fig. 2.12. Isolated modal resonances for scattering by an aluminum tube with $b/a = 0.2$ in water [2.14, Fig. 3]

The isolated modal resonances, so clearly distinguished in the right-hand column in Fig. 2.13, can be summed. The result of the summation corresponds to pure re-radiation, because the background is here completely excluded. In Fig. 2.14, the result of such a summation is presented. First, the real and the imaginary parts of the resonance components of all the partial-wave modes are summed separately, and then the modulus of the sum is computed. Each maximum on the curve corresponds to one of the modal resonance. The latter is labeled by two numbers (n, l). In the summation of the series (2.60) over n, the first eleven modes ($n = 0$–10) were taken into account.

The same resonances (n, l) are labeled on the curve in Fig. 2.15 where the total form function is presented. Here, in contrast to Fig. 2.14, thirty one terms were taken into account ($n = 0$–30).

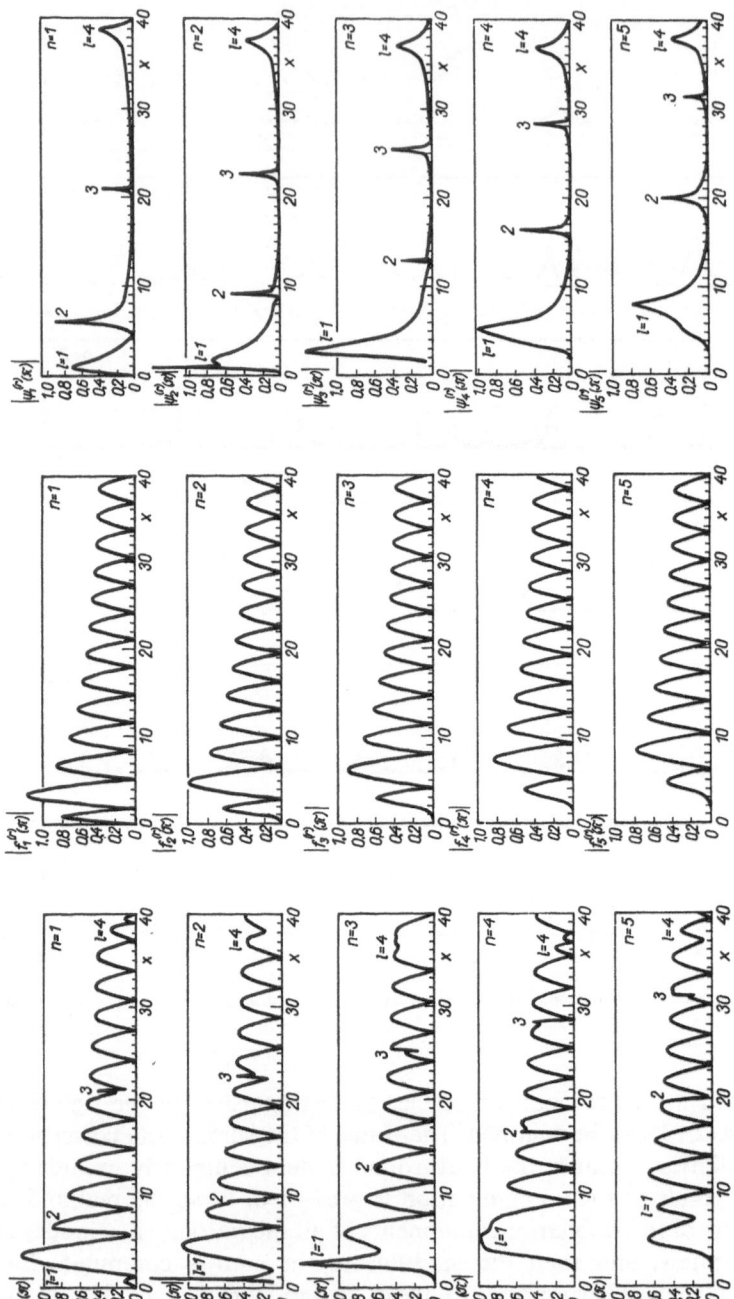

Fig. 2.13. Partial-wave form function $|f_n(\pi)|$ for scattering by an acoustically rigid cylinder (central column) and the isolated modal resonances $|\psi_n^{(\nu)}(\pi)|$ (right column). The central column represents the background component for each partial wave. Numbers l in the left- and right-hand columns indicate the family of the resonances; $l = 1$ corresponds to the Rayleigh wave and $l = 2, 3, 4$ to the Whispering Gallery waves [2.16, Figs. 12–14]

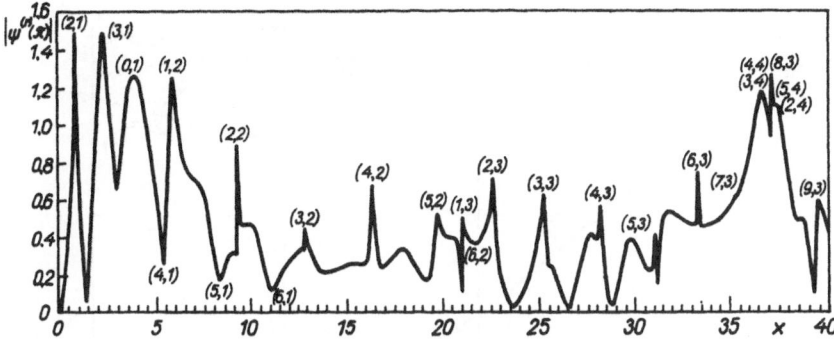

Fig. 2.14. Re-radiated form function $|\psi^{(r)}(\pi)|$ for scattering by an aluminum tube with $b/a = 2/3$ in water. Here the contribution of eleven ($n = 0$–10) partial waves is taken into account. Two numbers (n, l) identify the resonance [2.16, Fig. 15]

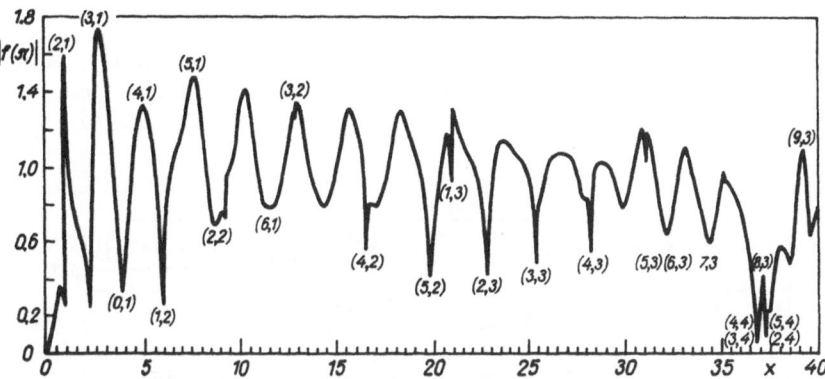

Fig. 2.15. Form function for scattering by an aluminum tube with $b/a = 2/3$ in water. Here the contribution of thirty one ($n = 0$–30) partial waves is taken into account. Two numbers (n, l) identify the resonance [2.16, Fig. 11]

In Fig. 2.16, the position of the poles in the complex x plane is shown. The poles are found from the equation $D_n = 0$, see (2.55). The index n marks the order of the mode along each branch l. Moving away from the origin, it takes on increasing values. For each peripheral wave, using a figure similar to the one presented in Fig. 2.16, the phase and group velocities can be calculated and the decay of the wave can be retraced. These poles are situated near the real x axis. They generate slowly decaying peripheral waves revolving around the tube. Besides this subsystem of poles, there is another one which, in the considered case of a very thickwalled tube (with $b/a = 2/3$), practically does not differ from the system of poles found for the scattering from an acoustically rigid cylinder. The latter poles are determined from the equation $H_n^{(1)\prime}(x) = 0$. They possess a large imaginary part and generate the "creeping" (Franz-type) waves, which

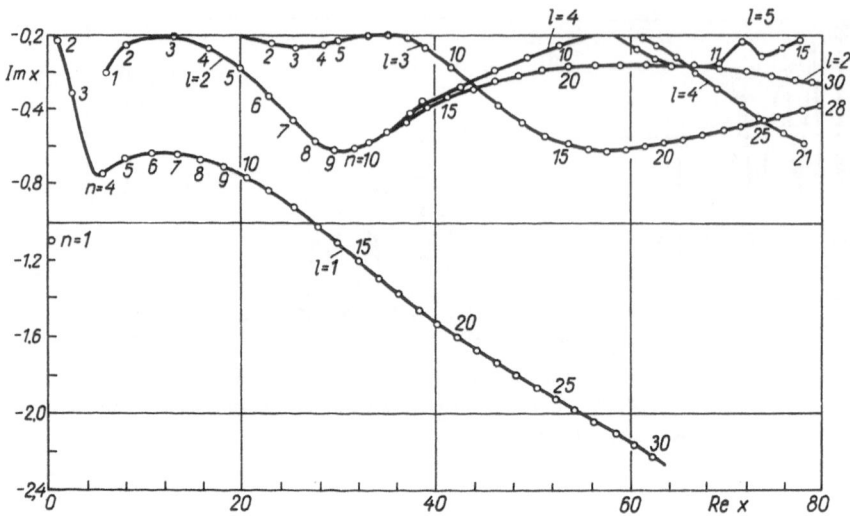

Fig. 2.16. Positions of the poles in the complex x plane for scattering by an aluminum tube with $b/a = 2/3$ in water. The poles are determined by the equation $D_n = 0$ (see (2.55)). The difference between the layer of poles $l = 1$ generating the Rayleigh wave and the other layers ($l \geq 2$) can clearly be observed. Along each layer the index n denotes the order of the mode [2.16, Fig. 10]

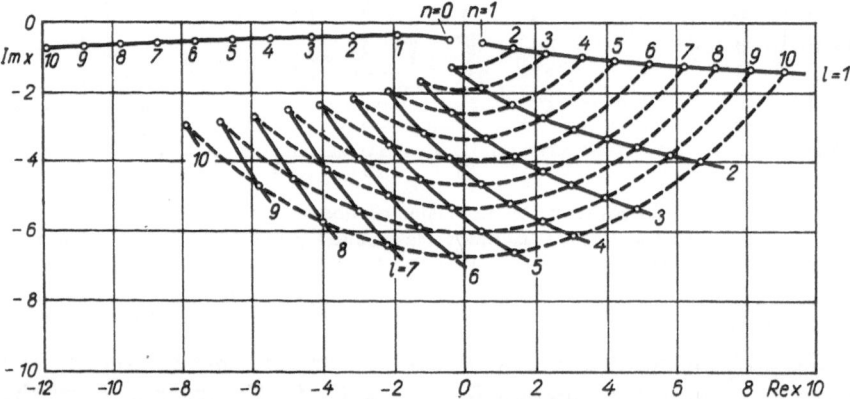

Fig. 2.17. The roots of the equation $H_n^{(1)\prime}(x) = 0$ in the restricted domain of the complex x plane [2.16, Fig. 1]

propagate on the outer surface of the tube in the liquid. Their influence on the total form function becomes apparent only at $x \lesssim 10$. The poles corresponding to the Franz-type waves on the acoustically rigid cylinder are shown in Fig. 2.17.

Using either the knowledge of isolated modal resonance (Fig. 2.13), or the poles (Fig. 2.16), an acoustic spectrogram can be obtained, which is a distinctive tag of the scatterer. In Fig. 2.18, the computed spectrogram is given. It is in good agreement with the results of experiments (Fig. 2.19). The latter were obtained using the method of resonance isolation and identification (Chap. 5). The measured spectrogram is contained as an element in the family of spectrograms shown in Fig. 5.11. In the acoustic spectrogram of Fig. 2.18, the calculated positions of resonances are labeled by arrows and the measured ones by circles. The unconfirmed positions of resonances in the experiment do not give rise to any doubts. The acoustic system used in the experiment did just not permit us to

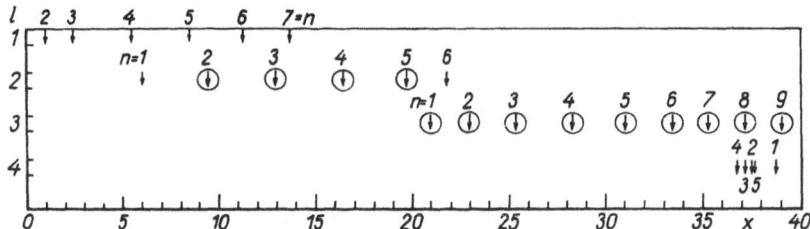

Fig. 2.18. Acoustic spectrogram for scattering by an aluminum tube with $b/a = 2/3$ in water. The computed positions of resonances are indicated by arrows and the measured positions by circles [2.16, Fig. 16]

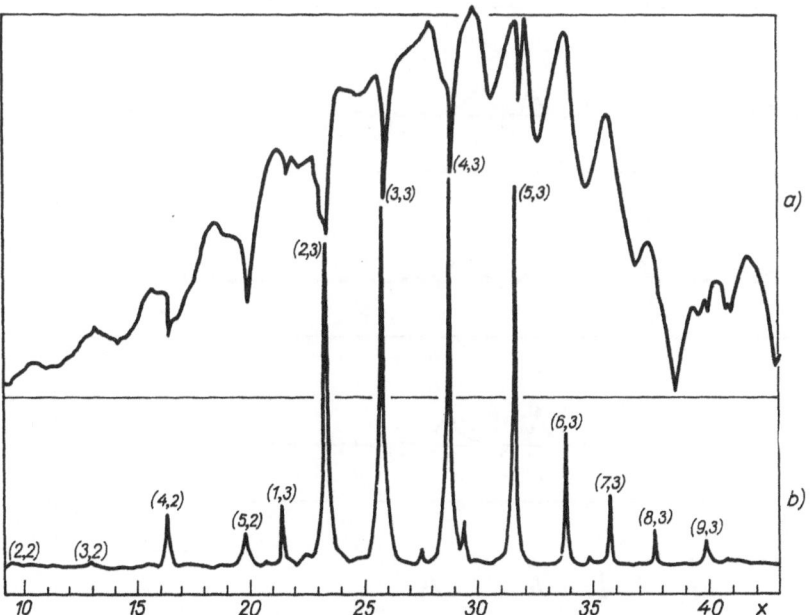

Fig. 2.19. Measured form functions of the acoustic pressure scattered by an aluminum tube with $b/a = 2/3$ in water. Curve **a** corresponds to steady-state scattering and curve **b** to transient scattering. Two numbers (n, l) identify the resonance [2.14, Fig. 3]

register them. It should be pointed out that due to the features of the experimental approach, in Fig. 2.19 on the ordinate axis a logarithm scale with an indefinite scale factor is used. Therefore the curves shown in Figs. 2.18 and 2.19 can be compared only at the resonance positions. Such a comparison is just what is needed to verify the correctness of the computed acoustic spectrogram.

The behavior of isolated resonances when the relative thickness of the shell is decreased is illustrated by the graphs of Figs. 2.20–22. In the case of shells with $b/a = 0.7$ and $b/a = 0.9$ the rigid background was subtracted; in the case of $b/a = 0.995$ ($h = 1/200$) the soft background was used. At fixed $n = n_0$ and in a restricted domain $x_a \leqslant x \leqslant x_b$ with h decreasing, the number of existing

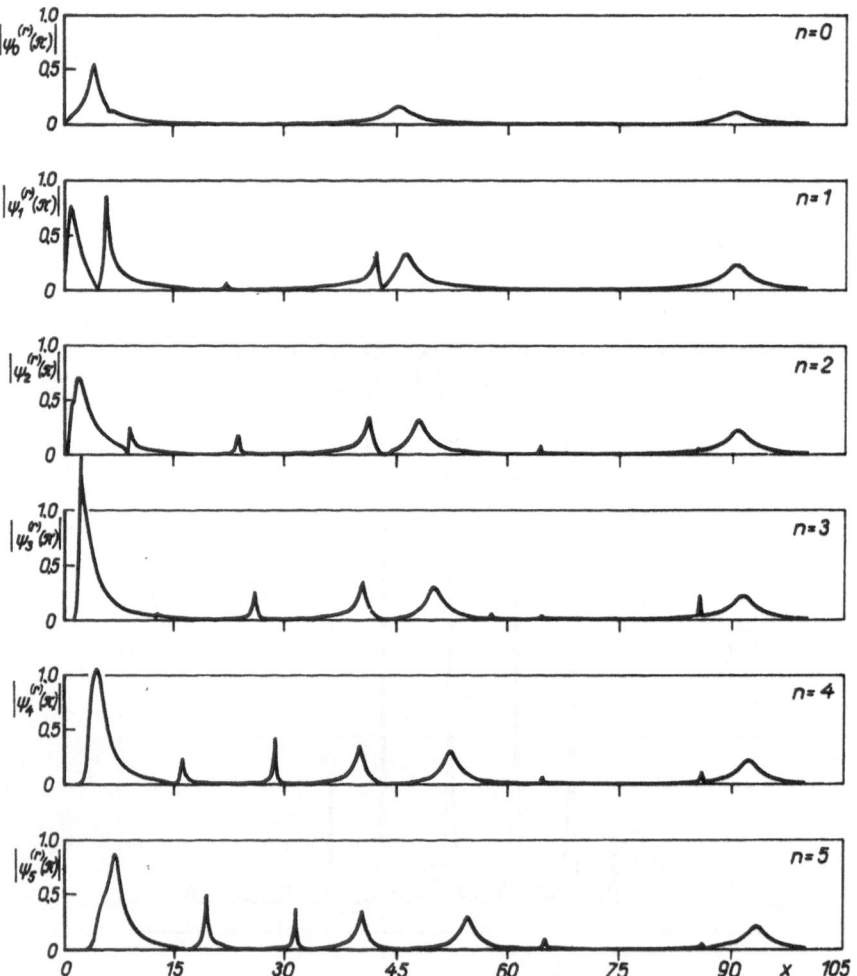

Fig. 2.20. Isolated modal resonances for scattering by an aluminum tube with $b/a = 0.7$ in water [2.14, Fig. 4]

resonances decreases, as can be seen from the presented plots. This fact can also be observed from the plots of the Regge poles (see Chap. 5). In the considered example of a very thin shell (with $h = 1/200$) in the range $0 \leqslant x \leqslant 105$, only one type of resonance remains. It corresponds to the zero-order symmetric Lamb-type wave S_0.

The range of the relative thickness of the shell $1/200 \lesssim h \lesssim 1/10$ is an intermediate one in the sense of the background (non-resonant) component. In the x range considered ($0 \leqslant x \leqslant 100$) its magnitude changes from that of the rigid background at $h = 1/10$ to that of the soft one at $h \sim 1/200$. Note that, just for this reason, the soft background was chosen in the case of the shell with

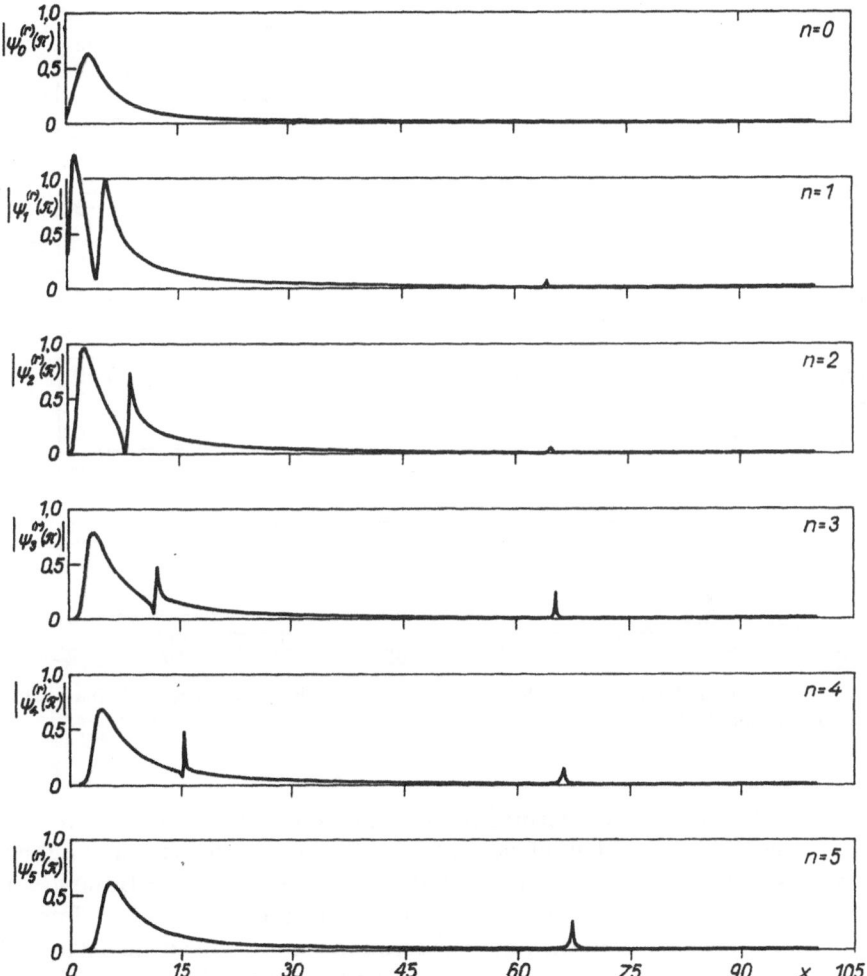

Fig. 2.21. Isolated modal resonances for scattering by an aluminum tube with $b/a = 0.9$ in water [2.14, Fig. 5]

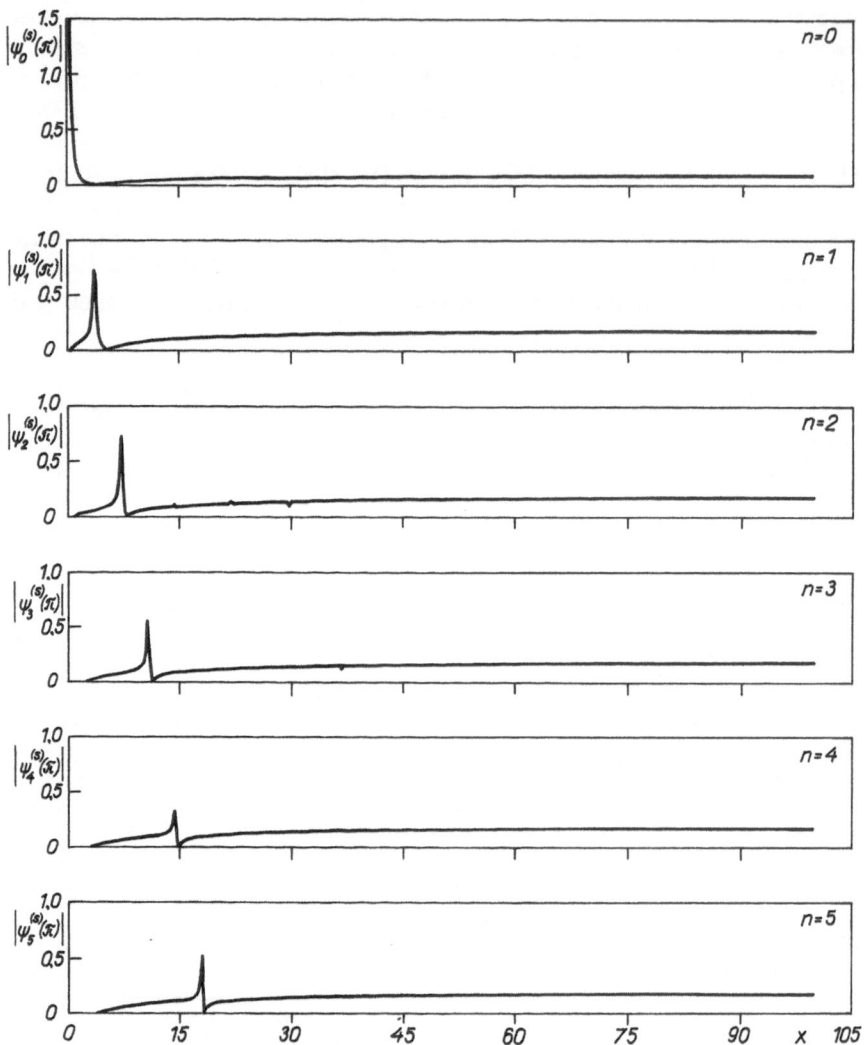

Fig. 2.22. Isolated modal resonances for scattering by an aluminum tube with $b/a = 0.995$ ($h = 1/200$) in water [2.14, Fig. 6]

$h = 1/200$. This fact can be interpreted physically in the following way. At such a small value of h, the incident plane wave perceives the empty shell as a bubble which behaves like an acoustically soft cylinder. For the modes $n = 0$ and $n = 1$, the change in the character of the background is illustrated by Fig. 2.23. For the case of a shell with $h = 1/10$ the rigid background is not quite adequate; for the shell with $h = 1/40$ the background coincides with neither a rigid nor a soft one, and for the shell with $h = 1/200$ the background nearly coincides with the soft one.

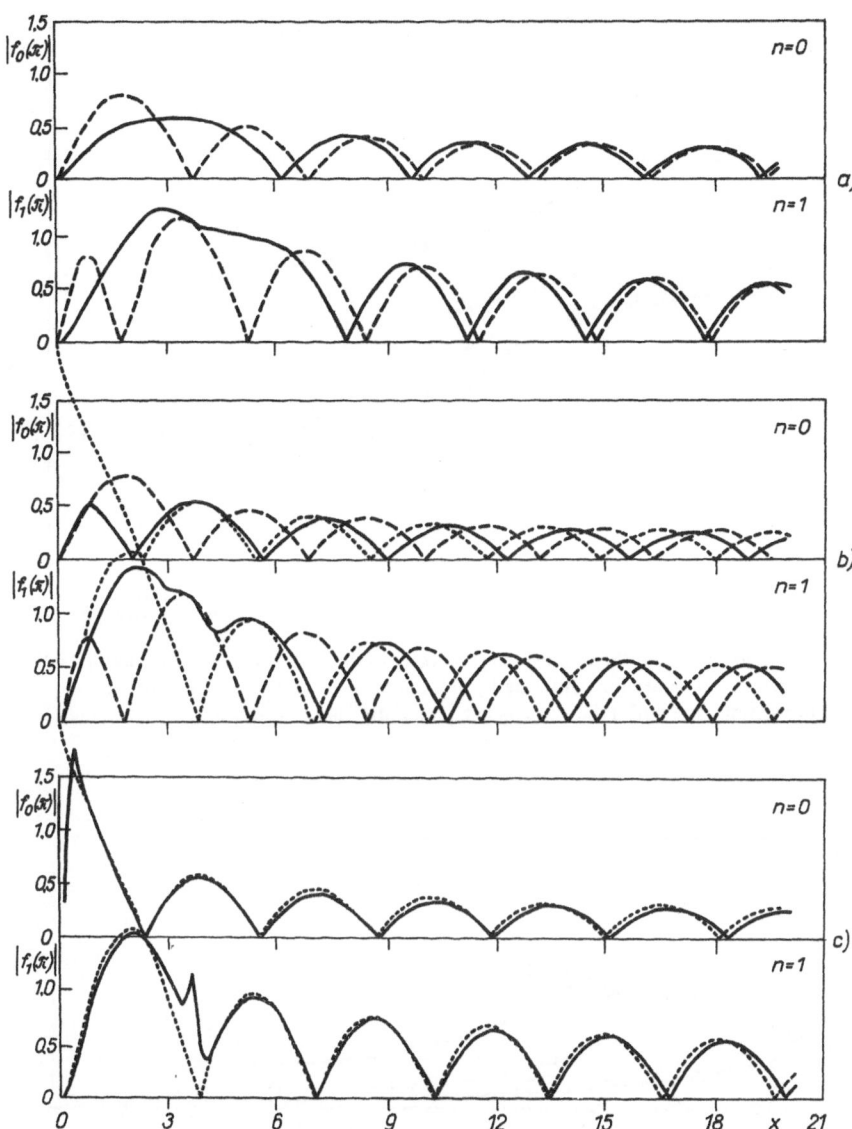

Fig. 2.23. Partial-wave form functions for scattering by aluminum shells (solid lines), by acoustically hard (dashed lines) and soft cylinders (points). The curves are shown for $n = 0$ and $n = 1$, and for three different relative thicknesses of the shell: $h = 1/10$ (**a**), $h = 1/40$ (**b**), and $h = 1/200$ (**c**) [2.14, Fig. 7]

From Fig. 2.23, it can be seen that in the intermediate region of thickness of the shell (as, for example, at $h = 1/40$), there is nevertheless a non-resonant background present which coincides with neither the rigid nor the soft one, but which is still smooth and quite distinct. Such a background was not considered

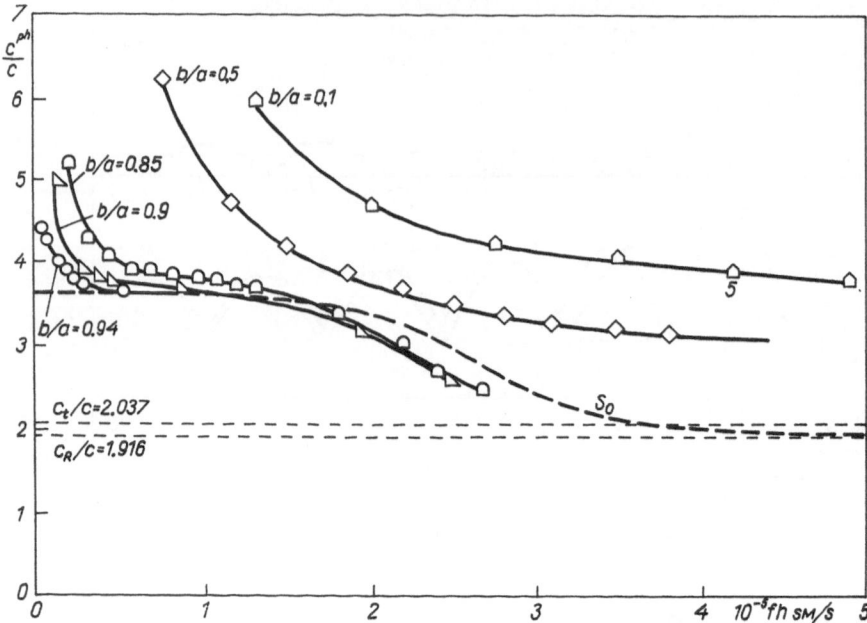

Fig. 2.24. Dispersion curves of the relative phase velocity of the first transverse Whispering Gallery wave T_1 at different b/a values for scattering by aluminum shells in water. The dispersion curve of the zero-order symmetric Lamb-type wave S_0 in a plane "dry" layer is shown as a dashed line [2.17, Fig. 9]

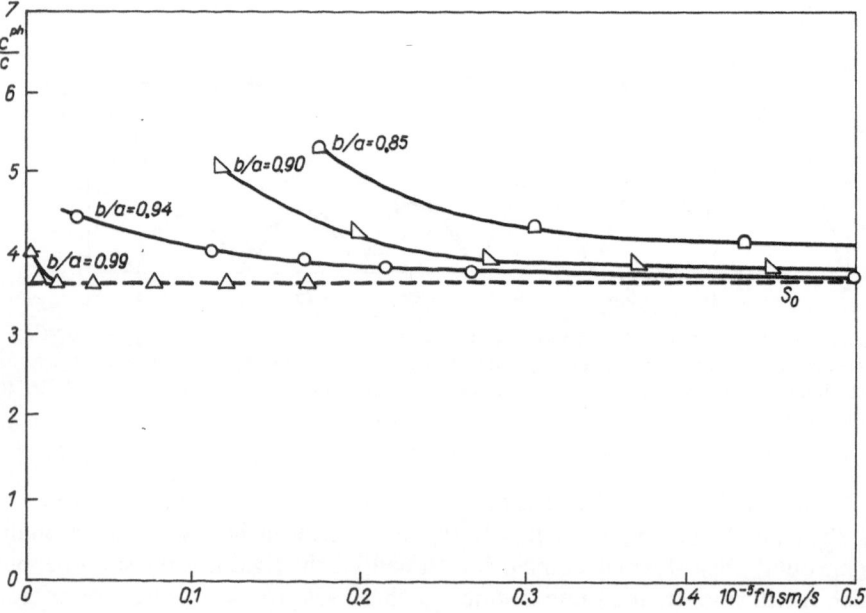

in Sect. 2.3 and we shall not dwell on this matter, referring instead to [2.14] where such an intermediate background is introduced and described.

In Fig. 2.24, the influence of the parameter b/a on the phase velocity of the first transverse Whispering Gallery wave T_1 is shown. The computation was carried out for the case of an aluminum shell in water. The dispersion curve of the zero-order symmetric Lamb-type wave S_0 in a plane "dry" layer is shown by a dashed line. In contrast to the above plots, the scale used here on the abscissa axis is

$$fh = \frac{xc}{2\pi}\left(1 - \frac{b}{a}\right). \tag{2.62}$$

In Fig. 2.25, the extreme left-hand portion of this plot on the fh axis is shown on a larger scale. At small values of fh, the dispersion curve of the S_0 wave bounds from below the dispersion curves of the T_1 wave for the different values of b/a. At large b/a (small h), the dispersion curve of the T_1 wave practically does not differ from that corresponding to the S_0 wave. This fact has been known for a long time, but here it obtains its graphical quantitative confirmation.

The author is grateful to Professor H. Überall for reprints of papers on the resonance scattering theory.

Fig. 2.25. Dispersion curves of the relative phase velocity of the first transverse Whispering Gallery wave T_1 at different b/a values for scattering by an aluminum shells in water. The dispersion curve of the zero-order symmetric Lamb-type wave S_0 in a plane "dry" layer is shown as a dashed line [2.17, Fig. 10]

3 Application of the Resonance Scattering Theory to Problems of Acoustic Wave Scattering by Elastic Spheres

An elastic body of spherical shape is the only one of all three-dimensional objects for which the exact solution of acoustic wave scattering can be obtained. Therefore, the problem of scattering from an elastic sphere has attracted the attention of investigators for a long time. The pioneering work by *Faran* [3.1] and *Hickling* [3.2–5] was followed by an impetuous stream of publications in which different, but very important, aspects of this problem were analyzed. The main attention was paid to the investigation of the form function and time dependence of the scattered acoustic pressure. The identification of the elastic body using information hidden in the scattered field was one of the ultimate aims.

A large amount of numerical results was obtained by summation of the *Rayleigh series* over eigenfunctions, and were interpreted mainly by applying the powerful tool of the *Sommerfeld–Watson transformation*. Both of these approaches were developed independently and some authors occasionally competed with each other, expressing critical, not always justified remarks concerning the method and the results obtained by supporters of the other orientation. In the so-called *Resonance Scattering Theory* (RST) the elements of both methods are present. The solution, as before, has a series form, but each term of the series is carefully analyzed. In particular, the resonance component and the background contained in it are isolated. The search for the resonance frequency of partial-wave modes from a characteristic equation, and conversion of the summation index n into a complex variable (theory of Regge poles) are closely connected with basic operations carried out in the Sommerfeld–Watson transformation. The advantages of both approaches are combined in the resonance scattering theory, and therefore the results obtained from it are so meaningful.

For each partial wave, using the resonance scattering theory, one can find the resonance frequency, determine the magnitude and the Q factor, relate the spectral lines of partial-wave modes to the resonances of peripheral (or "circumferential") waves, and find their phase and group velocities. By applying the resonance scattering theory it is possible to explain many of the phenomena observed in the form function and the time dependence of the secondary (scattered) acoustic field, including the angular diagram, and to indicate beforehand their properties. The resonance scattering theory can be used to obtain an acoustic spectrogram. The material presented below is consistent with that given in Chap. 2.

3.1   Scattering by a Solid Elastic Sphere

The problem of scattering of a plane acoustic pressure wave by a solid elastic sphere can be investigated similarly to that of scattering by a solid elastic cylinder. In fact, both of these problems have been considered in the same paper [3.6].

The acoustic pressure in a plane incident wave

$$p_i = p_* \exp[i(k\xi - \omega t)] \tag{3.1}$$

can be presented in the series form

$$p_i = p_* \exp(-i\omega t) \sum_{n=0}^{\infty} i^n (2n + 1) j_n(kr) P_n(\cos\theta) , \tag{3.2}$$

and the secondary (scattered) pressure can be written in a similar form [3.1, 2]

$$p_s = p_* \exp(-i\omega t) \sum_{n=0}^{\infty} i^n (2n + 1) R_n h_n^{(1)}(kr) P_n(\cos\theta) , \tag{3.3}$$

where

$$R_n = B_n/D_n . \tag{3.4}$$

The determinants of the third rank, B_n and D_n, and their elements are given in [3.1, 2, 7, 8]

$$B_n = \begin{vmatrix} d_{10} & d_{12} & d_{13} \\ d_{20} & d_{22} & d_{23} \\ 0 & d_{32} & d_{33} \end{vmatrix}, \qquad D_n = \begin{vmatrix} d_{11} & d_{12} & d_{13} \\ d_{21} & d_{22} & d_{23} \\ 0 & d_{32} & d_{33} \end{vmatrix}, \tag{3.5}$$

$$d_{11} = \frac{\rho}{\rho_1} x_2^2 h_n^{(1)}(x), \qquad d_{12} = [2n(n+1) - x_2^2] j_n(x_1) - 4x_1 j_n'(x_1) ,$$

$$d_{13} = 2n(n+1)[x_2 j_n'(x_2) - j_n(x_2)], \qquad d_{21} = -xh_n^{(1)\prime}(x) ,$$

$$d_{22} = x_1 j_n'(x_1), \qquad d_{23} = n(n+1) j_n(x_2) ,$$

$$d_{32} = 2[j_n(x_1) - x_1 j_n'(x_1)] ,$$

$$d_{33} = 2x_2 j_n'(x_2) + [x_2^2 - 2n(n+1) + 2] j_n'(x_2) ,$$

$$d_{10} = -\frac{\rho}{\rho_1} x_2^2 j_n(x), \qquad d_{20} = xj_n'(x), \qquad x_j = n_j x \quad (j = 1, 2) ,$$

$$n_1 = \frac{c}{c_l}, \qquad n_2 = \frac{c}{c_t} . \tag{3.6}$$

The partial-wave scattering function S_n and scattering phase shift δ_n are introduced according to the definition

$$S_n = 2R_n + 1 \equiv \exp(2i\delta_n) . \tag{3.7}$$

Introducing now the form function $f(\theta)$, which consists of partial form functions

$$f(\theta) = \sum_{n=0}^{\infty} f_n(\theta) , \tag{3.8}$$

$$f_n(\theta) = \frac{2}{x}(2n + 1) \exp(i\delta_n) \sin \delta_n P_n(\cos \theta) , \tag{3.9}$$

and using the asymptotic (at $kr \gg n$) representation of the spherical Hankel function

$$h_n^{(1)}(kr) \sim \frac{1}{kr} i^{-(n+1)} \exp(ikr) , \tag{3.10}$$

one can obtain the asymptotics of the scattered pressure in the far field

$$p_s \sim p_* \frac{a}{2r} \exp[i(kr - \omega t)] f(\theta) . \tag{3.11}$$

In (3.9) and (3.11) the notation

$$x \equiv ka, \qquad k = \omega/c \tag{3.12}$$

is used, where ω is the angular frequency, c is the sound velocity in the liquid, k is the wave number in the liquid, $x \equiv ka$ is the wave radius, a is the radius of the sphere, and p_* is a constant of dimension of pressure.

The partial scattering function S_n' , (3.7), can be represented either in the form

$$S_n = S_n^{(s)} \frac{L_n - z_n^{(2)}}{L_n - z_n^{(1)}}, \qquad S_n^{(s)} = -\frac{h_n^{(2)}(x)}{h_n^{(1)}(x)} \equiv \exp(2i\xi_n^{(s)}) , \tag{3.13}$$

or in the form

$$S_n = S_n^{(r)} \frac{L_n^{-1} - [z_n^{(2)}]^{-1}}{L_n^{-1} - [z_n^{(1)}]^{-1}}, \qquad S_n^{(r)} = -\frac{h_n^{(2)\prime}(x)}{h_n^{(1)\prime}(x)} \equiv \exp(2i\xi_n^{(r)}) , \tag{3.14}$$

where the following notations are used:

$$L_n = -\frac{\rho}{\rho_1} x_2^2 \frac{D_n^{(1)}}{D_n^{(2)}}, \qquad D_n^{(1)} = \begin{vmatrix} d_{22} & d_{23} \\ d_{32} & d_{33} \end{vmatrix}, \qquad D_n^{(2)} = \begin{vmatrix} d_{12} & d_{13} \\ d_{32} & d_{33} \end{vmatrix},$$

$$z_n^{(1)} = x\frac{h_n^{(1)\prime}(x)}{h_n^{(1)}(x)}, \qquad z_n^{(2)} = x\frac{h_n^{(2)\prime}(x)}{h_n^{(2)}(x)} , \tag{3.15}$$

$$\tan \xi_n^{(s)} = \frac{j_n(x)}{y_n(x)}, \qquad \tan \xi_n^{(r)} = \frac{j_n'(x)}{y_n'(x)} . \tag{3.16}$$

The quantities $S_n^{(s)}$ and $S_n^{(r)}$ [see (3.13) and (3.14)] are called partial-wave scattering functions for soft and rigid spheres, respectively. In these formulae L_n is a real quantity depending on spherical Bessel functions with arguments x_1 and x_2 (see (2.6)) and proportional to the quotient ρ/ρ_1, where ρ is the density of the ambient liquid and ρ_1 is the density of the material of the sphere. For $\rho \gg \rho_1$ there follows from L_n a limiting case of the partial-wave scattering function S_n , namely that corresponding to the acoustically soft sphere; for $\rho_1 \gg \rho$ we have the limiting case corresponding to the acoustically rigid sphere.

Since the expressions (3.13) and (3.14), obtained for the case of scattering by a sphere, are formally equivalent to the expressions (2.18) and (2.19) for the scattering problem by a cylinder, all the relations of the resonance scattering theory from Chap. 2 can be applied here with only one amendment, connected with the difference of functions $z_n^{(1)}$ and $z_n^{(2)}$. For scattering by a sphere, instead of (2.28) and (2.29) one should use either

$$\Delta_n^{(s)} = x \frac{j_n(x)j_n'(x) + y_n(x)y_n'(x)}{j_n^2(x) + y_n^2(x)} ,$$

$$s_n^{(s)} = \frac{1}{x} \frac{1}{j_n^2(x) + y_n^2(x)} ,$$

(3.17)

or

$$\Delta_n^{(r)} = \frac{1}{x} \frac{j_n(x)j_n'(x) + y_n(x)y_n'(x)}{[j_n'(x)]^2 + [y_n'(x)]^2} ,$$

$$s_n^{(r)} = -\frac{1}{x^3} \frac{1}{[j_n'(x)]^2 + [y_n'(x)]^2} .$$

(3.18)

Continuing the resonance formalism in these terms and expanding the new function $L_n(x)$ in a Taylor series, as was done in (2.31), one finds in the case of a soft background

$$f_n(\theta) = \frac{2}{x} (2n + 1) \exp(2i\zeta_n^{(s)}) \left(\sum_{l=0}^{\infty} \frac{\frac{1}{2}\Gamma_{nl}^{(s)}}{x_{nl}^{(s)} - x - \frac{1}{2}i\Gamma_{nl}^{(s)}} \right. $$
$$\left. + \exp(-i\zeta_n^{(s)}) \sin \zeta_n^{(s)} \right) P_n(\cos \theta) ,$$

(3.19)

and in the case of a rigid background

$$f_n(\theta) = \frac{2}{x} (2n + 1) \exp(2i\zeta_n^{(r)}) \left(\sum_{l=0}^{\infty} \frac{\frac{1}{2}\Gamma_{nl}^{(r)}}{x_{nl}^{(r)} - x - \frac{1}{2}i\Gamma_{nl}^{(r)}} \right. $$
$$\left. + \exp(-i\zeta_n^{(r)}) \sin \zeta_n^{(r)} \right) P_n(\cos \theta) .$$

(3.20)

Thus, for scattering by an elastic sphere, the partial-wave scattering function $f_n(\theta)$ consists of isolated resonances, superimposed on a background corresponding to the sound-impenetrable sphere: either acoustically soft or acoustically rigid. Therefore, all the consequences following from this fact and described above in the case of scattering by a cylinder can be fully applied in the case of scattering by a sphere.

In computations, the exact expressions are generally used instead of the approximate asymptotic formulae (3.19), (3.20). In the far field, the difference between the acoustic pressure scattered by the elastic sphere p_s and the acoustically soft sphere $p_s^{(s)}$ can be represented as

$$p_s - p_s^{(s)} = p_* \frac{a}{2r} \exp[i(kr - \omega t)] \psi^{(s)}(\theta) , \tag{3.21}$$

where

$$\psi^{(s)}(\theta) = \sum_{n=0}^{\infty} \psi_n^{(s)}(\theta) , \tag{3.22}$$

$$\psi_n^{(s)}(\theta) = -\frac{2i}{x}(2n + 1)\left[R_n + \frac{j_n(x)}{h_n^{(1)}(x)}\right]P_n(\cos\theta) . \tag{3.23}$$

Analogously, in the far field the difference between the pressure scattered by the elastic sphere p_s and the acoustically rigid sphere $p_s^{(r)}$ can be represented in the form

$$p_s - p_s^{(r)} = p_* \frac{a}{2r} \exp[i(kr - \omega t)] \psi^{(r)}(\theta) , \tag{3.24}$$

where

$$\psi^{(r)}(\theta) = \sum_{n=0}^{\infty} \psi_n^{(r)}(\theta) , \tag{3.25}$$

$$\psi_n^{(r)}(\theta) = -\frac{2i}{x}(2n + 1)\left[R_n + \frac{j_n'(x)}{h_n^{(1)\prime}(x)}\right]P_n(\cos\theta) . \tag{3.26}$$

Isolated modal resonances are investigated either according to $\psi_n^{(s)}$, (3.23), or according to $\psi_n^{(r)}$, (3.26), computed at fixed $\theta = \theta_0$.

3.2   Numerical Results for Scattering by a Solid Elastic Sphere

In Fig. 3.1, the form function modulus $|f(\pi; x)|$ is presented. The computation was carried out for the case of an aluminum sphere immersed in water. The physical parameters are defined by (2.49). In Fig. 3.1, the resonances are labeled by two numbers (n, l). The first determines the ordinal number (order) of the resonance and the second defines its type, i.e. it indicates the type of peripheral wave revolving around the sphere; $l = 1$ corresponds to the Rayleigh wave, and $l = 2, 3, 4, \ldots$ to the Whispering Gallery waves.

In Fig. 3.2, the dependence of $|f_n(\pi)|$ on x is given for the first six partial modes ($n = 0-5$). Here the number l determines the position of the resonance frequency of the partial wave.

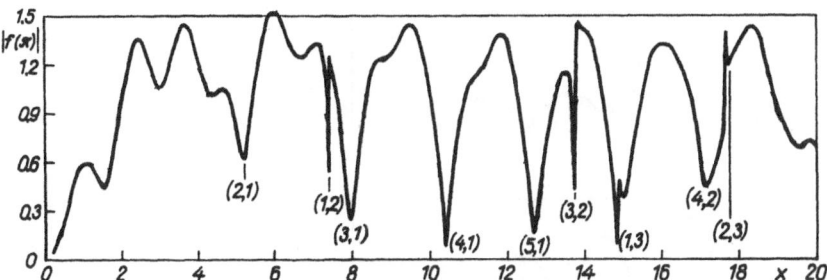

Fig. 3.1. The form function of the acoustic pressure backscattered by a solid aluminum sphere in water. Two numbers (n, l) denote the resonances of the peripheral waves [3.6, Fig. 1b]

Figures 3.1 and 3.2 are mainly presented to clearly show the difference in the total form function and partial-wave form function for scattering by a cylinder and a sphere (compare Fig. 2.1 with Fig. 3.1 and Fig. 2.2 with Fig. 3.2).

In Fig. 3.3, the form function modulus $|f(\pi; x)|$ (see (3.8)) is shown for scattering by a tungsten carbide sphere. The computation was carried out with the following parameters:

tungsten carbide: $\rho_1 = 13.8 \times 10^3 \, \text{kg/m}^3$, $c_l = 6860 \, \text{m/s}$, $c_t = 4185 \, \text{m/s}$,

water: $\rho = 1 \times 10^3 \, \text{kg/m}^3$, $c = 1476 \, \text{m/s}$.   (3.27)

The information on the resonances, hidden in Fig. 3.3, is obtained using the formalism of the resonance scattering theory. To this end the plots of the isolated modal resonances $|\psi_n^{(r)}(\pi; x)|$ [see (3.26)] are computed. In Fig. 3.4, the separation of each partial mode (left-hand column) into background (central column) and resonances (right-hand column) for the first seven partial modes $(n = 0–6)$ is shown. In the left-hand column the position of the resonance is marked by an arrow and the two numbers (n, l); in the right-hand column the position of the partial mode resonances is marked by the index l. In the left-hand column one can see for each mode the resonance irregularities ("notches"). In the central column, such irregularities are absent, as was supposed previously. Therefore, the graph in this column can be used as a background. The coherent differences between the complex quantities whose moduli are presented in the left-hand and central columns, are shown by an absolute value in the right-hand column. Here the resonances can be clearly observed, while in the left-hand column they can only be guessed. Each isolated modal resonance is located at some specific frequency and can be characterized by its magnitude and Q factor. For each resonance the position, magnitude and Q factor are connected with the inner structure and the external shape of the elastic scatterer.

The positions of isolated modal resonances in the right-hand column in Fig. 3.4 correspond exactly to the "notches", marked by arrows in the left-hand column. They are defined by the real parts of the coordinates of the poles, which are found from the equation $D_n = 0$, see (3.5). These poles are given in Table 3.1.

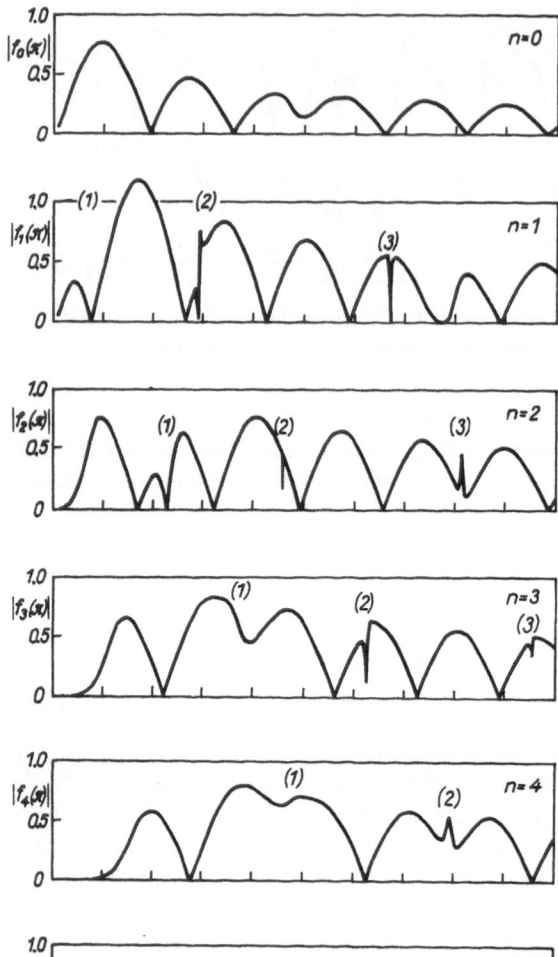

Fig. 3.2. The form functions of partial-wave modes for backscattering by a solid aluminum sphere in water, for the first six modes ($n = 0$–5). The number l labels the position of the eigenfrequencies of the partial waves [3.6, Fig. 2b]

In the limiting case of $\rho_1 \gg \rho$, the equation $D_n = 0$ can be represented in the form

$$d_{21}(d_{12}d_{33} - d_{32}d_{13}) = 0 . \tag{3.28}$$

This equation splits into two equations

$$d_{21} = 0, \quad \text{i.e. } h_n^{(1)\prime}(x) = 0 , \tag{3.29}$$

$$d_{12}d_{33} - d_{32}d_{13} = 0 . \tag{3.30}$$

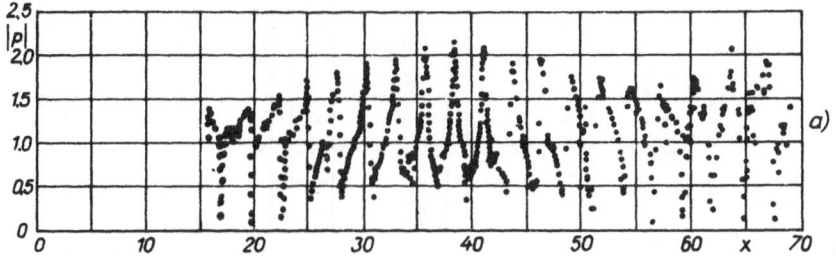

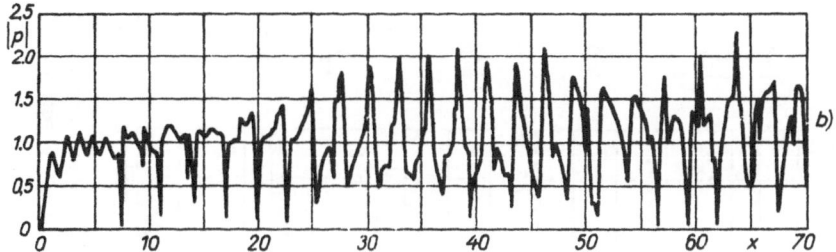

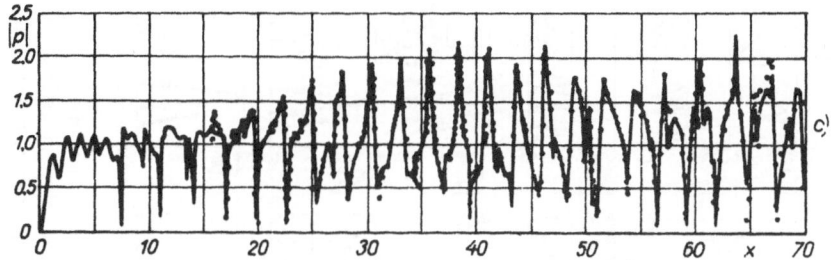

Fig. 3.3. The form function of the acoustic pressure backscattered by a solid tungsten carbide sphere in water: **a** result of measurements, **b** result of computation, **c** comparison of results [3.11, Figs. 2, 4, 5]

At fixed real x, only complex n values can be the roots of (3.29). These roots correspond to the Franz-type waves that propagate in the liquid surrounding the sphere. At fixed real (integer) n, only complex roots of (3.29) can exist. In Fig. 3.5, the position of the roots of (3.29) is shown in the complex x plane.

Equation (3.30) defines the eigenfrequencies of the elastic sphere. It was first obtained by Love in [3.9], not for the problem of acoustic wave scattering, but for the problem of mechanical vibrations of a sphere. It describes only one type of vibration in the scattering problem considered, namely the spheroidal vibrations. In Table 3.2 the first eight roots of (3.30) are given (the fundamental frequency and seven overtones). The computation is carried out for the case of

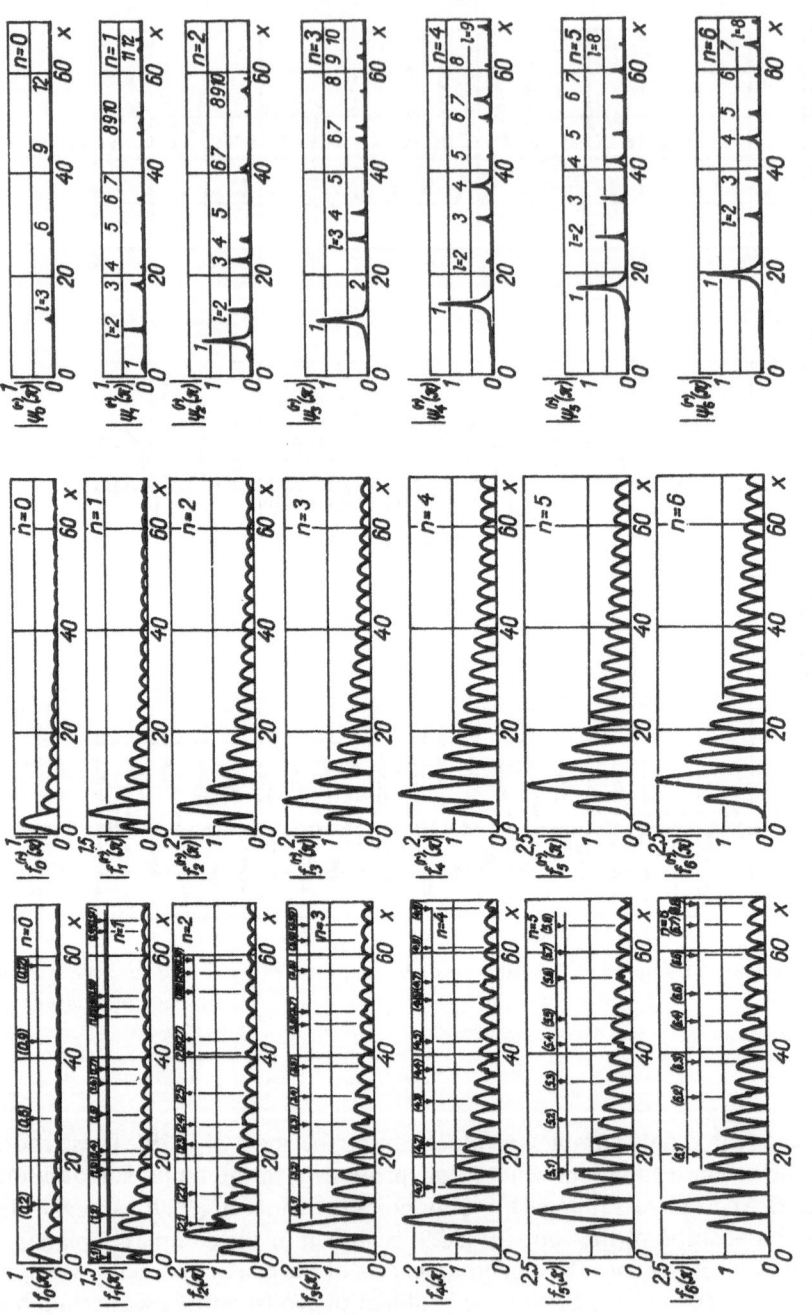

Fig. 3.4. Separation of the partial-wave modes into the background and the resonance components. The computation is carried out for backscattering by a solid tungsten carbide sphere in water for the first seven ($n = 0$–6) partial-wave modes. The total contribution of the partial-wave mode $|f_n(\pi; x)|$ is shown in the left-hand-column [3.8, Fig. 3]

Table 3.1. The roots of the equation $D_n = 0$ (see (3.5)) computed for the problem of scattering by a solid tungsten carbide sphere in water. The material parameters are given by (3.27). These roots are connected with the resonances of a solid sphere immersed in a liquid [3.8, Table IV]

l	$n = 0$	$n = 1$	$n = 2$	$n = 3$
1			7.454188 − i 0.080989	11.031209 − i 0.011226
2	6.971354	9.413069 − i 0.019219	13.405029 − i 0.004907	17.873797 − i 0.000017
3	11.459866 − i 0.114176	18.270501 − i 0.081086	22.982332 − i 0.054967	26.984220 − i 0.038497
4	16.897056	21.634283 − i 0.012550	26.852404 − i 0.042597	32.114726 − i 0.063452
5	26.116453	30.268030 − i 0.013539	34.358590 − i 0.000305	38.449662 − i 0.000235
6	28.028774 − i 0.077151	34.944002 − i 0.075475	41.111731 − i 0.070004	46.340030 − i 0.041577
7	35.177081	39.520633 − i 0.000534	43.936059 − i 0.008210	48.857674 − i 0.039020
8	44.175834	50.139928 − i 0.069096	52.557137 − i 0.001698	56.682811 − i 0.000341

l	$n = 4$	$n = 5$	$n = 6$	$n = 7$
1	14.086893 − i 0.134169	16.950044 − i 0.152263	19.723304 − i 0.168743	22.447117 − i 0.184422
2	22.422393 − i 0.003919	26.903325 − i 0.012046	31.259397 − i 0.022432	35.458506 − i 0.033807
3	30.769056 − i 0.028428	34.492196 − i 0.020451	38.223306 − i 0.013148	42.005683 − i 0.006611
4	37.120882 − i 0.074038	41.786274 − i 0.075450	46.116010 − i 0.069402	50.181108 − i 0.059958
5	42.615506 − i 0.003756	46.915162 − i 0.013521	51.351800 − i 0.028974	55.860119 − i 0.045438
6	50.720763 − i 0.018071	54.794431 − i 0.007272	58.781370 − i 0.002791	62.757778 − i 0.000267
7	54.295274 − i 0.064070	59.743283 − i 0.073189	64.960245 − i 0.071475	69.814976 − i 0.058342
8	60.786423 − i 0.000088	64.924286 − i 0.002346	69.181709 − i 0.010623	73.662435 − i 0.028144

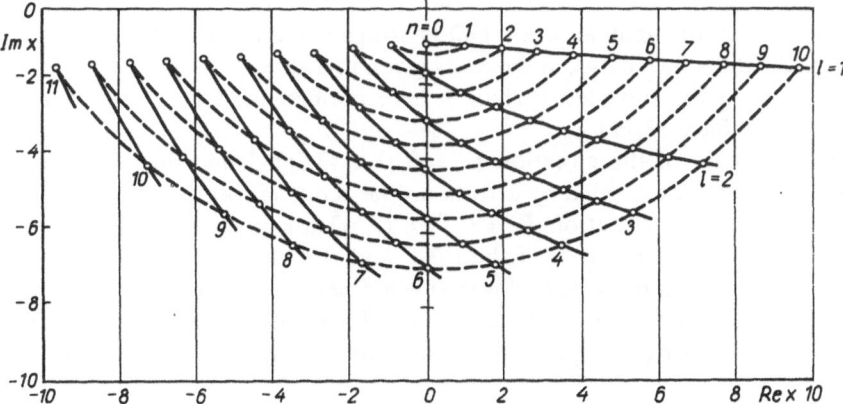

Fig. 3.5. Positions of the roots of the equation $h_n^{(1)\prime}(x) = 0$ in the complex x plane [3.17, Fig. 1]

a tungsten carbide sphere, see (3.27). Some of these roots are presented in the early paper [3.10].

For each root, as can be seen from Table 3.1, the real part is much larger than the imaginary one. The comparison of data from Tables 3.1 and 3.2 shows that the real part of each root of the equation $D_n = 0$ [see (3.5)] is close to the root of (3.30). A similar situation is typical for the scattering by metallic spheres

Table 3.2. The roots of (3.30), computed for a solid tungsten carbide sphere. They are purely real and correspond to the spheroidal vibrations of an elastic sphere [3.8, Table II]

l	$n = 0$	$n = 1$	$n = 2$	$n = 3$
1			7.466222	11.042275
2	6.971354	9.415125	13.405441	17.873798
3	11.469727	18.274795	22.454267	26.985318
4	16.897056	21.635142	26.854539	32.117101
5	26.116453	30.268062	34.358598	38.449670
6	28.031617	34.946245	41.113313	46.340369
7	35.177080	39.520652	43.936434	48.859091
8	44.175834	50.141564	52.557149	56.682916

l	$n = 4$	$n = 5$	$n = 6$	$n = 7$
1	14.097086	16.959475	19.732098	22.455370
2	22.422602	26.903800	31.260254	35.459658
3	30.769789	34.492697	38.223623	42.005843
4	37.123095	41.788080	46.117305	50.181985
5	42.615629	46.915602	51.352663	55.861310
6	50.720864	54.794504	58.781406	62.757782
7	54.296769	59.744615	64.961262	69.814489
8	60.786425	64.924344	69.182006	73.663181

in water. In the example considered, the sphere material is tungsten carbide, commonly used for manufacturing ball-bearings being a rather hard material, therefore the difference between the roots given in Tables 3.1 and 3.2 is so small. For this reason the tungsten carbide sphere is often used in experiments as a calibration object [3.11].

The real parts of the equation $D_n = 0$ [see (3.5)] define the positions of resonances. These are labeled by arrows on the acoustic spectrogram (Fig. 3.6).

The roots of the equation $B_n = 0$ [see (3.5)] are presented in Table 3.3. Comparison of data given in Tables 3.1 and 3.3 shows that the roots of equations $D_n = 0$ and $B_n = 0$ are close to each other, with the possible exception of the roots $l = 1$, which are connected with the Rayleigh wave. According to (3.4), the coefficient R_n is a quotient of two determinants and can be written in the form

$$R_n = \frac{B_n}{D_n} = \frac{(x - \hat{x}_{n1})(x - \hat{x}_{n2}) \ldots (x - \hat{x}_{nl})}{(x - x_{n1})(x - x_{n2}) \ldots (x - x_{nl})}, \tag{3.31}$$

where x_{nl} are the roots of $D_n = 0$ (Table 3.1) and $\hat{x}_{nl}$ are the roots of $B_n = 0$ (Table 3.3), calculated at fixed n values. For all $l \neq 1$ the equality $x_{nl} \approx \hat{x}_{nl}$ applies. Therefore the quotient (3.31) can be simplified to

$$R_n \cong \frac{x - \hat{x}_{n1}}{x - x_{n1}}. \tag{3.32}$$

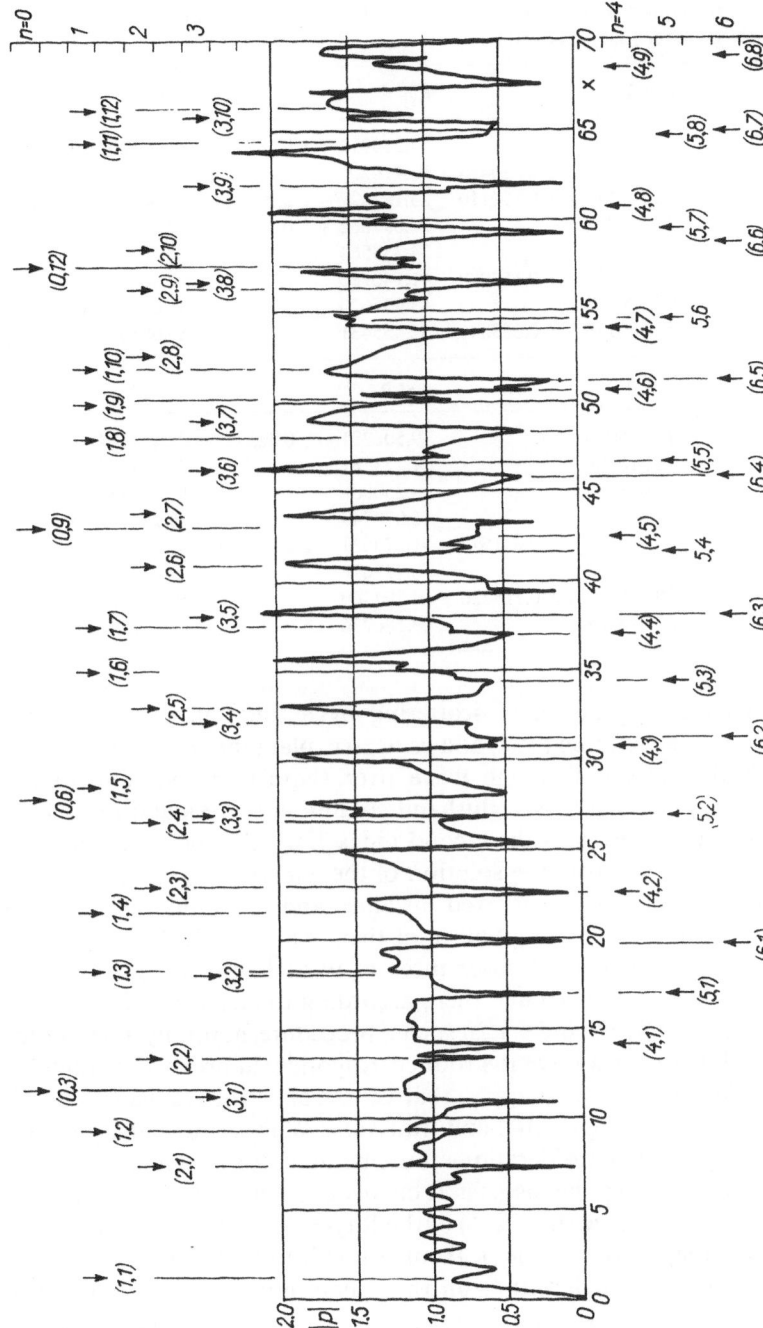

Fig. 3.6. The form function of the acoustic pressure and the acoustic spectrogram for backscattering of a solid tungsten carbide sphere in water. Two numbers (n, l) label the resonances of the peripheral waves–spectral lines [3.8, Fig. 4]

Table 3.3. The roots of the equation $B_n = 0$ (see (3.5)) computed for the problem of scattering by a solid tungsten carbide sphere in water [3.8, Table III]

l	$n = 0$	$n = 1$	$n = 2$	$n = 3$
1			7.375072 + i 0.288815	11.143492
2	6.971354	9.310693 − i 0.095453	13.416205	17.873824
3	11.193152 − i 0.190962	18.501618 + i 0.156110	23.187827 − i 0.105708	27.008103
4	16.897056	21.707322	26.697662 + i 0.135349	32.061206
5	26.116453	30.268735	34.358564	38.449490
6	28.010988	34.726717 − i 0.159080	41.140975	46.402945
7	35.177080	39.520806	43.936620	48.871284
8	44.175834	50.180119 − i 0.263045	52.555431	56.681593

l	$n = 4$	$n = 5$	$n = 6$	$n = 7$
1	13.851777	17.095015	19.508281 − i 0.363345	22.489914
2	22.427724	26.916126	31.273761	35.464029
3	30.760122	34.433913	38.250886	42.007645
4	37.097855	41.766048	46.070749	49.982849
5	42.608000	46.842766 − i 0.091676	51.374521 − i 0.170501	55.940225 + i 0.200978
6	50.735149	54.795235	58.779696	62.757123
7	54.529014 − i 0.07905	59.585117 + i 0.218726	64.894940	69.783835
8	60.787718	64.926607	69.186236	73.669591

For the considered problem of scattering by a tungsten carbide sphere immersed in water, the relation (3.32) allows us to explain the strong superiority of the contribution of the Rayleigh wave over those corresponding to the Whispering Gallery waves (cf. the right-hand column in Fig. 3.4 and Fig. 3.8A).

It should be mentioned that the second index l, which labels the family of resonances, was obtained from the solution of the equation $D_n = 0$ [see (3.5)]: the computation is carried out at fixed n values and the roots obtained are numbered according to increasing values of their real part. In fact, the same applies for the index l when the resonances of partial-wave modes are considered. First the n value is fixed and then, according to increasing x value, the resonances are labeled by the index l. In such a procedure, naturally, one cannot indicate to which peripheral wave the root corresponds. In order to distinguish the resonances for different wave types, that is, to isolate the resonances of the Rayleigh wave, and the longitudinal and transverse Whispering Gallery waves, some additional physical considerations must be used, for example, regarding the asymptotic behavior of the dispersion curves at small and large x values.

In Fig. 3.4, five resonances ($n = 2$–6) of the Rayleigh wave ($l = 1$) are shown. As is well known, the "dipole" mode (1, 1) corresponds to the motion of the body as a whole, and is generated neither on the sphere nor on the cylinder. The higher-order resonances of this wave are shown in Fig. 4.4 in a pure form, i.e. without any background.

In Fig. 3.7, the real parts of the roots of the equation $D_n = 0$ [see (3.5)] are given in the form of "Regge trajectories" (resonance frequencies versus mode

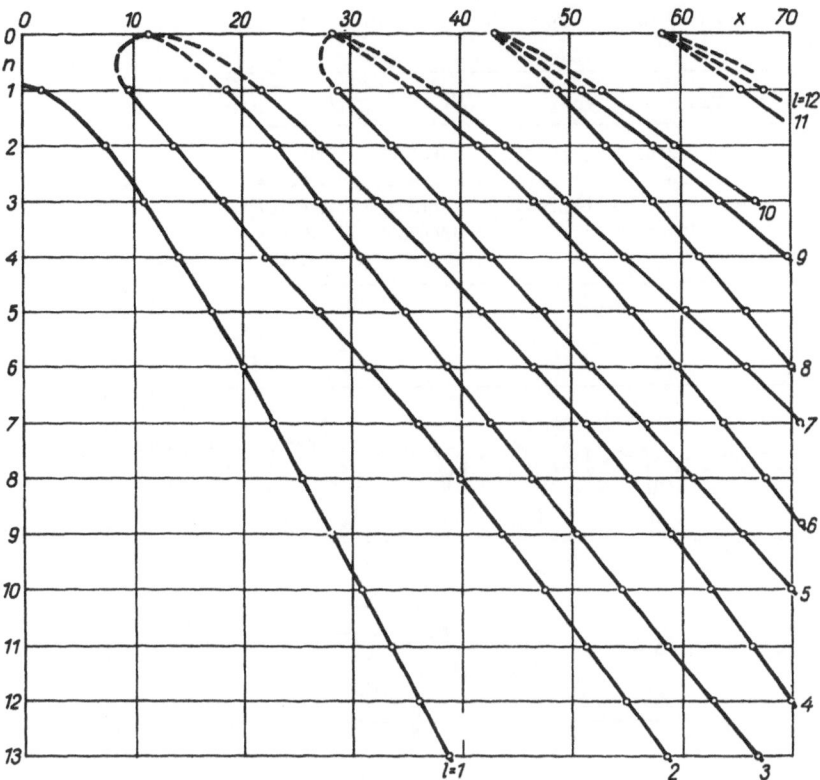

Fig. 3.7. The Regge trajectories of the peripheral waves in the backscattering from a solid tungsten carbide sphere [3.8, Fig. 7]

number). Using the Regge trajectories, one can find the relative phase and group velocities of each peripheral wave

$$\frac{c_l^{\mathrm{ph}}(x)}{c} = \frac{1}{\mathrm{Re}\{v_l + \frac{1}{2}\}}, \qquad \frac{c_l^{\mathrm{gr}}(x)}{c} = \frac{1}{\mathrm{Re}\left\{\dfrac{\partial v_l}{\partial x}\right\}}, \tag{3.33}$$

where (x, v_l) are the coordinates of a point along the lth Regge trajectory.

Form functions similar to those presented in Fig. 3.3 can be computed for θ values different from π. As indicated in [3.8], it is interesting to compute them at those observation points which coincide with zeroes of the Legendre polynomials given in Table 3.4. The observation angles $\theta = 140.7685°$, $\theta = 125.2644°$ and $\theta = 90°$ were used in the computation. In Fig. 3.8, the corresponding results are presented.

As can be seen from Table 3.4 and Fig. 3.8, in the domain $0 \leqslant x \leqslant 25$ and for the case B) (at $\theta = 140.7685°$) the contribution of the third mode has disappeared; on the graph the "notches" corresponding to the resonances (3, 1),

Table 3.4. The roots of the equation $P_n(\cos\theta) = 0$ for $n = 1$–5 [3.8, Table I]

$l \backslash P_n$	P_1	P_2	P_3	P_4	P_5
1	90°	54.7356°	90°	30.5556°	90°
2		125.2644°	39.2315°	149.4443°	29.2813°
3			140.7685°	70.1243°	150.7187°
4				109.8756°	58.7818°
5					121.2181°

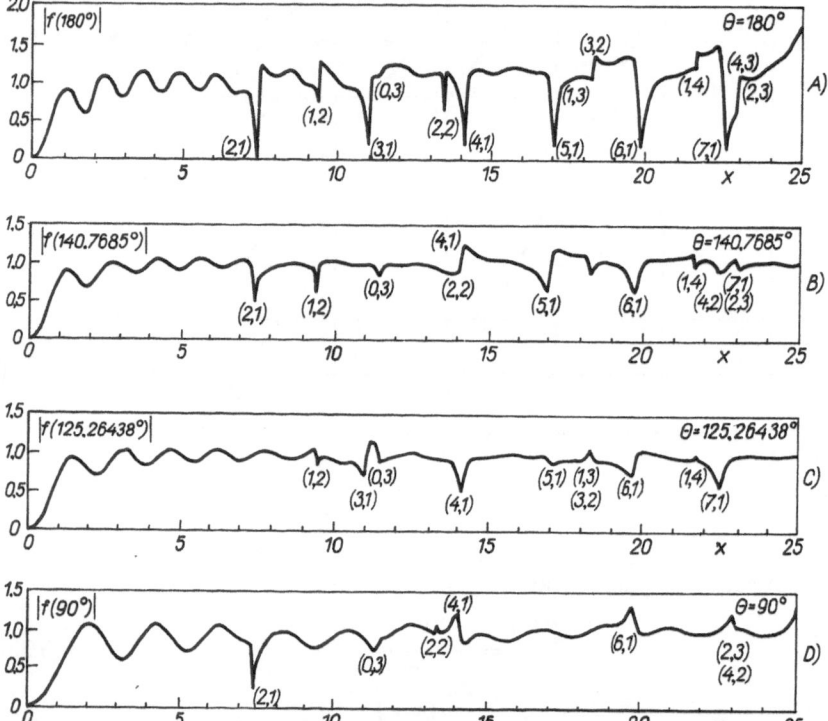

Fig. 3.8. The form functions of the acoustic pressure scattered by a solid tungsten carbide sphere in water at four fixed observation points: A) $\theta = 180°$ – all the modes are present; B) $\theta = 140.7685°$ – the third mode ($n = 3$) is absent; C) $\theta = 125.2643°$ – the second mode ($n = 2$) is absent: D) $\theta = 90°$ – all the odd modes are absent [3.8, Fig. 8]

(3, 2), ... are missing (compare A and B). In the case C (at $\theta = 125.2644°$) the resonances of the second mode ($n = 2$) are suppressed, i.e. the "notches" corresponding to the resonances (2, 1), (2, 2) ... vanish. The same can also be observed in the case D (at $\theta = 90°$). Here all the odd resonances are suppressed. An analogous situation occurs also at other observation angles (not given in Table 3.4), and for higher n values ($n > 5$), not given in Table 3.4.

In Fig. 3.8A the main resonances of successive modes, i.e. (2, 1), (3.1), (4, 1), . . . (n, 1) of the Rayleigh wave revolving around the sphere, can be clearly observed. They periodically repeat themselves in the form function and cause the deep and narrow dips. If one chooses such a resonance, say (5, 1), and computes the angular diagram at $x = 16.95$ corresponding to this resonance, then for backscattering ($\theta = 180°$) it will be of zero magnitude. This fact is illustrated by Fig. 3.9.

The Rayleigh wave has a sufficiently large Q factor, so that its resonances are fairly narrow (see Figs. 3.8A and 4.4). The resonance (5, 1) is one of them. The minimum in the direction $\theta = 180°$ on the angular diagram (Fig. 3.9) will be rapidly filled at a small (of the order of 1%) deviation from the resonance frequency. The resonance frequency $x = 16.95004$ is the root of (3.5), $D_n = 0$ (Table 3.1).

In Fig. 3.10, we show the modal resonances for scattering by water-immersed spheres of three different materials: tungsten carbide, aluminum and brass. As can be seen, with n increasing, the magnitudes of modal resonances of the Rayleigh wave (n, 1) diminish.

In terms of partial modes, the resonances (0, 1) and (0, 3) correspond to pure radial vibrations; both their magnitude and Q factor are small.

As was indicated in [3.12], for the mode (2, 1), the position on the x axis of the eigenfrequencies of free vibrations of the sphere is proportional to the velocity c_t. Using the resonance position (n, l) in the form function, one can find the phase velocity according to the formula

$$c_l^{\text{ph}} = c \frac{x_{nl}}{n + 1/2}, \qquad (3.34)$$

where x_{nl} is the coordinate of the resonance (n, l). In Table 3.5 we give the values of the phase velocity of the Rayleigh wave ($l = 1$) at the point corresponding to its second resonance ($n = 2$). These velocities are larger than c_t and, certainly, larger than the velocity of the Rayleigh wave on a half-space. It should be

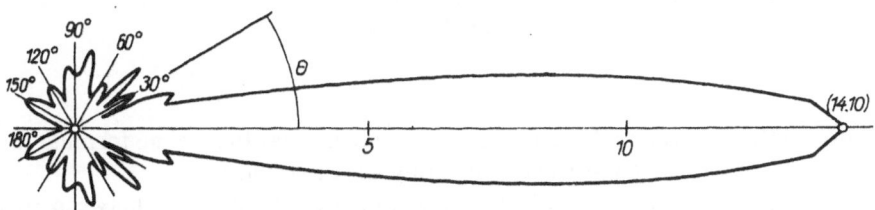

Fig. 3.9. The angular diagram for scattering by a solid tungsten carbide sphere in water. It corresponds to the resonance (5,1) at $x = 16.95$ (see Fig. 3.8A). The minimum at $0 = 180°$ can be observed in it [3.8, Fig. 9]

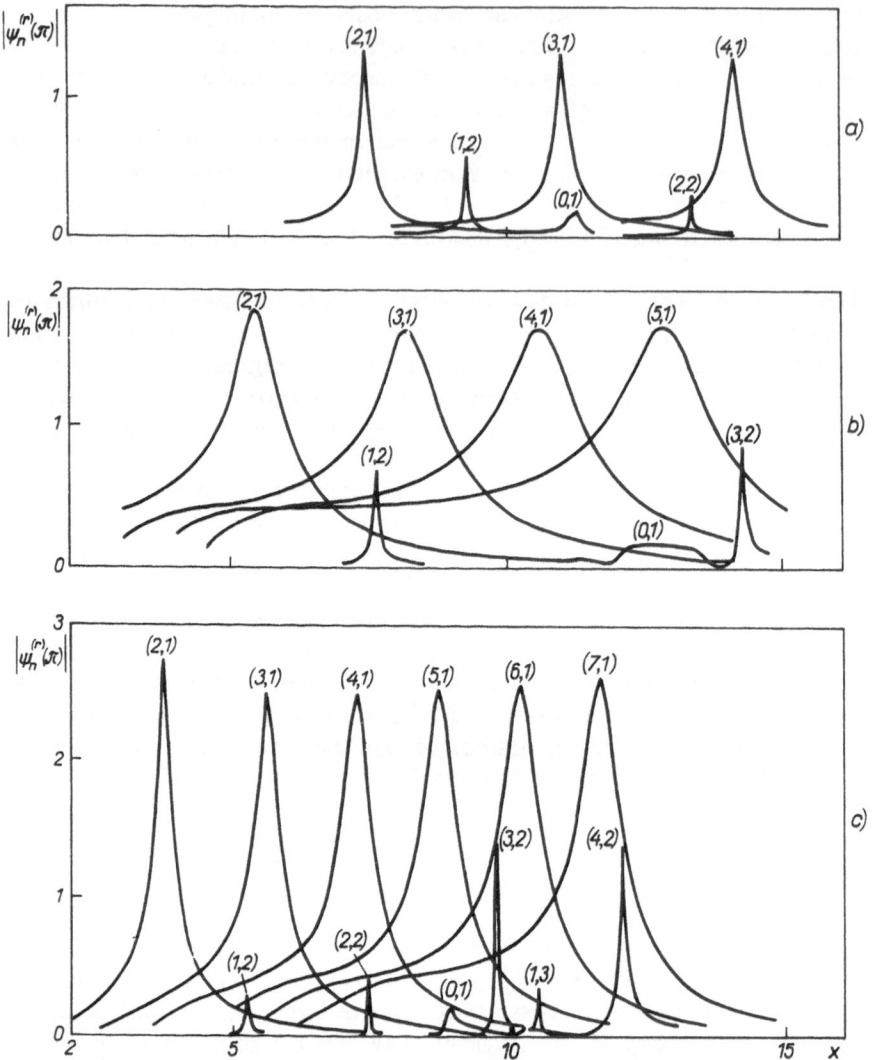

Fig. 3.10. The contribution of the resonance component $|\psi_n^{(r)}(\pi; x)|$ at backscattering: **a** by a solid tungsten carbide sphere; **b** by an aluminum sphere; **c** by a brass sphere. Two numbers (n, l) label the resonance positions of the peripheral waves [3.12, Fig. 1]

mentioned that in general, for an arbitrary x value, so far there is no formula for the velocity of the Rayleigh wave depending on the frequency.

The width of the resonance Δx is determined as the distance between two points situated at opposite sides from the maximum where the magnitude is $2^{-1/2}$ times the peak value. Analogously to the simple harmonic oscillator, using

Table 3.5. The phase velocity of the Rayleigh wave $(l = 1)$ at the frequency corresponding to its second resonance $(n = 2)$ computed for the spheres of different materials

Material	Tungsten carbide	Aluminum	Brass
c_t	4185 m/s	3120 m/s	2110 m/s
x_{21}	7.43	5.57	3.78
c^{ph}/c	2.972	2.228	1.512
c^{ph}/c_t	1.051	1.057	1.061
c^{ph}	4399 m/s	3297 m/s	2238 m/s

the width of the resonance, one can find the Q factor

$$Q = \frac{x_0}{\Delta x}, \tag{3.35}$$

where x_0 is the resonance frequency.

On the other hand, using either the computed or measured time dependences of the echo signal for a long, but finite harmonic incident pulse with a frequency $x = x_0$, the same quantity can be obtained from

$$Q = -\frac{\pi N}{\ln \delta}, \qquad \delta = \frac{p(\tau)}{p(\tau_0)}, \tag{3.36}$$

where $p(\tau_0)$ is the magnitude of the echo signal at the beginning of the decay and $p(\tau)$ is the magnitude after n full turns.

In the case of a tungsten carbide sphere, the width of the resonance $(2, 10)$ is 0.15 and the Q factor obtained from (3.35) is 49. These two values are in correspondence with each other.

Using the measured data in [3.12] for the resonances $(3, 1)$, $(4, 1)$, $(5, 1)$ and $(6, 1)$ in the case of a tungsten carbide sphere, the following values of the Q factor were obtained: 60, 50, 50 and 60, respectively. If the widths of the resonances are used, then the following values of Q will be obtained: 60, 60, 60 and 60, respectively.

As can be seen from Fig. 3.10, the widths of the Rayleigh wave resonances $(n, 1)$ almost do not change in the case of a tungsten carbide sphere, while they increase with increasing n for aluminum and brass spheres.

Figure 3.10 can be considered as the totality of three acoustic spectrograms. As can be seen from this figure, the acoustic spectrogram really takes the role of an acoustic tag, or measure, of the scatterer. Since here the shape of the body is unchanged, the spectrogram now contains information on its material. The information on the elastic scatterer is presented in Fig. 3.10 in a more visible and handy fashion than in the form function (see Figs. 3.1 and 3.3). This, however, is not always the case. In some frequency ranges the form function transmits the information about the scatterer rather well.

To date, as far as we know, there does not exist any procedure to relate the spectrogram to either the physical parameters of the sphere (ρ, c_1 and c_t) or some of their invariants.

3.3 Scattering by an Elastic Spherical Shell

The solution of the scattering problem of a plane acoustic pressure wave by an elastic spherical shell with liquid inside can be presented in a series form:

$$p_s = p_* \exp(-i\omega t) \sum_{n=0}^{\infty} i^n(2n+1) R_n h_n^{(1)}(kr) P_n(\cos\theta) , \qquad (3.37)$$

where

$$R_n = \frac{B_n}{D_n} . \qquad (3.38)$$

The elements of the determinants of the sixth rank B_n and D_n are given in [3.13–15].

Using considerations similar to those presented above for the cases of scattering by a solid elastic sphere and by a circular cylindrical shell, and following the procedure given in [3.13], we can introduce the partial scattering function

$$S_n = 2R_n + 1 . \qquad (3.39)$$

It can be presented in two different forms

$$S_n = S_n^{(s)} \frac{F_n - z_n^{(2)}}{F_n - z_n^{(1)}}, \qquad S_n = S_n^{(r)} \frac{F_n^{-1} - [z_n^{(2)}]^{-1}}{F_n^{-1} - [z_n^{(1)}]^{-1}} , \qquad (3.40)$$

where

$$F_n = \frac{\rho}{\rho_1} \frac{D_n^{(1)}}{D_n^{(2)}} . \qquad (3.41)$$

The functions $S_n^{(s)}$ and $S_n^{(r)}$ are defined by (3.13) and (3.14); the functions $z_n^{(1)}$ and $z_n^{(2)}$ are determined by (3.15). The determinants $D_n^{(1)}$ and $D_n^{(2)}$ in (3.41) are of the fifth rank [3.14]

$$D_n^{(1)} = \det(a_{ij}) \quad (i \neq 1, j \neq 1) ,$$
$$D_n^{(2)} = \det(a_{ij}) \quad (i \neq 2, j \neq 1) . \qquad (3.42)$$

In the far field the magnitude of the acoustic pressure scattered by the shell can be presented in the form given by (3.11) in which the form function $f(\theta)$ defined by (3.8) is contained. The modal resonance component is isolated by subtracting from f_n the rigid background $S_n^{(r)}$ in the case of a thick-walled shell, and by subtracting of the soft background $S_n^{(s)}$ in the case of a thin-walled shell.

When either the relative thickness of the shell is $1/100 \lesssim h \lesssim 1/10$ or the densities of liquid ρ, ρ_2 and elastic ρ_1 media are commensurable, it is appropriate to introduce the intermediate background [3.13]. However, a general background formula for shells of arbitrary thickness has recently been given in [3.16].

As the results of computation have shown, $F_n(x)$ [see (3.41)] changes with x rather slowly, with the exception of the regions around the frequencies at which $D_n^{(2)}$ passes through zero. The magnitude of the function $F_n(x)$ on the slowly changing "plateau" is rather large in the case of thin-walled (almost acoustically soft) shells, and is near zero in the case of thick-walled (almost acoustically rigid) shells. Therefore, when $F_n(x_0)$ is subtracted from $F_n(x)$, where x_0 is an arbitrary point on the plateau of the function $F_n(x)$, the graph of the difference would look like that of $F_n(x)$ in the case of an acoustically rigid sphere. According to this, the following difference expressions are introduced:

$$\bar{F}_n(x) \equiv F_n(x) - F_n(x_0) \tag{3.43}$$

and

$$\bar{z}_n^{(1)}(x) = z_n^{(1)}(x) - F_n(x_0) ,$$
$$\bar{z}_n^{(2)}(x) = z_n^{(2)}(x) - F_n(x_0) . \tag{3.44}$$

Using these definitions, the form function S_n , (3.40), can be presented in the form

$$S_n = S_n^{(s)} \frac{\bar{z}_n^{(2)} - F_n}{\bar{z}_n^{(1)} - F_n} , \tag{3.45}$$

$$S_n = S_n^{(i)} \frac{[\bar{z}_n^{(2)}]^{-1} - \bar{F}_n^{-1}}{[\bar{z}_n^{(1)}]^{-1} - \bar{F}_n^{-1}} , \tag{3.46}$$

where the partial scattering function of the intermediate background is introduced

$$S_n^{(i)} = S_n^{(s)} \frac{\bar{z}_n^{(2)}}{\bar{z}_n^{(1)}} = - \frac{h_n^{(2)}(x)}{h_n^{(1)}(x)} \frac{z_n^{(2)}(x) - F_n(x_0)}{z_n^{(1)}(x) - F_n(x_0)} , \tag{3.47}$$

$$S_n^{(i)} = - \frac{xh_n^{(2)\prime}(x) - \dfrac{\rho}{\rho_1} h_n^{(2)}(x) \left[\dfrac{D_n^{(1)}}{D_n^{(2)}}\right]_{x=x_0}}{xh_n^{(1)\prime}(x) - \dfrac{\rho}{\rho_1} h_n^{(1)}(x) \left[\dfrac{D_n^{(1)}}{D_n^{(2)}}\right]_{x=x_0}} . \tag{3.48}$$

It is easy to prove that the function $S_n^{(i)}$ satisfies the unitary condition $|S_n^{(i)}| = 1$. Besides, in one of the limiting cases, when $\rho_1/\rho \to 0$, it turns into $S_n^{(s)}$, and in the other limiting case, when $\rho/\rho_1 \to 0$, it turns into $S_n^{(r)}$.

According to the definition

$$2R_n = S_n - 1 , \tag{3.49}$$

this expression can be written either as

$$2R_n = (S_n - S_n^{(s)}) + (S_n^{(s)} - 1) ,$$ (3.50)

or as

$$2R_n = (S_n - S_n^{(i)}) + (S_n^{(i)} - 1) .$$ (3.51)

In (3.50) and (3.51) the first term defines the isolated resonance component and the second term defines the harmonic background.

Using the relation (3.46) in (3.51), we obtain

$$2R_n = S_n^{(i)} \frac{[\bar{z}_n^{(2)}]^{-1} - [\bar{z}_n^{(1)}]^{-1}}{[\bar{z}_n^{(1)}]^{-1} - \bar{F}_n^{-1}} + (S_n^{(i)} - 1) .$$ (3.52)

The resonance frequencies x_{nl} are found from the characteristic equation

$$\mathrm{Re}\{[\bar{z}_n^{(1)}(x)]^{-1} - \bar{F}_n^{-1}(x)\} = 0 ,$$ (3.53)

i.e. from the condition that the real part of the denominator in (3.52) should be equal to zero. Using the procedure described above in the case of scattering by elastic bodies of cylindrical shape, the coefficient R_n can be presented in a form describing the resonances separately

$$2R_n = S_n - 1 = \exp(2\mathrm{i}\xi_n)\left[\sum_{l=0}^{\infty} \frac{\mu_{nl}}{x - x_{nl} + \frac{1}{2}\mathrm{i}\Gamma_{nl}} + 2\mathrm{i}\exp(-\mathrm{i}\xi_n)\sin\xi_n \right], \quad (3.54)$$

where

$$\mu_{nl} = \{[\bar{z}_n^{(2)}(x)]^{-1} - [\bar{z}_n^{(1)}(x)]^{-1}\}/\beta_{nl} ,$$

$$\tfrac{1}{2}\Gamma_{nl} = \alpha_{nl}/\beta_{nl} ,$$

$$\alpha_{nl} = (\mathrm{Im}\{[\bar{z}_n^{(1)}(x)]^{-1} - [\bar{F}_n(x)]^{-1}\})_{x=x_{nl}} ,$$ (3.55)

$$\beta_{nl} = \left(\frac{\mathrm{d}}{\mathrm{d}x} \mathrm{Re}\{[\bar{z}_n^{(1)}(x)]^{-1} - [\bar{F}_n(x)]^{-1}\} \right)_{x=x_{nl}} ,$$

$$S_n^{(i)} = \exp(2\mathrm{i}\xi_n) .$$

The relation (3.54) was obtained by expansion of (3.53) in a Taylor series in the vicinity of the resonance frequency x_{nl}. Narrow resonance peaks show up near $x = x_{nl}$ when the functions μ_{nl} and Γ_{nl} changes slowly with x and the width of the resonance Γ_{nl} is small. Strictly speaking, the maximum of the function $\mu_{nl}/(x - x_{nl} + \frac{1}{2}\mathrm{i}\Gamma_{nl})$ is not situated exactly at $x = x_{nl}$, but only near it, because the function μ_{nl} and Γ_{nl} are (slowly) changing with x.

The fixed point $x = x_0$ in (3.43) and (3.44) is in some sense arbitrary, but as soon as it is chosen from the domain where the graph of the function $F_n(x)$ is a horizontal plateau, the difference in choice of the value $x = x_0$ will exert only small changes in the functions $F_n(x)$, $\bar{z}_n^{(1)}(x)$, $\bar{z}_n^{(2)}(x)$, μ_{nl} and Γ_{nl}, as well as in the value of the resonance frequency x_{nl}.

From (3.54) one can obtain

$$S_{\hat{n}} = \exp(2i\xi_n) \sum_{l=0}^{\infty} \frac{x - x_{nl} - \frac{1}{2}i\Gamma_{nl}}{x - x_{nl} + \frac{1}{2}i\Gamma_{nl}} \tag{3.56}$$

with poles

$$x = x_{np} = x_{nl} - \tfrac{1}{2}i\Gamma_{nl} \tag{3.57}$$

and zeroes

$$x = x_{nz} = x_{nl} + \tfrac{1}{2}i\Gamma_{nl} \tag{3.58}$$

in the complex plane x.

3.4 Numerical Results for Scattering by Thick-Walled Spherical Shells

Some plots of isolated modal resonances will be presented in this section. They are computed for scattering by thick-walled ($b/a = 0.2$–0.9) spherical shells [3.13]. The physical parameters of the problem are defined by (2.49). The observation point is situated in the far field. Only the case of backscattering ($\theta = \pi$) is considered. The plots are computed for the first few values of n in the domain $0 \leqslant x \leqslant 100$.

The total magnitude of the nth partial wave is defined by f_n , (3.9), which contains the partial scattering function S_n , (3.39) [with R_n defined by (3.38)]. If in (3.7) instead of S_n , (3.39), the expressions $S_n^{(s)}$, (3.13), $S_n^{(r)}$, (3.14), and $S_n^{(i)}$, (3.47), are used, then the nonresonant soft, $f_n^{(s)}$, rigid, $f_n^{(r)}$, and intermediate, $f_n^{(i)}$, backgrounds will be obtained. The difference between the total magnitude f_n and the background determines the resonance component

$$\psi_n^{(s)} = f_n - f_n^{(s)}, \qquad \psi_n^{(r)} = f_n - f_n^{(r)}, \qquad \psi_n^{(i)} = f_n - f_n^{(i)}$$

corresponding to the soft, rigid, and intermediate backgrounds, respectively.

As to be expected, it actually turned out that $|\psi_n^{(r)}| \equiv |\psi_n^{(i)}|$ for spheres with $b/a = 0.2$ and $b/a = 0.7$.

In Fig. 3.11, we represent the plot of $|\psi_n^{(r)}(\pi; x)|$ computed for the case of a sphere with $b/a = 0.2$ at $n = 0, 1, 2, 3$ in the domain $0 \leqslant x \leqslant 20$. The plot of $|\psi_n^{(i)}(\pi; x)|$ practically coincides with the latter. In Fig. 3.12, we give the plots of isolated modal resonances of the same sphere, but in the broader frequency band $0 \leqslant x \leqslant 100$.

The rigid background is fully adequate for a sphere with $b/a = 0.7$. This fact is illustrated in Figs. 3.13 and 3.14. When the rigid background is subtracted, the modal resonance frequencies will be found from

$$F_n^{-1}(x) = \text{Re}\left([z_n^{(1)}(x)]^{-1} \right). \tag{3.59}$$

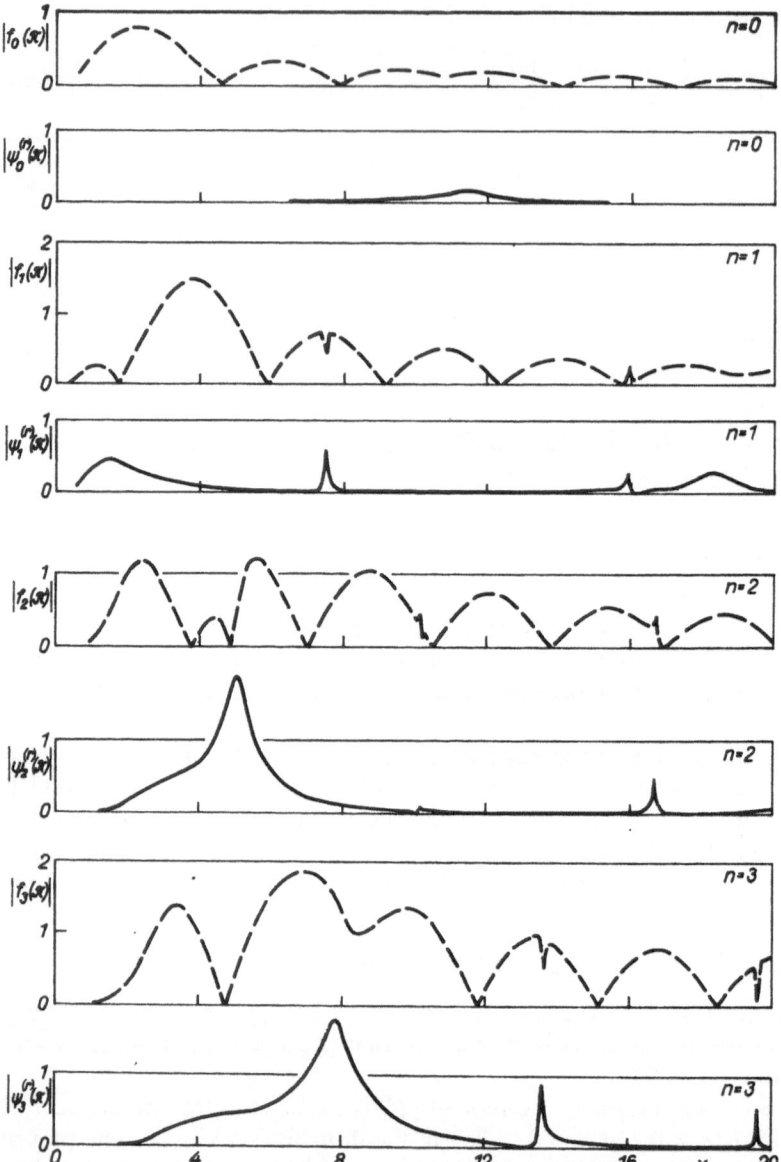

Fig. 3.11. The total form function $|f_n(\pi; x)|$ (dashed line) and the form function of the resonance component $|\psi_n^{(r)}(\pi; x)|$ (solid line) for the partial mode $n = 0, 1, 2, 3$ for backscattering by an aluminum sphere with $b/a = 0.2$ in water [3.13, Fig. 2]

Fig. 3.13. a Graphical solution of the characteristic equation (3.59); here F_1^{-1} is shown as solid line and Re $([z_1^{(1)}(x)]^{-1})$ by a dot-and-dash line. **b** The form function $|\psi_1^{(i)}(\pi; x)|$ computed at $x_0 = 0.5$ in the case of a sphere with $b/a = 0.7$. The dashed lines show the correspondence between the solution obtained in the case when the rigid background was used and the positions of the resonance peaks obtained in the case of the intermediate background [3.13, Fig. 4]

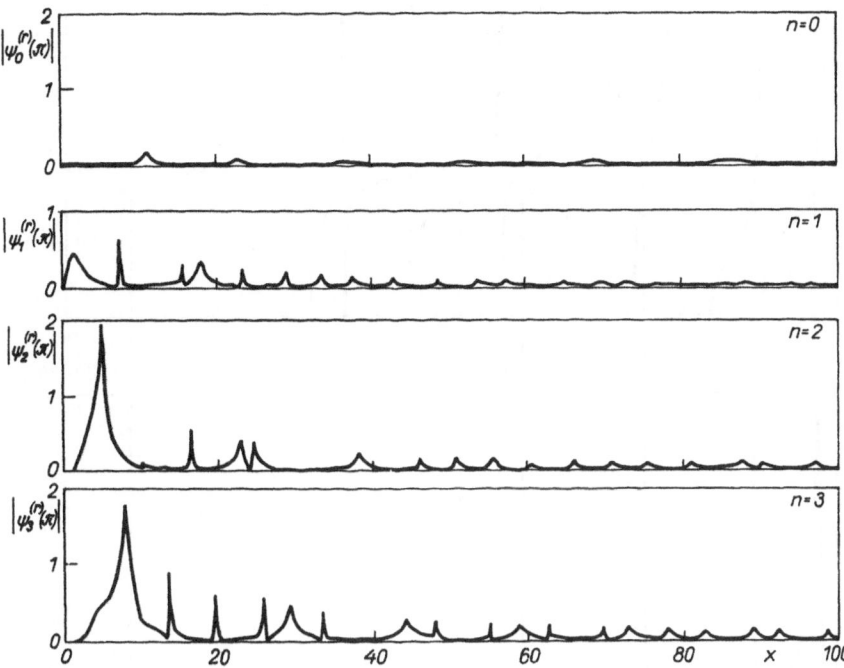

Fig. 3.12. The form function of isolated modal resonance $|\psi_n^{(r)}(\pi; x)|$ for $n = 0, 1, 2, 3$ for back-scattering by an aluminum sphere with $b/a = 0.2$ in water [3.13, Fig. 3]

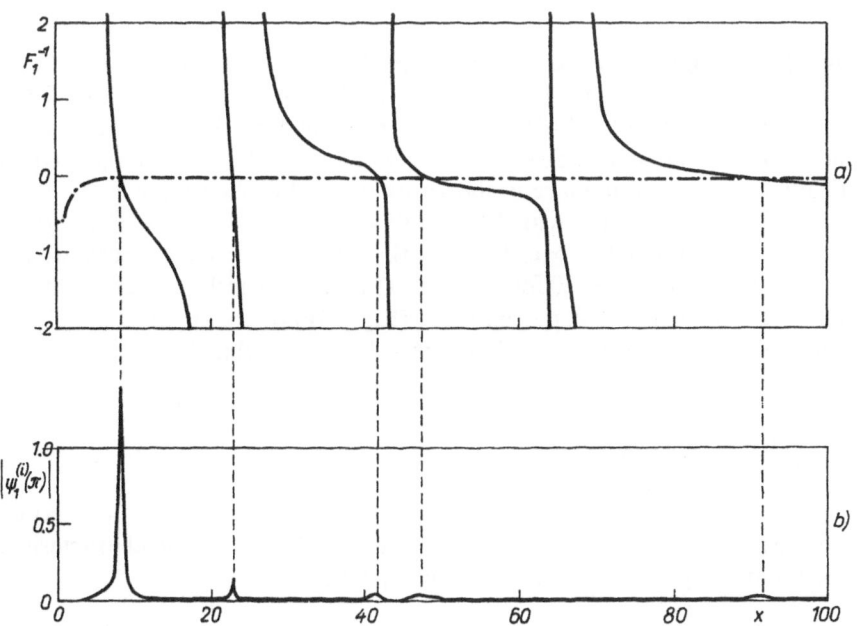

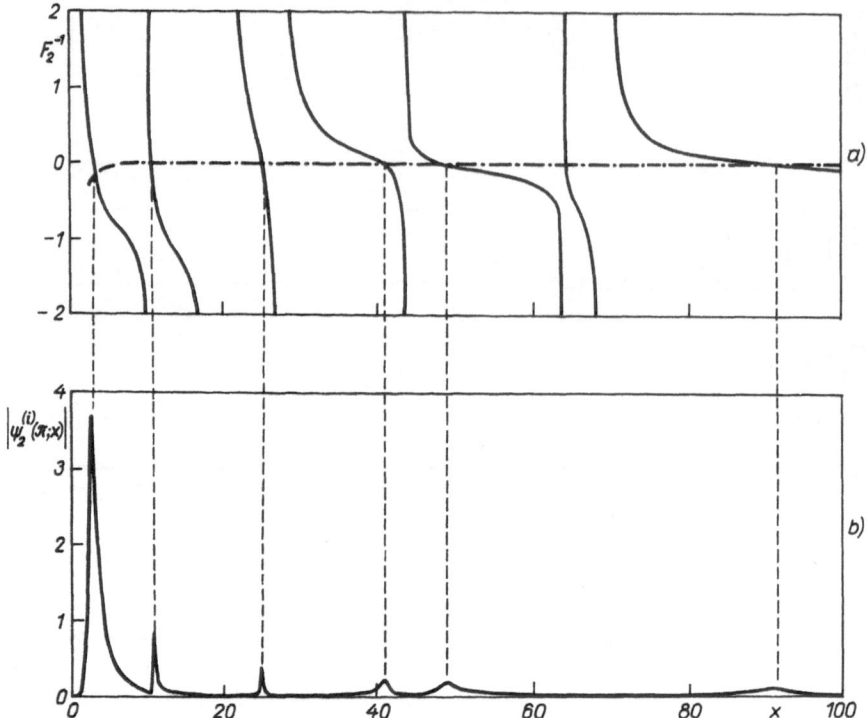

Fig. 3.14. Same as Fig. 3.13 but for the mode $n = 2$ [3.13, Fig. 5]

The left- and right-hand sides of (3.59) are shown in Figs. 3.13a and 3.14a for the partial modes $n = 1$ and $n = 2$, respectively. As can be seen from Figs. 3.13b and 3.14b, where the plots of $|\psi_n^{(i)}(\pi; x)|$ are presented for the same modes, the resonance peaks exactly correspond to the solution of the characteristic equation (3.59). This correspondence is shown by vertical dashed lines.

In Fig. 3.15, for a sphere with $b/a = 0.9$, we present plots of $|\psi_n^{(i)}(\pi; x)|$ for $n = 1, 2, 3$ in the domain $0 \leqslant x \leqslant 100$. For the intermediate background we use $x_0 = 0.5$ for $n = 1$ and $x_0 = 20$ for $n = 2, 3$. This choice can be explained by the example of the second ($n = 2$) partial mode (Fig. 3.16) where in part a the plot of the function $F_2(\pi; x)$ is presented, in part b the graphical solution of the equation

$$[\bar{F}_2(x)]^{-1} = \text{Re}\{[\bar{z}_2^{(1)}(x)]^{-1}\} \tag{3.60}$$

is shown, and in part c the plot of $|\psi_n^{(i)}(\pi; x)|$ is given. As before, the vertical dashed lines show the correspondence between the roots of the characteristic equation and the position of the resonance peaks.

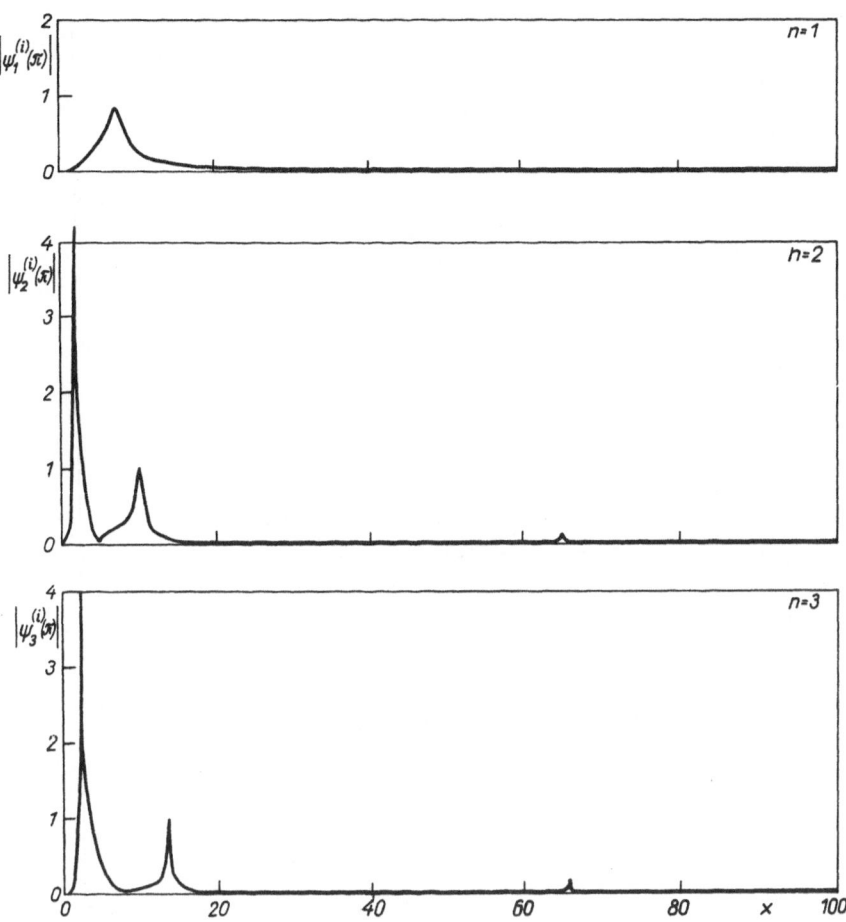

Fig. 3.15. Form functions of isolated modal resonances $|\psi_n^{(i)}(\pi; x)|$ for scattering by an aluminum sphere with $b/a = 0.9$ in water [3.13, Fig. 6]

The function $F_2(x)$, indeed, possesses the properties needed for choosing it as an intermediate background. Its plot has a form of a broad plateau separated by sharp cuts. If the x_0 value (in the example under consideration, $x_0 = 20$) is chosen in order that $F_2(x_0)$ be situated on the plateau of the function $F_2(x)$, then the difference $\bar{F}_n(x) = F_n(x) - F_n(x_0)$ will behave as the function $F_n(x)$ does in the case of an acoustically rigid sphere, as was needed in Sect. 3.3. Generally speaking, every value from the interval at the end points of which $F_n(x)$ has singularities can be chosen as x_0. The choice itself is not critical, but changing of the x_0 value will shift the zeroes of the function $\bar{F}_n$, and, respectively, the zeroes of $\bar{F}_n^{-1}$. Nevertheless, since $[\bar{z}_n^{(1)}]^{-1}$ is small, the positions of isolated modal resonances of the thick-walled sphere are defined by the zeroes of the $\bar{F}_n^{-1}$.

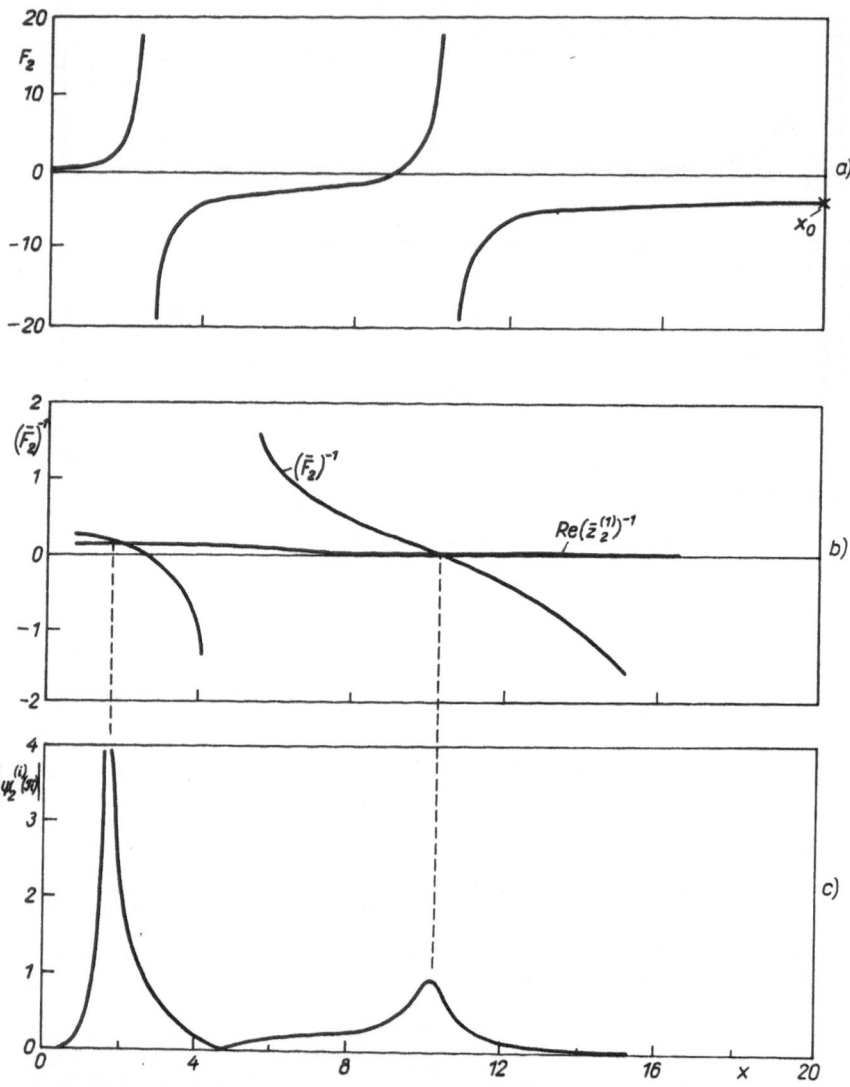

Fig. 3.16. a Plot of $F_2(x)$. **b** The graphical solution of (3.60). **c** The form function $|\psi_2^{(i)}(\pi; x)|$ for scattering by an aluminum sphere with $b/a = 0.9$ in water at $x_0 = 20$. The dashed lines show the correspondence between the solution of the characteristic equation in the case of the intermediate background and the positions of the resonance peaks [3.13, Fig. 7]

4 Synthesis of the Backscattering Form Function for a Solid Elastic Sphere Using the Sommerfeld–Watson Transformation

The material presented below is based on the results obtained by *Williams* and *Marston* in [4.1, 2].

The problem of a plane acoustic wave scattered by a solid elastic sphere has an exact solution in series form [4.3]. For given physical parameters of the material of the sphere ρ_1, c_1 and c_t and that of the surrounding liquid ρ, c, it is easy either to compute directly the secondary (scattered) field of the acoustic pressure in the series form, or to apply to it the Sommerfeld–Watson transformation, which transforms the series into an integral in the complex frequency plane, and then to calculate the integral. Changing in the proper way the initial contour, the calculation can be reduced to the evaluation of two contour integrals. One of them passes through the saddle point, and the other envelopes the poles. Usually, both these integrals are evaluated on the assumption of large x values (in the short-wavelength approximation). The first integral is evaluated analytically using the saddle point method and gives the contribution made by the specularly reflected wave. The second integral is evaluated numerically and gives the contributions made by the shock and slip waves generated by peripheral and creeping waves, respectively.

As far as we know, the authors of papers which use the Sommerfeld–Watson transformation limit themselves, as a rule, to analyzing the singularities of the integrand in the complex plane and to finding the coordinates of the poles and saddle points. The residues of the poles were seldom calculated. They furnish the peripheral waves revolving around on the sphere. The total form function, in which contributions of all waves generated in the elastic body are present, was not computed. Apparently this may be caused by the large amount of computation time required, because the coordinates of the poles and the residues of the poles have to be computed for each value of the wave radius x (with small step size l_x over a broad range). Only recently, in [4.2], the acoustic field scattered by a sphere was computed and analyzed by this method.

4.1 Approximate Formula for the Description of the Form Function

As is well known, in the short-wavelength approximation (at $x \gg 1$) the acoustic pressure scattered by a sphere can be represented as a sum of two components

$$p(x) = p_1(x) + p_2(x) , \tag{4.1}$$

where $p_1(x)$ is the contribution furnished by specularly reflected and refracted waves, and $p_2(x)$ is the contribution corresponding to peripheral and creeping waves. In (4.1), the geometrical divergence should be taken into account by an additional spreading factor.

In [4.2], using the information obtained from an experimental investigation of the same problem [4.1], the assumption was made that the contribution of the refracted waves is small in comparison with that corresponding to the specularly reflected wave and therefore it may be neglected. The contribution of the specularly reflected wave was obtained using the saddle-point method, which was proposed in [4.4] for the scattering problem by an elastic cylinder. The formulae for this contribution are given in [Appendix A of 4.1]. At back-scattering, for numerical calculations the contribution of the specularly reflected wave can be described by the simple formula

$$p_1(x) = v, \qquad v = \frac{1 - \zeta}{1 + \zeta}, \qquad \zeta = \frac{\rho c}{\rho_1 c_1}. \tag{4.2}$$

Here v is the reflection coefficient of a plane wave from a plane boundary of two liquid media at normal incidence. It is the limit to which the outer spherical reflection coefficient tends for $x \to \infty$. As noted in [4.1], in the numerical example considered there, the spherical and plane reflection coefficients practically coincide in all the x range considered. Even at the bottom boundary 0 and throughout this range (at $x = 20$) the difference in modulus of the indicated coefficients does not exceed 10^{-3}.

The contribution of the slip waves, caused by the creeping waves, is small. It is considerable only at small x (at $x \lesssim 15$); with growing x it diminishes and can be neglected.

The contribution of the shock waves, caused by the peripheral waves, can be presented in the form

$$p_2(x) = \sum_{l=1}^{q} f_l(x), \tag{4.3}$$

where $f_l(x)$ determines the contribution of each peripheral wave (index l) and q defines the number of waves taken into account.

According to the concept of the geometrical theory of diffraction, at back-scattering the contribution $f_l(x)$ can be written in the form

$$f_l(x) = q_{0l}(x)\varepsilon_{0l}(x) \exp[ix\gamma_l(x)] \frac{-1}{1 - \varepsilon_l(x)\exp[ix\delta_l(x)]}, \tag{4.4}$$

where

$$\varepsilon_l(x) = \exp[-2\pi\kappa_l(x)], \qquad \delta_l(x) = \frac{2\pi}{v_l^{ph}(x)},$$

$$\gamma_l(x) = 2\left\{[1 - \cos\theta_l(x)] + \frac{\pi - \theta_l(x)}{v_l^{ph}(x)}\right\},$$

$$\varepsilon_{0l}(x) = \exp\{-2[\pi - \theta_l(x)]\kappa_l(x)\} ,$$

$$\theta_l(x) = \arcsin \frac{1}{v_l^{ph}(x)} . \tag{4.5}$$

In (4.4) and (4.5) the following notations are used: $q_{0l}(x)$ is the "landing" complex magnitude of the peripheral wave, $\varepsilon_l(x)$ is the damping coefficient of the peripheral wave for one full turn, $\delta_l(x)$ is the increase of the phase for one full turn, $\varepsilon_{0l}(x)$ is the damping coefficient of the peripheral wave on the path from the critical entrance point to the critical re-radiation point, $\gamma_l(x)$ is the increase of the phase over the path from the critical entrance point to the critical re-radiation point, $\theta_l(x)$ is the angle determining the critical entrance point, $v_l^{ph}(x)$ is the relative (divided by c) phase velocity of the wave, and $\kappa_l(x)$ is the damping coefficient per unit length of propagation along the arc.

The magnitude $g_{0l}(x)$ represents the total contribution by the left- and right-hand peripheral waves generated on a circle which is cut from the sphere by a sector of angle $\theta_l(x)$

At backscattering, in addition to the "landing" magnitude $g_{0l}(x)$ it is convenient to introduce the complex "launching" magnitude $g_l(x)$ of the wave which has made one incomplete turn around the sphere

$$g_l(x) = g_{0l}(x)\varepsilon_{0l}(x) . \tag{4.6}$$

Then (4.4) takes the form

$$f_l(x) = -g_l(x) \exp[ix\gamma_l(x)] \frac{1}{1 - \varepsilon_l(x) \exp[ix\,\delta_l(x)]} . \tag{4.7}$$

The functions $\delta_l(x)$ and $\kappa_l(x)$ are defined by

$$\delta_l(x) = \frac{2\pi}{x} \operatorname{Re}\{v_l(x)\}, \qquad \kappa_l(x) = \operatorname{Im}\{v_l(x)\} , \tag{4.8}$$

where $v_l(x)$ is the coordinate of the pole in the complex plane of the Sommerfeld–Watson transformation relating to the lth peripheral wave.

In contrast with the scattering problem by a solid elastic cylinder where the relative (divided by c) phase velocity of the wave is defined by the formula

$$v_l^{ph}(x_n) = x_n/n , \tag{4.9}$$

in the scattering problem by a sphere it is given by the formula

$$v_l^{ph}(x_n) = x_n/(n + 1/2) . \tag{4.10}$$

At the resonance frequency the equality

$$\operatorname{Re}\{v_{ln}\} = n ,$$

$$2\pi a = (n + 1/2)\frac{2\pi c_l^{ph}}{\omega} \tag{4.11}$$

holds, i.e. exactly $(n + 1/2)$ wavelengths fit the meridian circle length. Propagating on the meridian arc the peripheral wave receives an additional phase shift $\exp(-i\pi/2)$ at each pole (at $\theta = 0$ and $\theta = \pi$) [4.5], and after making one full turn, taking into account the additional phase shift at the poles, it is superimposed on itself in phase and therefore is amplified.

4.2   Features of the Rayleigh Wave

In Fig. 4.1a, the form function of the acoustic pressure scattered by a solid elastic sphere is given. It was obtained by direct summation of the series of eigenfunctions and represents the exact solution of the scattering problem. The

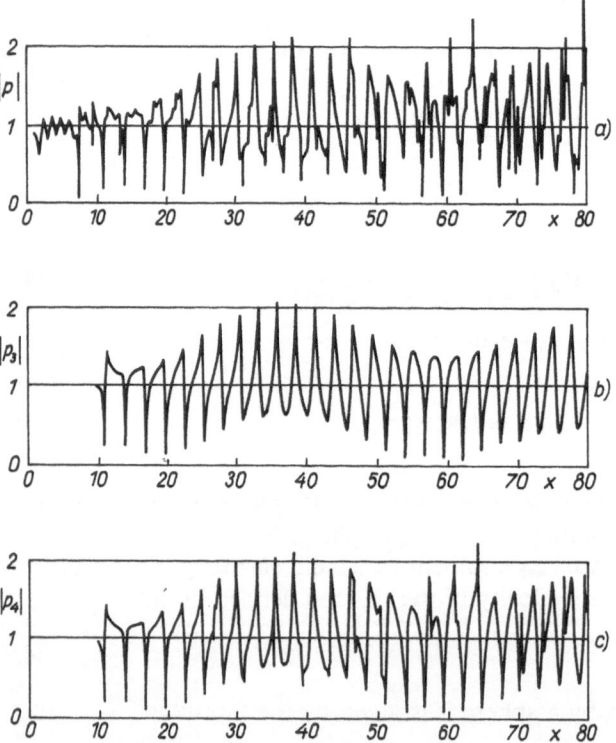

Fig. 4.1. The form function of the acoustic pressure scattered by a solid elastic sphere. Curve **a** presents the exact solution obtained in the series form; curve **b** corresponds to the approximation of the form function by (4.14), i.e. when the contributions furnished by the specularly reflected and the Rayleigh wave are taken into account; curve **c** corresponds to the approximation of the form function by (4.15), when the contributions of the specularly reflected, the Rayleigh wave, and the Whispering Gallery waves are taken into account [4.2, Figs. 4, 6]

computation was carried out in the case of a tungsten carbide sphere in water with the following parameters

tungsten carbide:

$$\rho_1 = 13.8 \times 10^3\,\text{kg/m}^3, \quad c_1 = 6860\,\text{m/s}, \quad c_t = 4185\,\text{m/s}, \qquad (4.12)$$

water: $\rho = 1 \times 10^3\,\text{kg/m}^3, \qquad c = 1476\,\text{m/s}$.

From the distinctive extrema of this curve it is easy to recognize that in the domain considered ($1 \lesssim x \lesssim 80$) at least three types of waves revolving around the sphere contribute to the form function.

At $x \lesssim 10$ the extrema corresponding to the Franz-type wave can be clearly observed. Up to $x \lesssim 7.5$ they furnish the main contribution to the form function. In this domain the curve of the form function looks like the corresponding curve in the scattering problem by an acoustically rigid sphere. Therefore, this domain of the form function is occasionally called the "quasi-rigid region".

At $x \gtrsim 7.5$ the main contribution to the form function is furnished by the shock wave caused by the Rayleigh peripheral wave. It is the resonances of this wave that determine the appearance of the form function curve. On this background the resonances of another peripheral wave are visible which provide the fine structure of the form function curve. This peripheral wave is the first (the slowest) Whispering Gallery wave (WG1).

In [4.2] the landing g_{0l} and the launching g_l magnitudes of two types of waves were calculated, namely the Rayleigh wave (index $l = 1$) and a Whispering Gallery wave (index $l = 4$). In Fig. 4.2, the dependence of their magnitudes on x is shown.

It can be seen from the presented plots that for the physical parameters given by (4.12), the modulus of magnitude g_{01} is much larger than that of g_{04}. With x increasing, the modulus of magnitude g_1 slowly increases and is of order 0.5. Just for this reason the form function is mainly governed by the contributions of two types of waves – by the specularly reflected wave, with magnitude modulus 0.9693, and the shock wave caused by the Rayleigh wave.

In Table 4.1, the dependence on x of the pole coordinate v_1 generating the Rayleigh wave in such a way that $\text{Re}\,v_1 = n$ ($n = 6, 7, 8, \ldots, 30$) is given. The

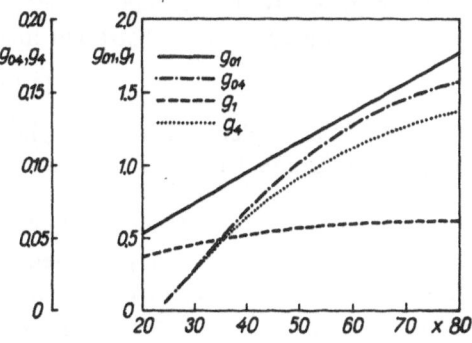

Fig. 4.2. The modulus of complex "landing" g_{0l} and "launching" g_l magnitudes of the peripheral waves revolving around on a solid elastic sphere. The Rayleigh wave is marked by the index $l = 1$ and the first Whispering Gallery wave is marked by $l = 4$ [4.2, Fig. 2]

Table 4.1. The Rayleigh wave characteristics for scattering by a solid elastic sphere: ν_1 is the pole position in the complex plane, v_1^{ph} is the relative phase velocity, ε_1 is the damping factor for one full turn [4.1, Table I]

n	x	ν_1	v_1^{ph}	ε_1
6	19.7231	5.9996 + i 0.0616	3.0343	0.679
7	22.4472	5.9995 + i 0.0679	2.9930	0.653
8	25.1401	7.9995 + i 0.0747	2.9577	0.625
9	27.8130	8.9995 + i 0.0805	2.9277	0.603
10	30.4712	9.9995 + i 0.0864	2.9020	0.581
11	33.1196	10.9995 + i 0.0923	2.8800	0.560
12	35.7583	11.9995 + i 0.0981	2.8607	0.540
13	38.3910	12.9995 + i 0.1040	2.8438	0.5202
14	41.0190	13.9996 + i 0.1090	2.8289	0.5042
15	43.6421	14.9996 + i 0.1145	2.8156	0.4870
16	46.2612	15.9996 + i 0.1207	2.8037	0.4684
17	48.8774	16.9996 + i 0.1262	2.7930	0.4525
18	51.4907	17.9988 + i 0.1316	2.7833	0.4374
19	54.1021	18.9988 + i 0.1371	2.7745	0.4226
20	56.7114	19.9988 + i 0.1418	2.7664	0.4103
21	59.3189	20.9988 + i 0.1473	2.7590	0.3963
22	61.9253	21.9996 + i 0.1527	2.7522	0.3831
23	64.5298	22.9988 + i 0.1582	2.7459	0.3701
24	67.1333	23.9991 + i 0.1635	2.7403	0.3580
25	69.7353	24.9991 + i 0.1688	2.7347	0.3462
26	72.3364	25.9991 + i 0.1741	2.7297	0.3349
27	74.9370	26.9991 + i 0.1794	2.7250	0.3239
28	77.5366	27.9991 + i 0.1847	2.7206	0.3133
29	80.1353	28.9991 + i 0.1900	2.7165	0.3031
30	82.7329	29.9988 + i 0.1948	2.7126	0.2941

small difference of Re ν_1 from the integer can be explained by the computational accuracy. The values of the relative phase velocity of the Rayleigh wave $v_1^{ph}(x)$ and the damping factor $\varepsilon_1(x)$ for one full turn are entered in this table.

In Fig. 4.3, plots of $v_1^{ph}(x)$ and $\varepsilon_1(x)$ are given. With x increasing, $v_1^{ph}(x)$ diminishes monotonically. The curve $v_1^{ph}(x)$ is situated above the straight line

$$v_1^{ph}(x) = 2.5847 \tag{4.13}$$

corresponding to the relative phase velocity of the Rayleigh wave on a plane elastic half-space (without any contact with the liquid). The latter is shown by a dashed line.

The modulus of the contribution furnished by the shock wave which is caused by the Rayleigh wave at backscattering has the form presented in Fig. 4.4. Maxima of this curve correspond to the resonances of the Rayleigh wave and its minima correspond to the antiresonances. With x increasing, both the magnitude of the Rayleigh wave and the Q factor diminish.

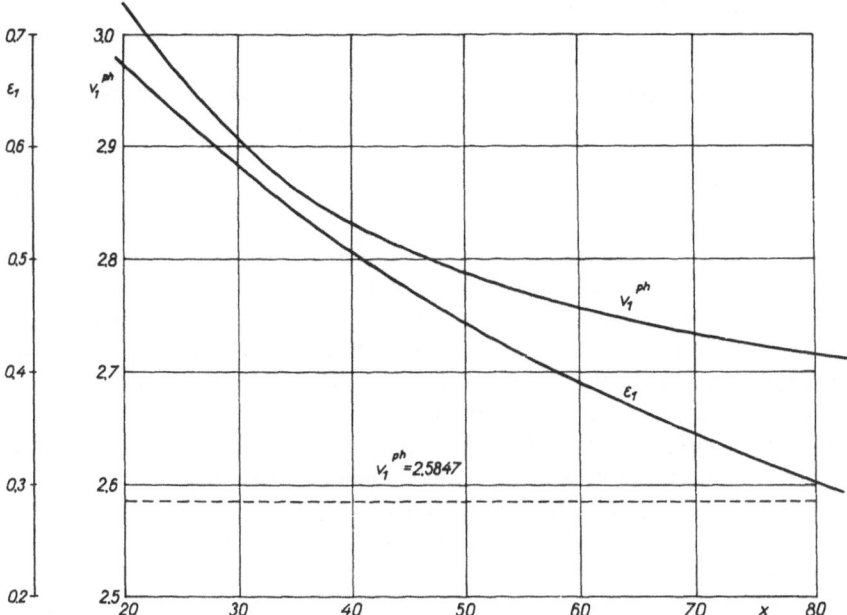

Fig. 4.3. The Rayleigh wave characteristics for scattering by a solid elastic sphere: $v_1^{\mathrm{ph}}(x)$ is the relative phase velocity, $\varepsilon_1(x)$ is the damping factor over one full turn. The relative phase velocity of the Rayleigh wave on a plane dry elastic half-space is shown by the dashed line

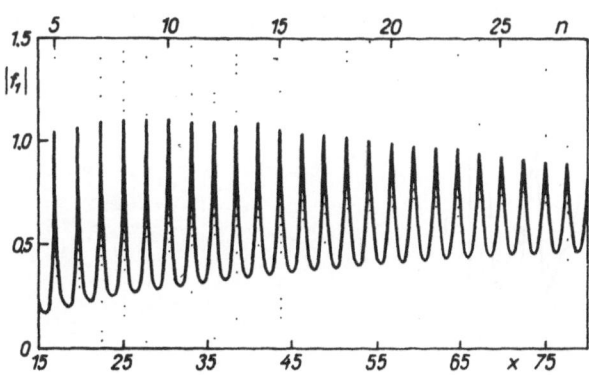

Fig. 4.4. The modulus of the Rayleigh wave contribution to the backscattering form function [4.2, Fig. 3]

Knowing the quantity $f_l(x)$ is useful no matter how it was obtained. In particular, using the quality concept [4.6, Chap. 4] it is easy to find the damping coefficient from Fig. 4.4. The numerical values obtained in this way are in good agreement with those presented in Table 4.1.

4.3 Synthesis of the Form Function

On the assumption that the form function is generated by the specularly reflected wave and by the shock wave caused by the Rayleigh peripheral wave, the backscattering pressure in the far field can be written in the form

$$p_3(x) = v + f_1(x) . \tag{4.14}$$

The modulus $|p_3(x)|$ is presented in Fig. 4.1b. On the whole, without the fine structure, curve b is a sufficiently good approximation of the exact solution, curve a in Fig. 4.1. In a more restricted x range ($30 \leqslant x \leqslant 50$), a comparison of the exact solution and its approximation is performed in Fig. 4.5a where the scale is enlarged both on the abscissa and the ordinate. The exact solution is shown by the solid line and its approximation by the dashed line.

The physical interpretation of curve b in Fig. 4.1 is similar to that given in [4.6, pp. 78, 79], and therefore we shall not consider this question here.

When the contribution of the shock wave caused by the Whispering Gallery wave is taken into account, the following approximation of the form function will be obtained

$$p_4(x) = v + f_1(x) + f_4(x) . \tag{4.15}$$

The modulus $|p_4(x)|$ is shown in Fig. 4.1c. The comparison of the exact solution and its approximation given by (4.15) is presented in Fig. 4.5b.

On the whole, the approximation given by (4.15) is better than that given by (4.14); it does not, however, describe well the form function structure in the whole x range considered. It is possible that besides the shock waves caused by

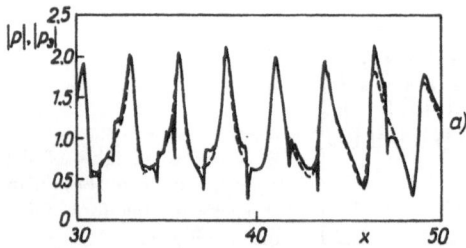

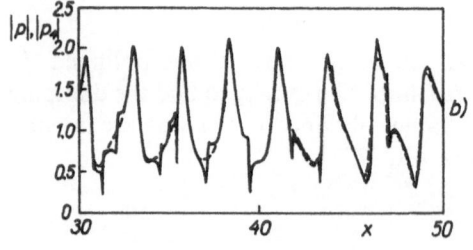

Fig. 4.5. The form function (solid line) and its approximation (dashed line); **a** the approximation defined by (4.14); **b** the approximation defined by (4.15) [4.2, Figs. 5, 7]

the Rayleigh wave and the first Whispering Gallery wave, the other peripheral waves add their contributions to the form function (see Chap. 7, Fig. 7.15).

4.4   Comparison with Experiment

To verify the computed results a model experiment was performed. The simplified scheme of the experiment and its methodology are presented in [4.1]. The results of a comparison of the computed and experimental dependence of the Rayleigh wave's magnitude on the wave radius x is presented in Fig. 4.6. The experimental data differ from the computed results to the extent of at most five per cent. The possible causes of this discrepancy are considered in [4.1]. It is typical that the experimental points show some structure.

We believe that one of the possible causes of the discrepancy between the measured and computed data is a methodological mistake. In the experiment, a short pulse with four periods in a sinusoid ($s = 4$) was used. The magnitudes of the specularly reflected wave and of the shock wave caused by the Rayleigh wave were measured for an incident pulse of this duration. As is well known [4.7, pp. 66–69], the spectrum of the incident wave is rather broad in this case and the incident wave "sees" the sphere in a rather broad range of the wave radius x. In such a case, in the time dependence the magnitude of the shock wave is determined not only by the point in the form function corresponding to the chosen frequency, but also by its neighborhood. At $s = 4$ and the physical parameters of the problem used (see (4.12)), the main lobe of the loading spectrum covers neighboring resonances of the Rayleigh wave, and the side lobes are covering approximately twelve successive resonances. Therefore, for a short pulse the magnitude of the shock wave caused by the Rayleigh peripheral

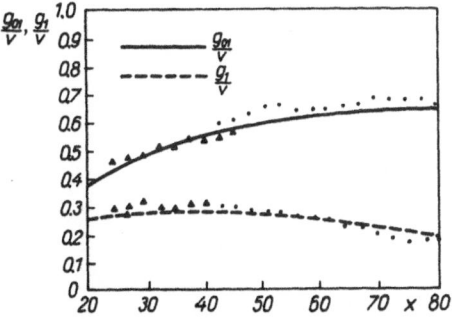

Fig. 4.6. Computed and measured results for the magnitude of the backscattered Rayleigh wave. The magnitudes are normalized to the magnitude of the specularly reflected wave. The computed results are shown by the solid line. The dots mark the experimental results for the 2.54×10^{-2} m diameter sphere, the triangles for the 1.27×10^{-2} m diameter sphere. The upper and lower curves are the results of computation for one incomplete and one complete turn, respectively [4.1, Fig. 8]

wave depends not on one, but on several resonances and becomes a washed-out characteristic. On the contrary, the computed amplitude is found on the assumption of infinite pulse duration (at $s \to \infty$). In this case the spectrum of the loading is narrow (sharp) and it "sees" the sphere only at one point of the wave radius x; here the computed magnitude is a local characteristic. Therefore, in principle, the measured amplitudes for a short pulse do not necessarily have to coincide with the computed ones in the case of a long pulse. (In this connection, see also [4.8].)

In order to compare the computed results with experiment, the latter should be carried out using a long, but finite, incident pulse and the magnitudes of B_1, C_1, C_{10} and C_{11} (see Fig. 5.1) should be measured. From these data the quasi-stationary level of the magnitude of the shock wave caused by the Rayleigh wave can be found and compared with the stationary one.

The author is grateful to Professor P.L. Marston for his kindness in sending him the reprints of papers concerning the Sommerfeld–Watson transformation for the problem of acoustic wave scattering by an elastic sphere.

5 Resonance Isolation and Identification Method

One of the difficulties in investigating the scattering of acoustic waves by elastic bodies is the separation of the background in the total form function. This problem is very acute when the method of the resonance scattering theory is applied to not very thick shells (with $h \leqslant 1/10$).

In the method of resonance isolation and identification, developed by a group of French acousticians [5.1–10], the separation of the background is accomplished in a very ingenious way.

5.1 The Essence of the Method

The method considered is an experimental one, and the separation of the background is realized instrumentally.

Usually the form function is obtained by registration of the time dependence for a sufficiently long but finite incident pulse, when the scattering process becomes totally settled (the B_1 level in Fig. 5.1). In the *Method of Resonance Isolation and Identification* (which is termed "MIIR"), in addition to obtaining the form function and sometimes instead of it, it is proposed to construct the re-radiation form function, which is obtained by registration of the point corresponding to the cutting-off of the long finite pulse (the C_1 level in Fig. ·5.1). In this way the contributions of the specularly reflected and refracted waves that form the background can be eliminated, which allows one to obtain only the contributions of the shock waves caused by the peripheral waves revolving around the elastic body. It should be noted that the idea of constructing several frequency dependences by using the time dependences computed at different values of the wave radius x_0 and duration τ_* of the incident pulse, was proposed by us [5.11, p. 164]. In [5.11] it was particularly pointed out that it is possible to construct the re-radiation form function. The authors of the resonance isolation and identification method came to this idea independently and added to it a very important feature, namely the angular diagram of re-radiation.

In Fig. 5.2, the computed, curve a, and measured, curve b, form functions for backscattering by a solid aluminum cylinder in water are presented. The computation was carried out with the following parameters

aluminum: $\rho_1 = 2.79 \times 10^3$ kg/m^3, $\quad c_1 = 6380$ m/s, $\quad c_t = 3100$ m/s ,

water: $\quad \rho = 1 \times 10^3$ kg/m^3, $\quad c = 1470$ m/s . $\hspace{3cm}$ (5.1)

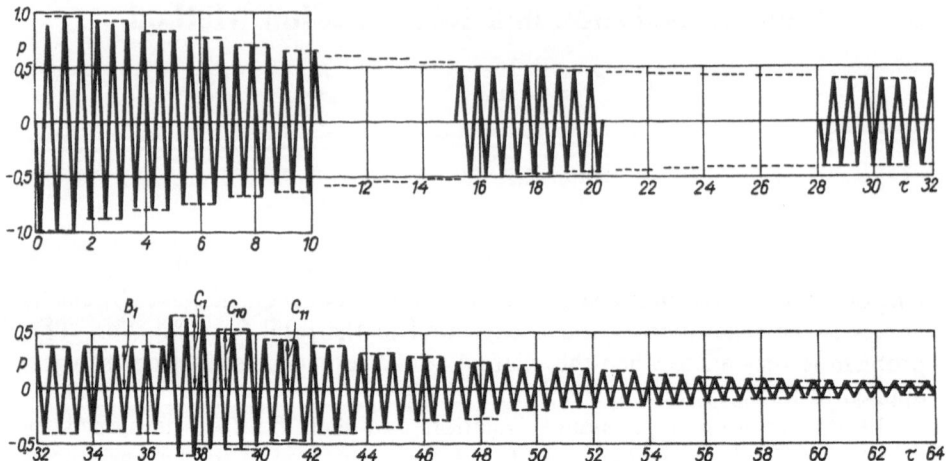

Fig. 5.1. The time dependence of the acoustic pressure scattered by a thin cylindrical shell with $h = 1/512$ at the resonance frequency $x = 10.97$ (corresponding to the third resonance of the S_0 wave) and for a large, but finite duration of the incident pulse $\tau_* = 36.66$, the number of wavelengths in the pulse being $s = 64$ [5.11, Fig. 3.16]

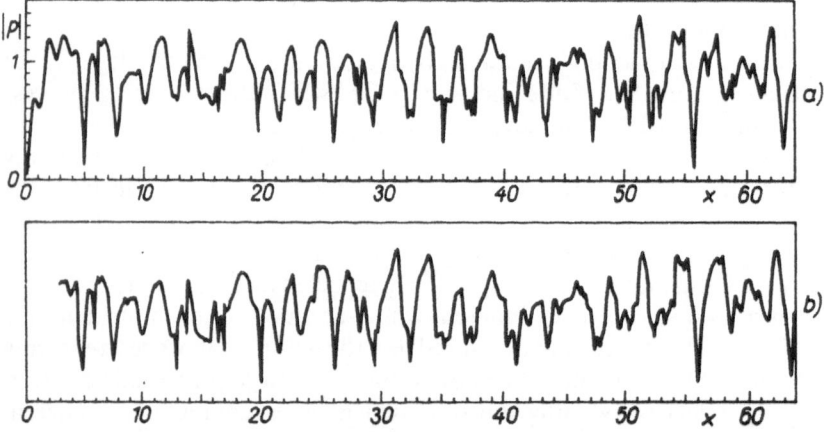

Fig. 5.2. The form function of the acoustic pressure scattered by a solid elastic cylinder. Curve **a** corresponds to the computed result and curve **b** to the measured one [5.5, Fig. 4]

The difference in normalization of the scattered acoustic pressure should be marked. We use the factor $[r(2 - a/r)]^{-1/2}$ and in papers on the resonance isolation and identification method it is chosen as $r^{-1/2}$. At large r ($r \to \infty$), as applies to the case considered below, the difference in normalization is $2^{-1/2}$.

In Fig. 5.3, curve a, the measured form function is presented on a larger scale than in Fig. 5.2, but in a more restricted range; the re-radiation form function,

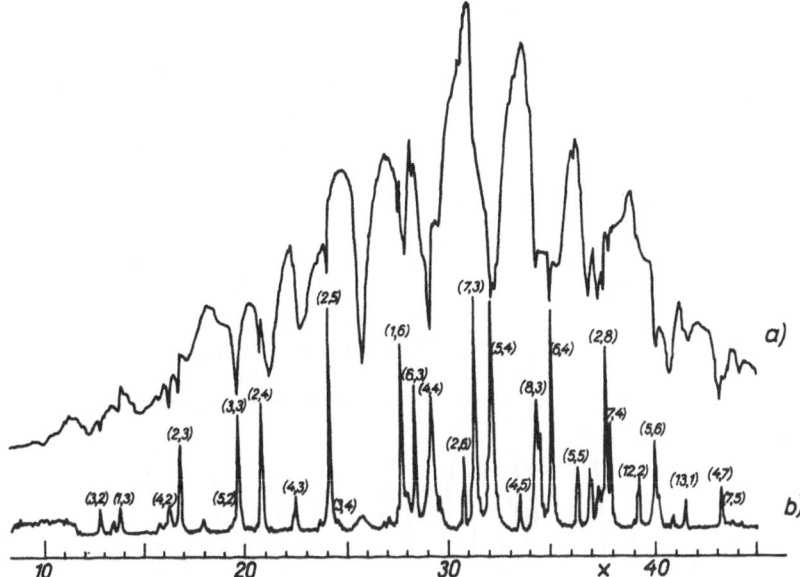

Fig. 5.3. The measured form function, curve **a** and the re-radiated form function, curve **b** for backscattering by a solid elastic cylinder [5.5, Fig. 7]

curve b, is shown together with it. The comparison of these curves shows that the peaks of the re-radiation form function correspond to the regions of rapid changes in the form function.

The plot of the form function can be considered as a "vibration" relative to the central portion of the "carrying" line (dashed line in Fig. 5.1). At $x \gtrsim 10$ this carrying line is formed by the contributions of the specularly reflected and refracted waves. It is significant that the extrema of the re-radiation form function are clearly distinguished by their position and their magnitude relative to the carrying line which is of small magnitude.

The measured curves of Fig. 5.3 are connected only by the abscissa axis (the ordinate axes are unrelated). This circumstance, apparently, is caused by the procedure used, and has no influence on the following considerations. The first stage of this method ends when the curve of the re-radiation form function is obtained. The maxima of this curve correspond to the resonance of the shock waves caused by the peripheral waves. In order to obtain the ordinal number n of the resonance the angular diagram of re-radiation is constructed for the x values corresponding to the maxima of the re-radiation form function. In Fig. 5.4, two such diagrams are given. One of them, curve a, corresponds to $x \sim 17$, and the other, curve b, to $x \sim 35$. The order of the resonance is equal to one half of the number of maxima (or minima) on the angular diagram of re-radiation. In the latter, on the ordinate axis the modulus of the acoustic pressure is presented and therefore the number of extrema on this curve is twice the number of wavelengths which fit the perimeter of the guiding circle. Curve a

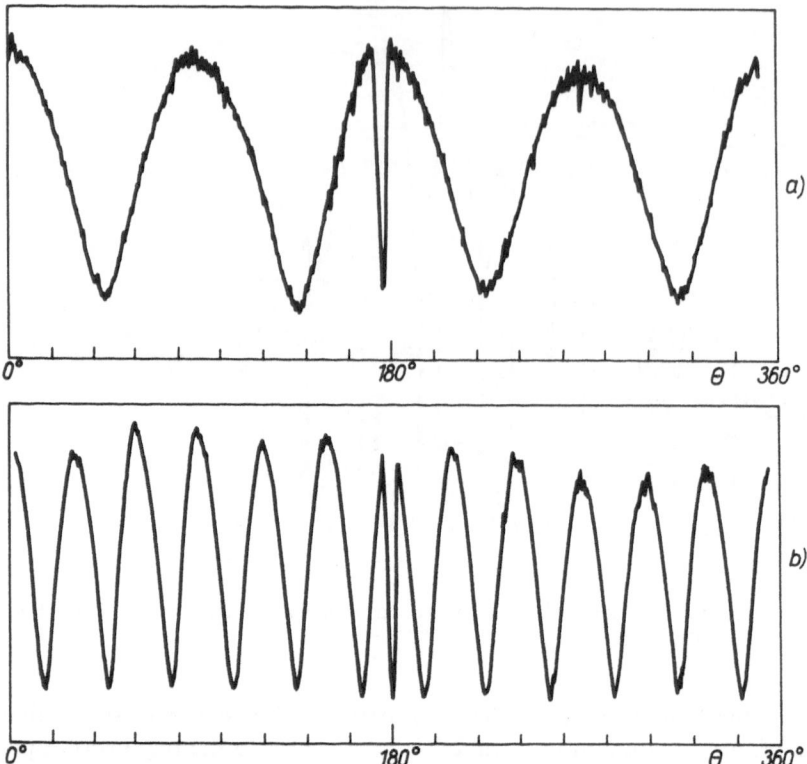

Fig. 5.4. The angular diagrams of re-radiation for two different resonances. Curve **a** corresponds to the resonance $(2, 3)$, and curve **b** to the resonance $(6, 4)$ [5.5, Fig. 8]

corresponds to the second resonance, and curve b to the sixth. The narrow dip on the angular diagram of re-radiation is connected with equipment realization of the method and should be disregarded. (At angle $\theta = \pi$ the receiver is situated between the transmitter and the scatterer and shadows the latter.)

Resonances of the same kind are gathered into families (see [5.12]). Two numbers (n, l) are used to label the resonance. The first of them indicates the order of the resonance and the second determines its family. Thus, for example, the resonances presented in Fig. 5.4 are labeled $(2, 3)$ and $(6, 4)$, respectively. In Fig. 5.3b all the resonances are labeled which lie in the considered range. At fixed x values corresponding to the resonance frequency, exactly n_* wavelengths of the peripheral wave fit the guiding circle, and, at this frequency, in the series on $\cos n\theta$ the main contribution is furnished by the term with $n = n_*$.

In Fig. 5.4 the ordinate scale is absent. Thus the magnitudes of neither the maxima nor the minima of this curve can be defined, but these are not needed for determining the ordinal number of the resonance.

The angular diagram of re-radiation corresponds to the steady-state regime (the elastic body radiates the energy stored up when the emitter is turned on).

Therefore, one would assume that the values of all the maximal magnitudes on the angular diagram of re-radiation should be equal and their envelope would be a straight line. Actually the envelope is not a straight line. The changeability of the maximal magnitudes on the angular diagrams of re-radiation can be connected either with the contributions of the waves of other families, or with equipment error. The latter can result from the fact that the measurement is carried out at a fixed delay after the end of the incident pulse and here it is easy to jump from the C_1 level to the C_{10} level (Fig. 5.1) which will lead to the error of the indicated type. The ordinal number n of a resonance, however, is not connected with the value of the maximal magnitude, but only with the number of the maxima and therefore is well determined.

5.2   Results

At present the results obtained by this method concern the scattering problems by solid elastic cylinder and thick-walled cylindrical shells.

5.2.1   Solid Elastic Cylinder

The results of the investigation of resonances of the peripheral waves for scattering by a solid aluminum cylinder in water [with parameters given by (5.1)] are gathered in the x–n plane. As a result the so-called Regge trajectories are obtained. The relative phase velocity of the wave

$$v_l^{\mathrm{ph}}(x) = c_l^{\mathrm{ph}}(x)/c \qquad (5.2)$$

can be easily obtained using the curves presented in Fig. 5.5

$$v_l^{\mathrm{ph}}(x) \equiv x/n . \qquad (5.3)$$

The less the Regge trajectory differs from a straight line, the weaker is the dispersion of the velocity of the wave.

As can be seen from the diagram of the Regge trajectories, two types of peripheral waves (with indices $l = 1$ and $l = 2$) can be generated beginning with $x = 0$, but the waves with $l > 2$ can be generated only above certain cutoff frequencies.

5.2.2   Thick-Walled Cylindrical Shell

The Regge trajectories strongly depend on the relative thickness of the shell. In Figs. 5.6–8, the Regge trajectories are shown at three different h values: $h = 1/3$ (more exactly at $b/a = 0.67$), $h = 1/5$ and $h = 1/10$. The physical parameters of the problem are given by (5.1).

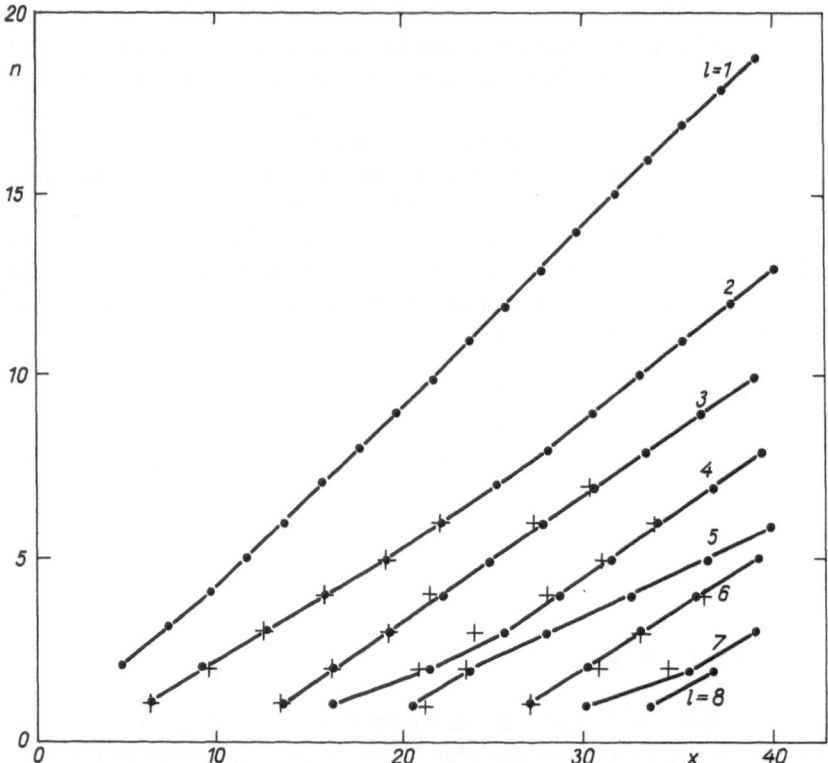

Fig. 5.5. The Regge trajectories of peripheral waves. Results of computation are marked by solid circles, and those of measurement by crosses [5.5, Fig. 14]

In each plot three trajectories are presented: index $l = 1$ denotes the Rayleigh wave, index $l = 2$ the S_0 wave, and index $l = 3$ the A_1 wave (more precisely, with diminishing values of h the peripheral Whispering Gallery waves of indices $l = 2$ and $l = 3$ are transformed into the Lamb-type waves: zero-order symmetric, S_0, and first-order antisymmetric, A_1, respectively). The smaller the value of h, the farther from each other the Regge trajectories of waves of different types are situated. The thinner the shell, the higher the cutoff frequency of the $l = 3$ wave. This experimental fact is in good agreement with what was found before by computation in [5.13].

The comparison of the Regge trajectories of the $l = 1$ wave at different values of h shows that with h diminishing, the trajectories are shifted into the domain of small x, i.e. at fixed x the phase velocity of this wave will be smaller for smaller h.

Fig. 5.7. The Regge trajectories of three families of resonances of peripheral waves ($l = 1-3$) in the case of backscattering by a thick-walled cylinder with $b/a = 0.80$. Results of computation are marked by solid circles, and those of measurement by crosses [5.5, Fig. 16]

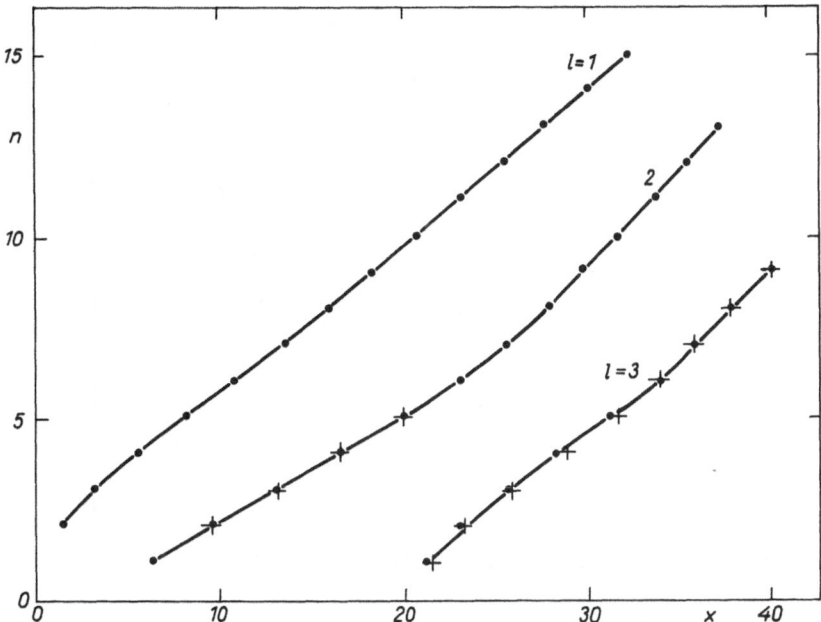

Fig. 5.6. The Regge trajectories of three families of resonances of peripheral waves ($l = 1$–3) in the case of backscattering by a thick-walled cylinder with $b/a = 0.67$. Results of computation are marked by solid circles, and those of measurement by crosses [5.5, Fig. 15)

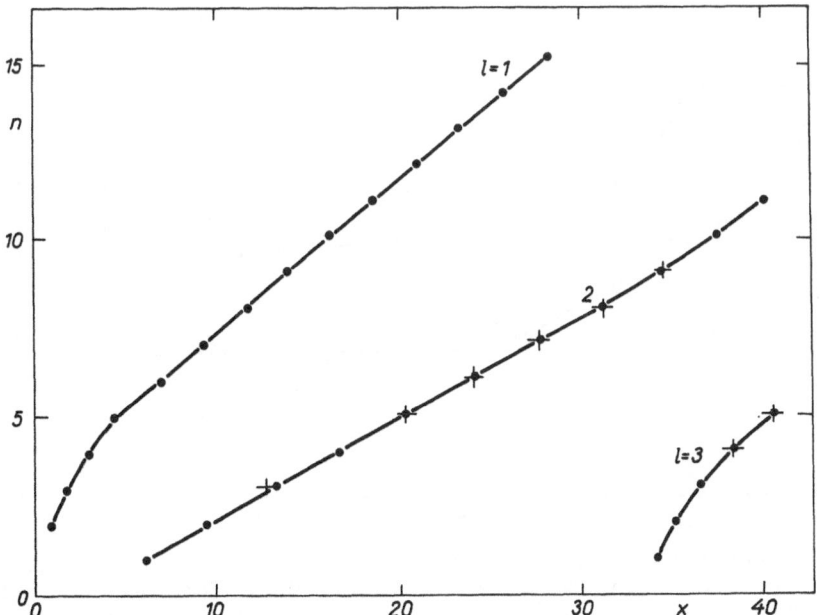

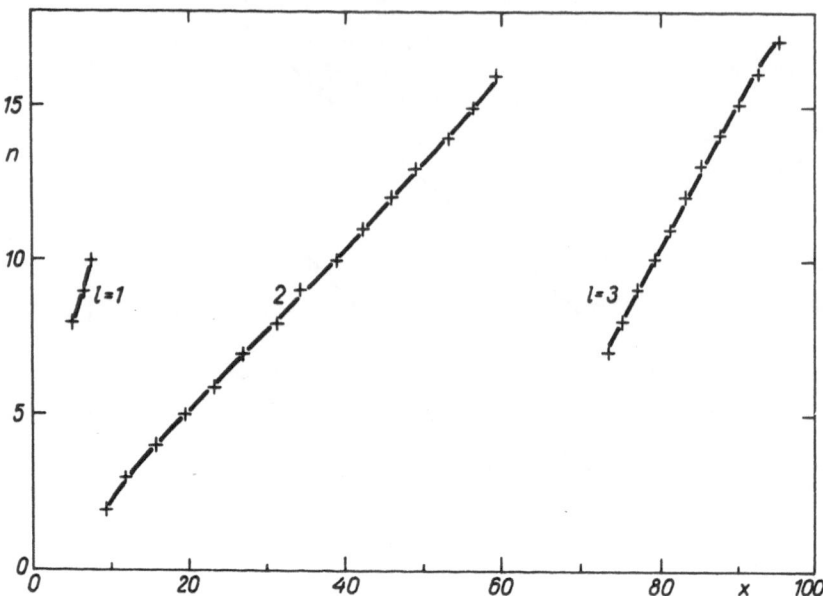

Fig. 5.8. The Regge trajectories of three families of resonances of peripheral waves ($l = 1$ corresponds to the Rayleigh wave, $l = 2$ to the S_0 wave and $l = 3$ to the A_1 wave) in the case of scattering by a (relatively) thick-walled cylinder with $b/a = 0.90$. Results of measurement are marked by crosses [5.5, Fig. 20]

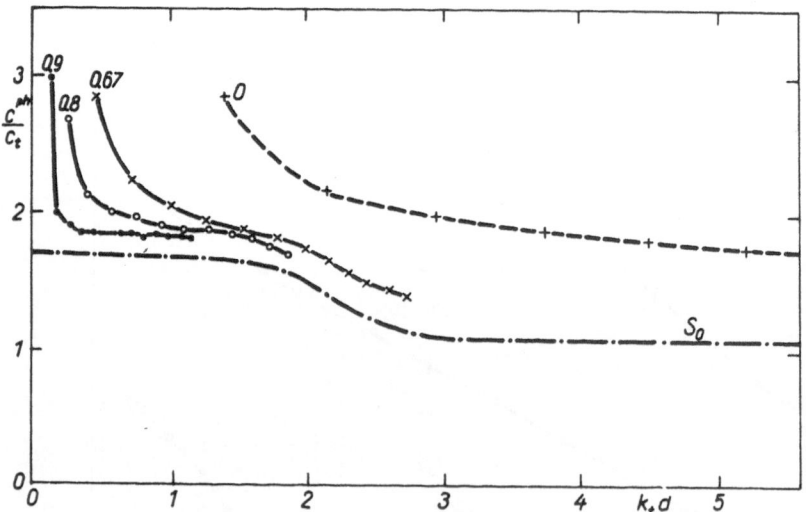

Fig. 5.9. The dispersion curves of the phase velocity of the $l = 2$ wave. Skew crosses correspond to $b/a = 0.67$, open circles to $b/a = 0.80$ and solid circles to $b/a = 0.90$. The solid cylinder result (with $b/a = 0$) is marked by dashed line, and the thin plate result by dot-and-dash line [5.5, Fig. 21]

At $h = 1/3$ the velocity of the $l = 2$ wave is dispersive; with h diminishing the dispersion decreases and beginning with $h = 1/5$ the wave becomes almost non-dispersive (the Regge trajectories do not differ from straight lines).

The thinner the shell, the bigger the velocity of the $l = 3$ wave at fixed x.

For comparison, instead of the wave radius on the abscissa axis the wave thickness (or half-thickness) is often used. This parameter is very natural for

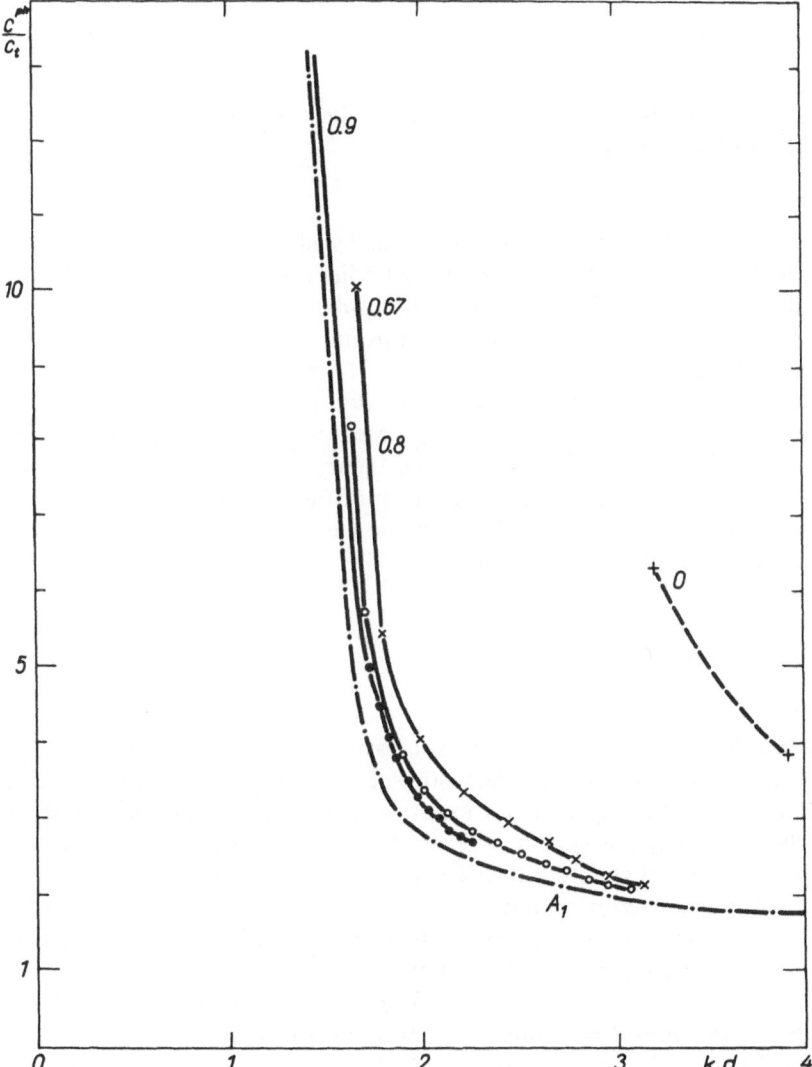

Fig. 5.10. The dispersion curves of the phase velocity of the $l = 3$ wave. Crosses correspond to $b/a = 0.67$, open circles to $b/a = 0.80$ and solid circles to $b/a = 0.90$. The solid cylinder result (with $b/a = 0$) is marked by a dashed line, and the thin plate result by a dot-and-dash line [5.5, Fig. 22]

a plane thin layer, from where it was transferred into the scattering problem. In the case of a "dry" layer, i.e. without any contact with liquid, instead of the wave number in the liquid the wave number of the transverse wave is used

$$k_t = \omega/c_t \ . \tag{5.4}$$

In the following figures the dependence of the relative phase velocity of the $l = 2$ wave (Fig. 5.9) and the $l = 3$ wave (Fig. 5.10) on the wave half-thickness of the transverse wave are presented

$$k_t d = \frac{\omega}{c_t} \frac{(a-b)}{2}, \qquad 2d = a - b \ . \tag{5.5}$$

In contrast with the velocity used before, here the phase velocities are normalized not by c, but by c_t

$$v_l^{ph} = c_l^{ph}/c_t \ . \tag{5.6}$$

Besides the curves relating to the thick-walled shells, the dispersion curves of the phase velocities of waves on two limiting bodies are given: for a solid elastic cylinder (dashed line) and a thin plate (dot-and-dash line). The comparison of the presented curves shows that with h diminishing, the $l = 2$ wave transforms into the S_0 wave, and that with $l = 3$ into the A_1 wave.

The experimentally obtained dispersion curves of the S_0 wave revolving around the cylindrical shell are in good agreement with independent numerical results obtained by the resonance scattering theory [5.14, Figs. 8–10]. On the level of the Regge trajectories the comparison of the measured and computed results is shown in Figs. 5.6 and 5.7, where the measured results are shown by circles, and the computed ones by crosses.

5.3   Acoustic Spectrogram

When the positions on the x axis of the successive resonances (index n) of each wave type (index l) are known, the acoustic spectrogram of the elastic scatterer can be constructed. In Fig. 5.11, an example of such spectrograms for three different elastic scatterers is shown: solid elastic cylinder (with $b/a = 0$) and two thick-walled cylindrical shells (with $b/a = 0.67$ and $b/a = 0.80$, respectively).

The acoustic spectrogram can be used as a distinctive tag, or marker, of the elastic scatterer. One can see, particularly from Fig. 5.11, that the acoustic spectrogram is rather sensitive to the relative thickness of the shell.

As was noticed by the authors of [5.7], with a diminishing value of h the resonances of the family $l = 2$ shift rather little on the x axis (this can also be observed in Figs. 5.6–8); on the other hand, the resonances of the family $l = 3$ shift noticeably in the range of higher frequencies.

As a whole, the acoustic spectrogram can be used for obtaining the properties, for example, the material and thickness, of the elastic scatterer. Using

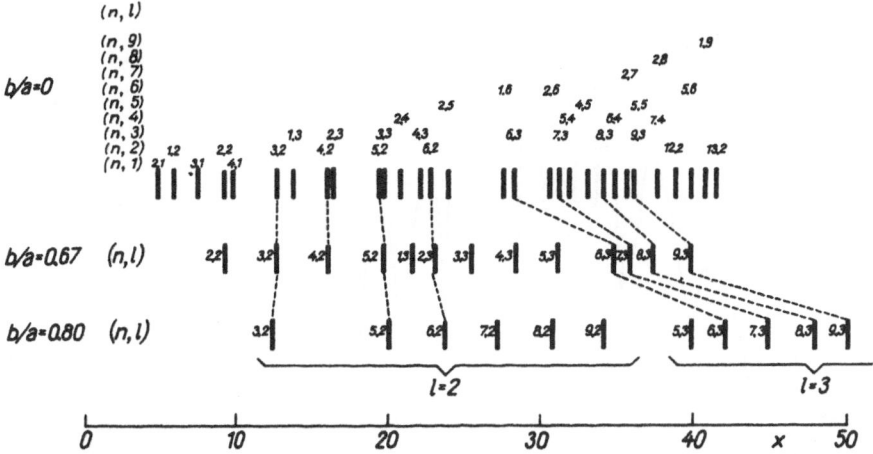

Fig. 5.11. Acoustic spectrograms of elastic cylinders in water for $b/a = 0$ (solid cylinder), $b/a = 0.67$ and $b/a = 0.80$ [5.7, Fig. 4]

graphs of the dependence of the (n, l) resonance position versus h constructed beforehand, the latter value can be obtained. Using the presented spectrogram, it is easy to find the thickness of the shell from the position on the x axis of one particular, say the sixth or the ninth, resonance of the $l = 3$ wave. Here it is assumed that the scatterer is a cylindrical shell with material parameters known beforehand.

5.4 Resonances of Guided Waves

A careful investigation of the re-radiation form functions shows that they contain resonances which cannot be described by the procedure of the resonance scattering theory in the case of normal incidence. In Fig. 5.12, such a family of resonances is shown as curve b. These resonances are labeled by arrows, and can be identified in the same way as the others. In Fig. 5.13, two examples of identification are shown. They concern the resonance with index $n = 2$ at $x = 6.5$ and the resonance with index $n = 3$ at $x = 8.8$.

The generation of these additional (or supplementary) resonances is connected with the angular diagram of the emitter. In fact, in the experiment the cylinder is actually insonified not by a plane wave with direction of propagation normal to the longitudinal axis of the cylinder, but by a wave at oblique incidence. Therefore in the cylinder the generated waves propagate along its longitudinal axis. To verify this supposition the following experiment was carried out. The receiver was placed at some distance from the emitter along the axis of the cylinder on the same straight line as the emitter (see scheme in

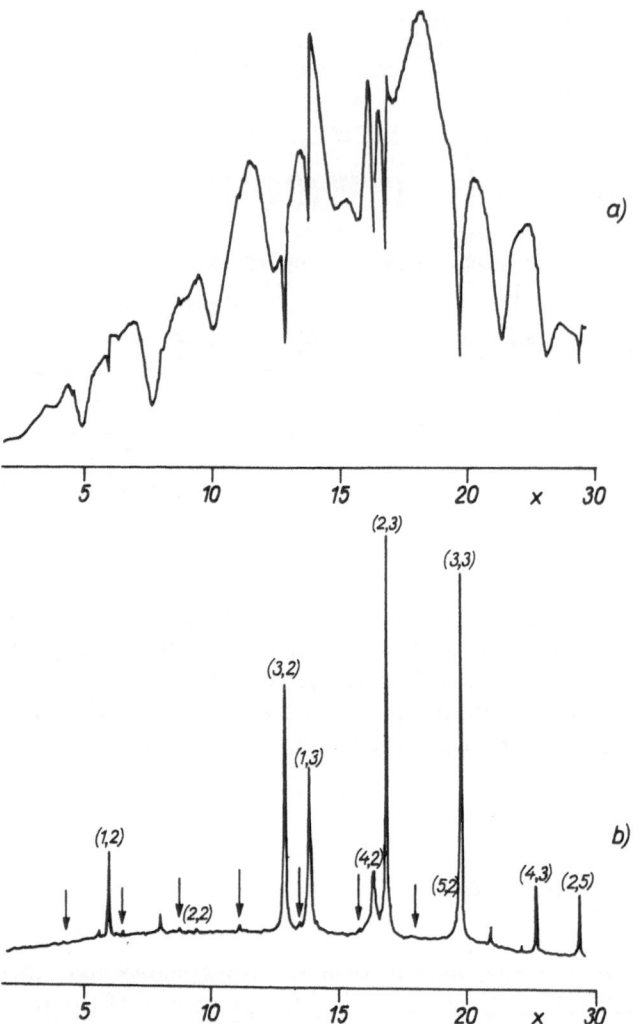

Fig. 5.12. The measured form function, curve **a** and the re-radiation form function, curve **b** for the case of scattering by a solid aluminum cylinder in water [5.6, Fig. 3]

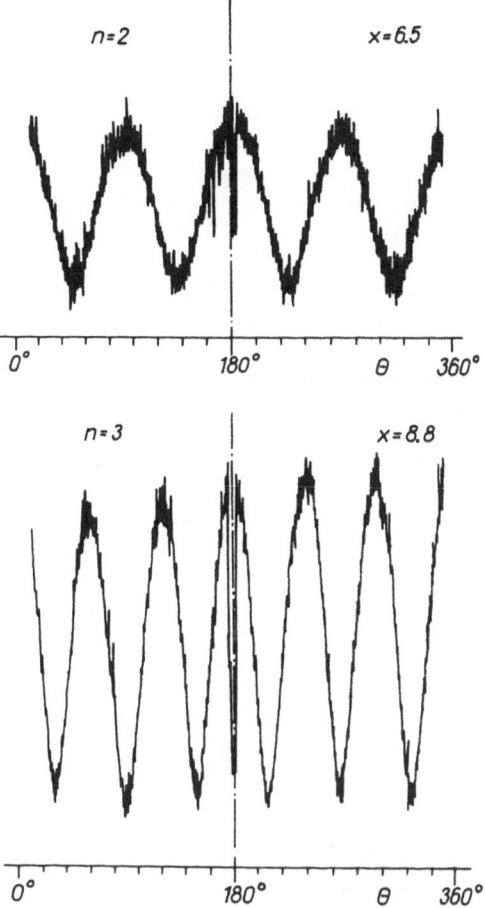

Fig. 5.13. The angular diagrams of re-radiation for the identification of two supplementary resonances $n = 2$ (at $x = 6.5$) and $n = 3$ (at $x = 8.8$) [5.6, Fig. 6]

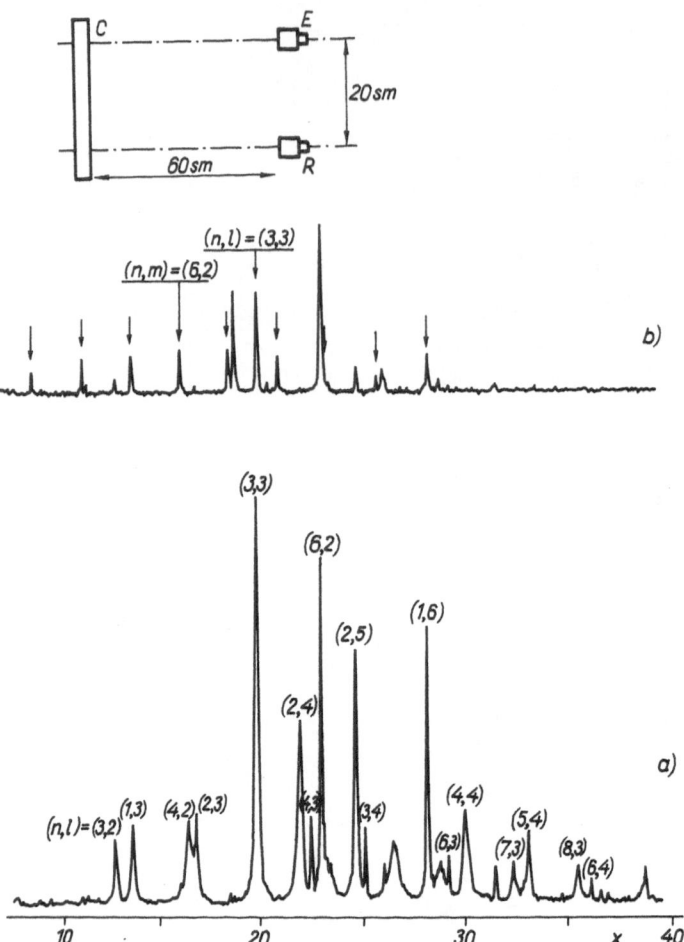

Fig. 5.14. The measured form functions of the acoustic pressure scattered by a solid aluminum cylinder. Curve **a** with one emitter, used as an emitter-receiver; curve **b** with emitter and receiver separated along the axis of the cylinder. On the top, the scheme of the experiment is shown: E – emitter, R – receiver, C – cylinder [5.6, Fig. 7]

Fig. 5.14). In such a case the shock waves, caused by the peripheral waves propagating on the cylinder along helical paths, arrive at the receiver. In Fig. 5.14, curve b, the received re-radiation form function is shown. For comparison, the re-radiation form function at the emitter (in the regime of receiver), curve a, is also given. It is significant that on curve a the resonances of waves propagating along helical paths can also be observed, but they are not so well observed as in curve b. In Fig. 5.15, the Regge trajectories of these supplementary resonances are shown. The measured positions of resonances (crosses

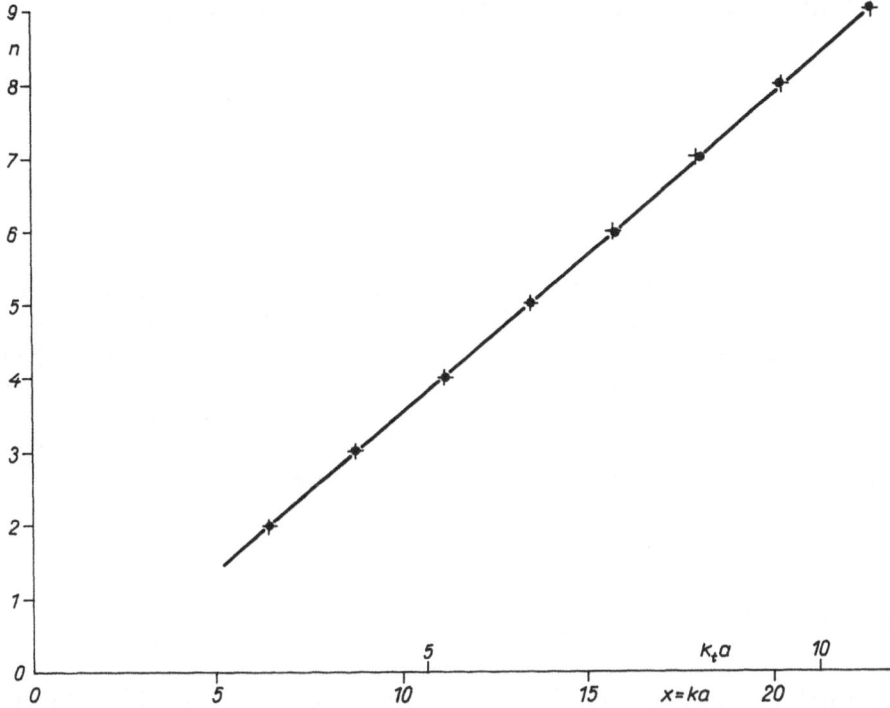

Fig. 5.15. The Regge trajectories of the supplementary resonances for the case of a solid cylinder. The numerical results are marked by solid circles and the measured ones by crosses [5.6, Fig. 9]

on the figure) are in good agreement with the computed ones (circles). The latter are found from the solution of the three-dimensional problem under the assumption of a small angle of incidence.

The author is grateful to Professor J. Ripoche for his kindness in sending him the preprints of papers on the method of resonance isolation and identification, and thus acquainting him with some of the results before their publication.

6 Pulsed Resonance Identification Method

Here, in addition to the method described in Chap. 5, an alternative experimental method is outlined for isolation of the resonance components of the peripheral waves. The distinguishing feature of this approach is the utilization of short in time (broadband) incident pulse. This ultrasonic method exhibits good capabilities for detection, characterization, and classification of resonances. This approach allows one to reduce the time needed for an experimental investigation, admittedly at the expense of the exactness of the result.

6.1 The Essence of the Method

As is well known, beginning from pioneering experimental [6.1, 2] and computed [6.3–5] results for a single incident pulse that is short compared to the physical dimension of the body, the echo structure of backscattered acoustic pressure from elastic bodies of cylindrical and spherical shape consists of multiple periodic pulses. The first of these is caused by the specular reflection and the subsequent ones are generated by the slip and shock waves corresponding to the creeping and peripheral waves revolving around the body, respectively. Depending on the frequency and length of the incident pulse and the inner structure of the elastic body, the successive pulses in the echo can either overlap or be separated. In Fig. 6.1, the time dependence of the acoustic (echo) pressure backscattered by a cylindrical shell with $h = 1/512$ is shown. The physical parameters of the problem are given by (9.3). The incident pulse is three periods in length. For such a short pulse the subsequent pulses in the time plot are clearly distinguished. Each of them is also three periods in length. In the example considered, the structure of the time dependence is very simple because it is caused only by two waves, namely the specularly reflected wave (the refraction in the thin cylindrical layer included) and the S_0 wave. At the chosen carrier frequency of the incident pulse $x_0 = 12.81$, the contribution of the creeping wave in the time plot is small and can be neglected. In the problem considered, the form function is also very simple (Fig. 6.2). The chosen carrier frequency $x_0 = 12.81$ corresponds to the 3.5 antiresonance. For a short incident pulse, this particular frequency practically does not affect the structure of the time dependence.

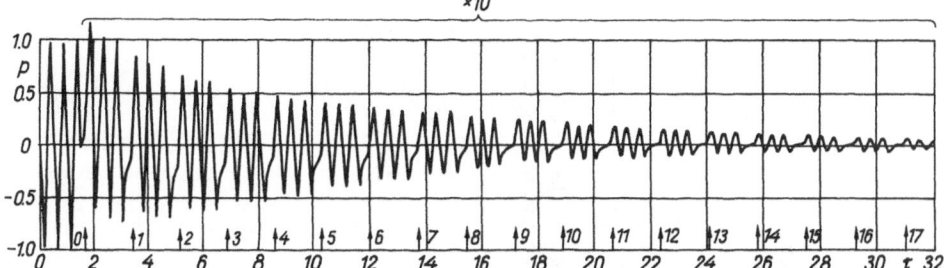

Fig. 6.1. The time dependence of the acoustic pressure backscattered by a thin empty Armco iron cylindrical shell with $h = 1/512$ in water. The incident pulse has a carrier frequency $x_0 = 12.81$ and is three periods in length ($s = 3$). The pulses corresponding to the successive circumnavigations of the S_0 wave are multiplied by factor 10. Arrow and number label the number of the full turns around the shell. Geometrical divergence of the rays should be taken into account by an additional spreading factor [6.18, Fig. 3.15]

We shall now turn to the description of the method, which consists of three stages. In the first stage, as used in the calibration technique for acoustic scattering measurements [6.6], the background is excluded; in the second stage, the re-radiation form function is obtained, and in the third stage the observed resonances are identified.

The background component is excluded by suppression of the specularly reflected pulse in the echo time dependence. This is achieved by filtration of the time dependence [6.7].

A broadband transducer is used in the pulse-echo mode. The target is insonified by ultrasonic short pulses, and oriented in order to get a maximum backscattered echo. The elastic body and the transducer, working as emitter-receiver, are immersed in a tank whose dimensions are very large in comparison with the wavelength of sound in water. The electronic chain includes an amplifier, a gate and a spectrum analyzer. The gate can be moved, and its length adjusted, to isolate the desired portion of echo. The backscattered signal gated in the temporal domain is transformed into the frequency domain by the spectrum analyzer. The frequency spectrum is then recorded on a plotter.

When the specular reflection is included in the time series (Fig. 6.3a) whose Fourier transform is calculated, we obtain the uncorrected form function. This spectrum includes the background contribution and a partial elastic response, because of the limitation of the gate. The minima observed in such plots correspond to resonances of the body.

A typical unnormalized form function is given in Fig. 6.4. The amplitude scale on the ordinate is logarithmic and the values of the dimensionless parameter $x \equiv ka$ are indicated on the abscissa. The exclusion of the specular signal (Fig. 6.3b) from the gated signal leads us directly to the re-radiation form function as shown in Fig. 6.5. The minima of the form function in Fig. 6.4 are replaced by maxima which indicate the pure resonances. The scale of amplitude in Fig. 6.4 is linear. For a given length of the gate the re-radiation form function

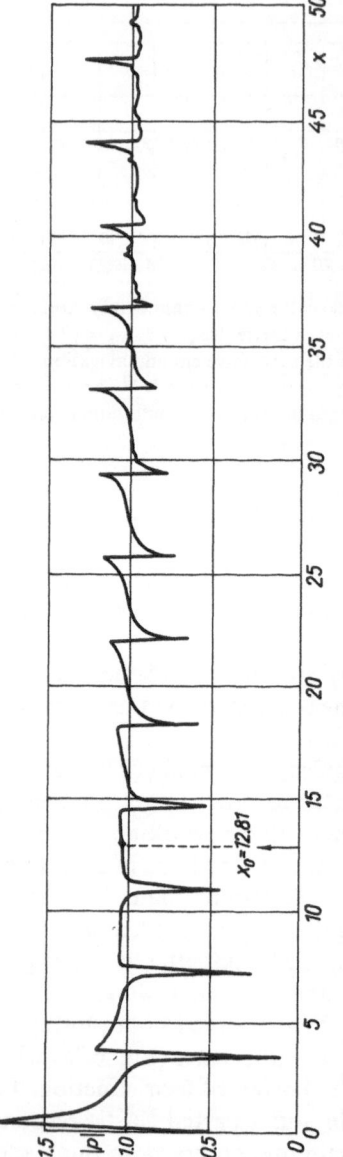

Fig. 6.2. The form function of the acoustic pressure backscattered by a thin empty Armco iron cylindrical shell with $h = 1/512$ in water. The carrier frequency $x_0 = 12.81$, which is labeled by the arrow, corresponds to the 3.5 antiresonance of the S_0 wave [6.18, Fig. 3.6]

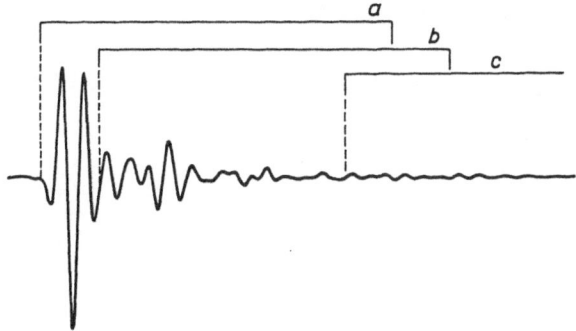

Fig. 6.3. Different positions of the gate on the echo time plot [6.7, Fig. 1]

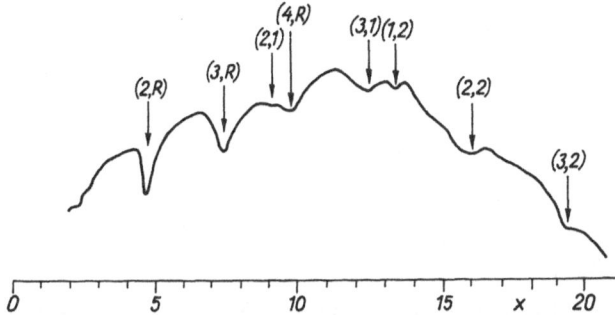

Fig. 6.4. Uncorrected form function. The identified resonances (n, l) are indicated by arrows [6.7, Fig. 2]

is modified as the delay increases. When the gate is sufficiently delayed above a "critical delay", the re-radiation form function remains unchanged (see Fig. 6.6).

Depending on the delay and length of the gate in the temporal domain, the resonances of some types of waves can be observed in the re-radiation form function. The wider the gate, the bigger the number of resonances distinguished in the re-radiation form function.

To determine the delay of the gate for observation of the complete elastic response, the author of [6.7] proceeded as follows. Initially the electronic gate was moved onto the specular echo and its length was adjusted until the form function (which is similar to that of the incident pulses) began to be distorted; then, the extreme position of the gate defines the position at which the gate has to be opened to isolate just the elastic response. The duration and the location of the gate are critical parameters.

If the length of the gate is too short, low resonances can be absent. It appears that a critical length of the gate exists for which any modification is visible in the shape of the form function. The length of the gate has to be larger or equal to this value for the measurements. This value varies with the typical dimension of the scatterer.

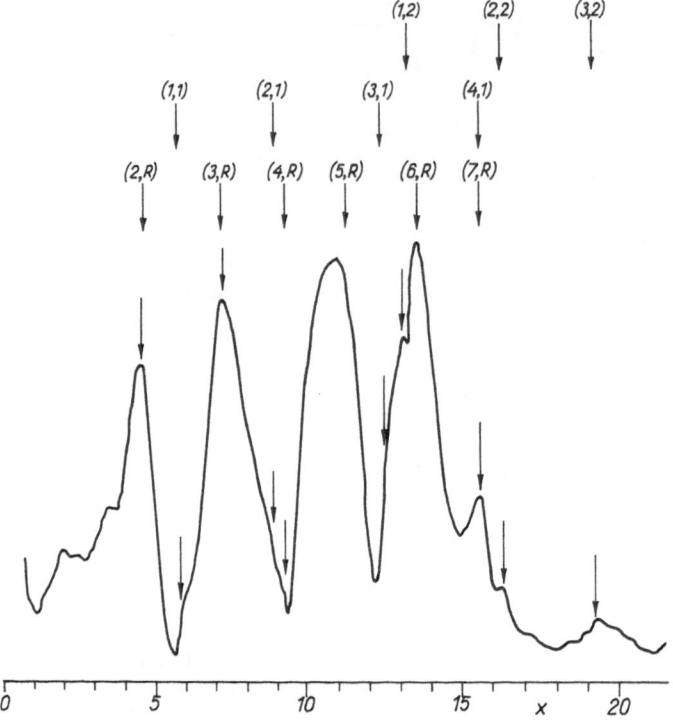

Fig. 6.5. Re-radiation form function observed when the gate is near the specular echo [6.7, Fig. 3]

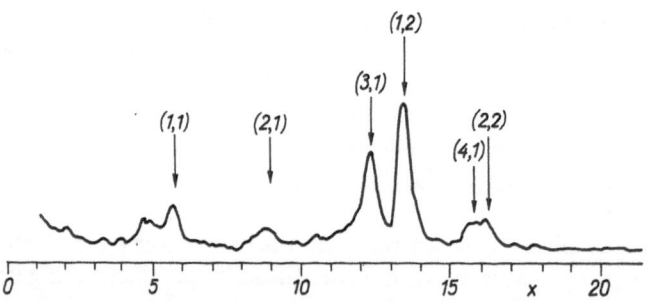

Fig. 6.6. Re-radiation form function observed when the gate is far from the specular echo [6.7, Fig. 4]

The order of the resonance and its family are not defined as yet. The arrows and numbers in Figs. 6.4–6.6 identifying the (n, l) values of the resonances have been obtained by comparison with theoretical results given in [6.8–10].

The impulse method, which allows the isolation of resonances and in addition the identification of each eigenmode of the scatterer, was independently proposed in [6.11]. Following [6.12], we shall describe this method for the case

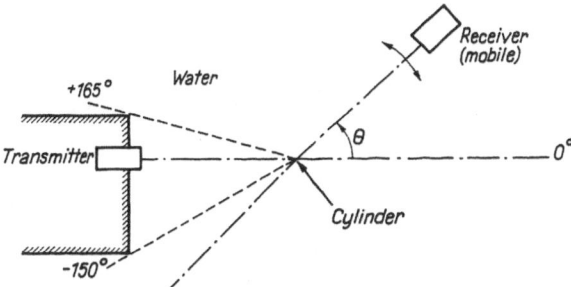

Fig. 6.7. Diagram of experimental setup [6.12, Fig. 1]

of scattering by a solid elastic cylinder. The scatterer is insonified by a short ultrasonic impulse. The transmitter is maintained steady at some distance from the cylinder. The scatterer is in the far field of the transducer in the whole bandwidth studied. The axis of the incident beam is perpendicular to the cylinder axis. The receiver is at the same distance from the target as the transmitter and is rotated around the scatterer's axis. The transmitter and receiver axis are both perpendicular to the cylinder's axis and cross one another (Fig. 6.7). The angle between the receiver and the transmitter was varied from $-150°$ to $+165°$ (in the experimental setup used it was impossible to achieve a wider range) because the receiver must not be placed between the transmitter and the scatterer. The angle θ is scanned in steps of 5 or 10 degrees with a precision of 0.02 degrees.

At each position of the receiver the signal is sent to a wideband amplifier and then sampled and recorded by a computer. Signal processing is very simple. Firstly, the part of the scattered signal which contains neither the direct signal radiation nor the specularly reflected signal was selected through a digital window. The following step is a Fourier transform of the time signal using the FFT algorithm and 2^{10} sampled points. The obtained re-radiation form function is recorded in the computer memory. If this procedure is applied to m values of the receiver angle θ, a number of re-radiated form functions m ($m = 32$ or 64) will be obtained.

Once this series of the re-radiated form functions is in the computer's memory, further processing is carried out. Firstly, the resonance frequencies $x_{n,l}$ are measured, then for each such frequency the magnitude of the re-radiation form function is displayed on a polar plot against θ. This can be achieved either manually or via the computer. In Fig. 6.8, as an example, two such angular diagrams are presented. The left-hand plot corresponds to the $(2, R)$ resonance at $x_{exp} = 3.84$ and the right-hand plot to the $(3, 1)$ resonance at $x_{exp} = 12.83$. The measurements are carried out for a solid copper cylinder immersed in water.

The order of the resonance n is equal to half the number of maxima on the angular diagram. The angle $\theta = 180°$ corresponds to backscattering and the relevant lobe is strongly distorted. It should be mentioned that in other directions the lobes of the angular diagrams have dents which are particularly

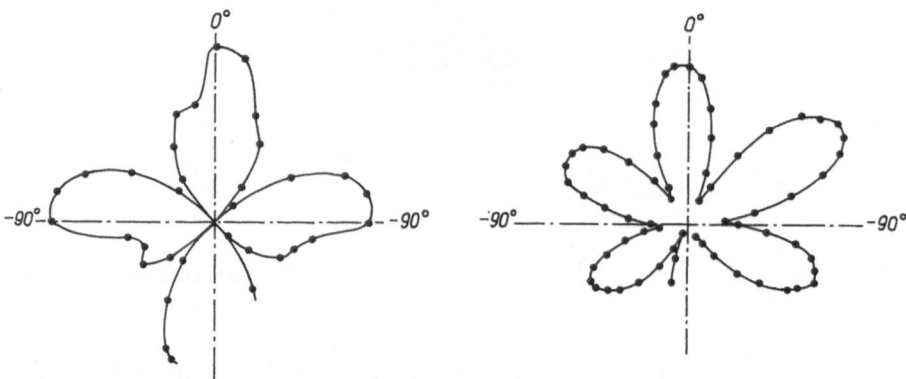

Fig. 6.8. Angular diagram for identification of the $(2, R)$ resonance (left) and the $(3, 1)$ resonance (right) for a solid copper cylinder in water [6.12, Figs. 3, 4]

distinguishable on the left-hand diagram of Fig. 6.8. Since the problem is symmetric relative to the axis $\theta = 0$, the angular diagrams of re-radiation have also to be symmetrical. In fact, the measured angular diagrams are asymmetric. The lobes are asymmetric with respect to their axes, and the diagram as a whole is asymmetric with respect to the $\theta = 0$ axis. However, the angular diagram answers its purpose completely as it permits us to determine unambiguously the order of the resonance.

When the third stage is carried out, the resonances have been identified on the re-radiation form function. The single-type resonances are thus grouped into families. As in any other method, the experiment does not allow us to determine the family of the resonance. For this purpose either the dispersion curves of the phase velocity or the Regge poles can be used.

The above-described method allows very quick isolation and identification of the resonances of an axisymmetric scatterer. The number of resonances studied depends on the range of values of x contained inside the bandwidth of the transmitted ultrasonic pulse. The accuracy of the method strongly depends on the position and the width of the gate and the discretization of the temporal domain used. The farther the gate is situated from the specular reflected pulse, the better the background in the re-radiation form function will be suppressed. The magnitude of the re-radiated pulses diminishes with increasing number of turns (and τ). Therefore the gate cannot be removed far enough.

6.2 Results

6.2.1 Solid Elastic Cylinder

In Figs. 6.4–6, the results of the experimental investigations of scattering by a solid elastic cylinder in the range $0 \leqslant x \leqslant 20$ have been presented. In the total form function, Fig. 6.4, the first three resonances of the Rayleigh wave $(2, R)$, $(3, R)$ and $(4, R)$, the two resonances of the first $(l = 1)$ Whispering Gallery wave $(2, 1)$ and $(3, 1)$, and the first three resonances of the second $(l = 2)$ Whispering Gallery wave $(1, 2)$, $(2, 2)$ and $(3, 2)$ can be observed. When the specularly reflected pulse is incompletely suppressed (Fig. 6.5), in addition to these resonances three more resonances of the Rayleigh wave $(5, R)$, $(6, R)$ and $(7, R)$ and two more resonances of the $WG1$ wave $(1, 1)$ and $(4, 1)$ are isolated and identified. The complete suppression of the specularly reflected pulse (see Fig. 6.6) permits us to distinguish clearly the resonances $(1, 1)$, $(2, 1)$, $(3, 1)$ and $(4, 1)$, and $(1, 2)$ and $(2, 2)$ which in Fig. 6.5 only appear as some irregularities of the curve. In Table 6.1, the experimental positions of the resonances (in the x domain) are compared with the theoretical predictions [6.9]. The agreement is generally quite good. In [6.7] the material parameters of aluminum and water are not given. In Chap. 7, the problem of scattering by an aluminum cylinder in water is considered in detail. The computation is carried out with parameters defined by (7.2). If the data presented in Tables 7.2–4 for the I variant are multiplied by a factor 0.966, they coincide with those given in Table 6.1. The computed form function (Fig. 7.1) is in good agreement with the measured ones (Figs. 6.4–6).

The influence of the elastic parameters of the solid cylinder immersed in water on both the form function and the re-radiation form function has been studied in [6.13] at low x values $(x < 30)$ for cylinders made of various materials (copper, iron, brass, stainless steel, bronze). All the resonances in the x range considered are identified. They form four families: R, $WG1$, $WG2$ and $WG3$. The upper orders of resonances have been observed experimentally for a copper cylinder, namely $(17, R)$, $(11, 1)$, $(6, 2)$ and $(6, 3)$. The difference between the measured and computed in [6.14] position of the resonances on the x axis, as a rule, is rather small and is approximately 0.2. In some cases this difference, however, is large (0.5). Using the results of the measurement, the Regge trajectories were calculated. They are similar to those presented in Chap. 5.

Table 6.1. The experimentally measured (first column) and the resonance positions computed according the resonance scattering theory (second column) of the Rayleigh wave and first two Whispering Gallery waves [6.7, Table I]

n	The Rayleigh wave		The $WG1$ wave		The $WG2$ wave	
	1	2	1	2	1	2
1			5.8	5.8	13.3	13.5
2	4.8	4.7	9.2	9.17	16.2	16.3
3	7.6	7.3	12.5	12.5	19.0	19.1
4	10.0	9.6	15.6	15.8		
5	11.1	11.7				
6	13.8	13.8				
7	15.6	15.8				

6.2.2 Cylindrical Shells

Systematic investigations of the backscattered acoustic pressure from thin shells were performed in [6.15]. Various materials (duraluminum, stainless steel, copper) and various relative thicknesses of the shells are considered. The shells are filled with air and immersed in water. The frequency band $f(a - b) < 1$ was considered (here a, b are the outer and inner radius of the shell, $f = \omega/2\pi$, ω is the circular frequency). In this frequency range the incident wave generates two peripheral waves in the shell, namely S_0 and A. As the results of measurements have shown, the A wave is generated only in a restricted frequency range depending on its relative thickness. It was demonstrated that the phase velocity of the A wave is maximal in this range. The properties of the S_0 and A wave are investigated in detail in Chap. 9, therefore we shall not discuss this matter here.

Backscattering by thicker shells (with $0.01 \leqslant h \leqslant 0.5$) was studied in [6.16]. Here, as in [6.7], the resonance scattering theory was used for identification of the resonances.

In the case of very thick shells (with $b/a = 0.85$) in the low frequency band it is shown that the incident plane wave generates both the A and A_0 waves in the shell. The resonances of the latter can be clearly observed beginning from $n = 10$. A similar situation appears in the upper part of Fig. 8.4, computed for a shell with $h = 1/10$.

In Fig. 6.9, the form function of the acoustic pressure backscattered by an aluminum shell with $h = 1/25$ in water is shown. The computation was carried out using the following parameters

duraluminum: $\rho_1 = 2.8 \times 10^3 \text{ kg/m}^3,$ $\quad c_l = 6370 \text{ m/s},$ $\quad c_t = 3130 \text{ m/s}$,

water: $\qquad \rho = 1 \times 10^3 \text{ kg/m}^3,$ $\quad c = 1480 \text{ m/s}$. $\qquad\qquad$ (6.1)

The computation was performed with step size $l_x = 1/256$. The resonances of two peripheral waves can be observed here. The S_0 wave resonances, as always, are labeled by $(n, 0)$, and those of the A wave by $[n[$. On a larger scale along the x axis, the same form function is shown in Fig. 6.10a. In the x range considered four resonances of the A wave can be clearly observed. The corresponding partial mode resonances are presented in Fig. 6.10b. With n increasing, the magnitude of the A wave slowly diminishes and the damping factor increases rapidly. The width Γ of the resonance is measured directly from the isolated modal resonance curve at an amplitude equal to $2^{-1/2}$ of its maximum. The results of experimental and theoretical investigations of the A wave are presented in Table 6.2, and are in good agreement with each other. With n (and x) increasing, the relative phase velocity of the A wave monotonically tends to unity.

The measured resonance positions of the S_0 wave are in rather good agreement with the computed ones (Table 6.3). It is true that with n (and x) increasing the difference in the resonance position increases. At $n = 30$ it is more than a unit.

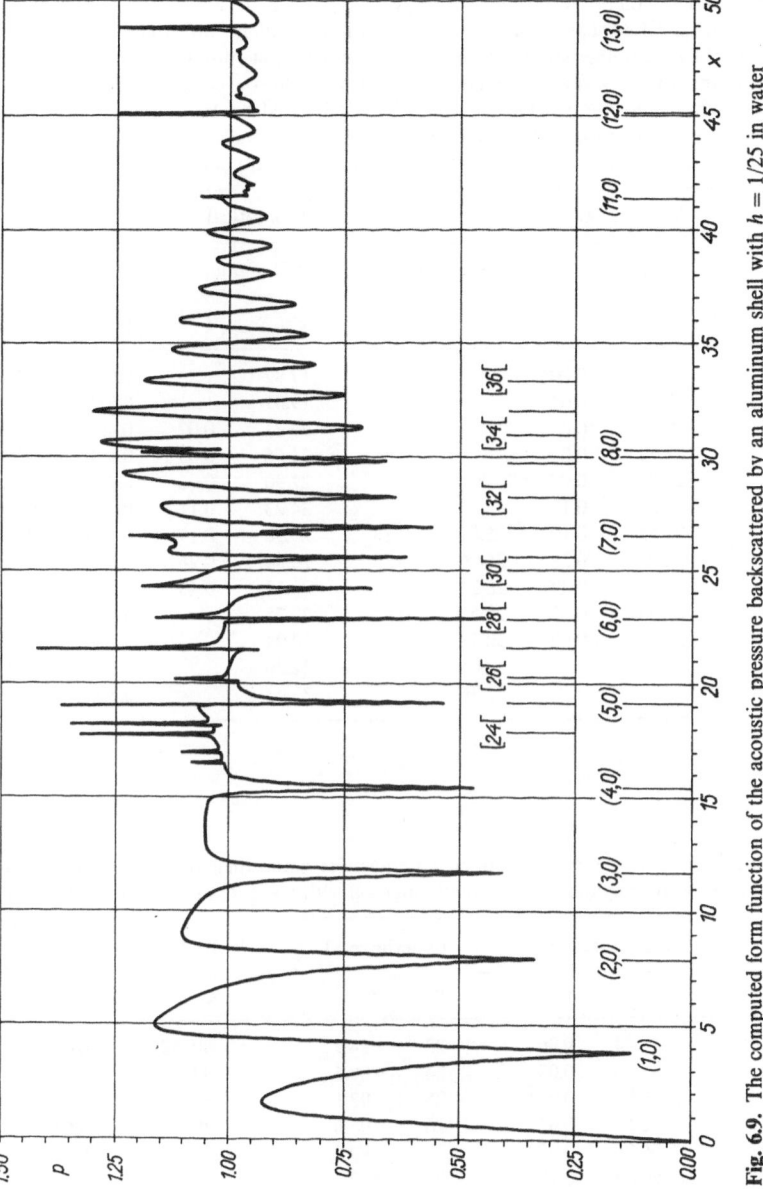

Fig. 6.9. The computed form function of the acoustic pressure backscattered by an aluminum shell with $h = 1/25$ in water

Table 6.2. Experimental and theoretical resonance frequencies x and their theoretical widths Γ (in units of x) for the A wave on a cylindrical duraluminum shell with $h = 1/25$ in water. The results obtained from a root search of the characteristic equation in the complex mode number plane (the Sommerfeld–Watson transformation, $n \rightarrow \nu$) are designated by SWT; those for the complex x plane (according to the procedure of the resonance scattering theory) are designated by RST analytical; and those obtained from the partial mode resonance components are designated by RST numerical. In the latter case the resonance magnitude and the relative phase velocity are given [Ref. 6.16, Table I]

Experimental		Theoretical				Theoretical			
		SWT		RST analytical		RST numerical			
n	x	x	Γ	x	Γ	x	Γ	ζ	c^{ph}/c
24		17.77	0.005			17.82		0.534	0.7425
25	19.0	19.09	0.01			19.07		0.515	0.7628
26	20.6	20.40	0.02			20.21	0.02	0.498	0.7773
27	21.9	21.72	0.04			21.56	0.03	0.486	0.7985
28	23.2	23.04	0.07			22.90	0.06	0.471	0.8179
29	24.5	24.35	0.12			24.23	0.11	0.458	0.8355
30	25.7	25.65	0.19	25.67	0.18	25.58	0.19	0.446	0.8527
31	27.0	26.93	0.28	26.96	0.32	26.84	0.26	0.436	0.8658
32	28.2	28.19	0.39	28.24	0.54	28.12	0.43	0.426	0.8788
33	29.5	29.43	0.53	29.51	0.78	29.73		0.414	0.9009
34	30.8	30.65	0.68	30.78		30.94		0.406	0.9100
35	31.9			32.06		32.20		0.398	0.9200
36	33.2			33.36		33.43		0.390	0.9286

Table 6.3. Experimental and theoretical resonance frequencies x and their theoretical width Γ (in unit of x) for the S_0 wave on a cylindrical duraluminum shell with $h = 1/25$ in water [6.16, Table II]

Experimental		RST analytical		Experimental		RST analytical	
n	x	x	Γ	n	x	x	Γ
11	41.1	41.44	0.06	23	84.9	85.58	0.44
12	44.8	45.15	0.03	24	88.5	89.21	0.51
13	48.5	48.84	0.004	25	92.1	92.82	0.57
14	52.2	52.55	0.03	26	95.6	96.41	0.64
15	55.9	56.25	0.08	27	99.2	100.00	0.72
16	59.5	59.89	0.10	28	102.7	103.57	0.79
17	63.2	63.63	0.15	29	106.2	107.12	0.88
18	66.9	67.31	0.19	30	109.7	110.65	0.96
19	70.4	70.97	0.23	31	113.2	114.17	1.05
20	74.1	74.65	0.29	32	116.6	117.67	1.15
21	77.7	78.30	0.34	33	120.1	121.15	1.25
22	81.4	81.95	0.39				

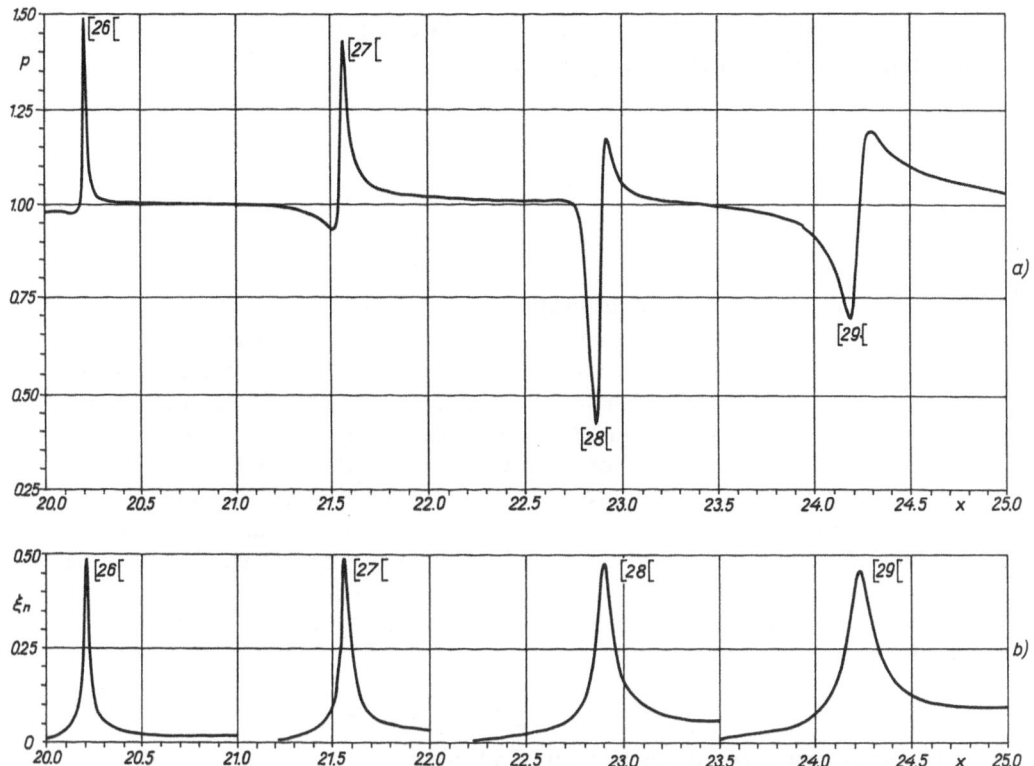

Fig. 6.10. The computed form function, (a), and the partial mode resonance components, (b), in a restricted x range. The scale on the abscissa is enlarged ten times compared with that used in Fig. 6.9.

6.2.3 Spherical Shells

The pulse method was used in [6.17] for an investigation of backscattering from spherical shells, one and three layered. The relative thickness of the shell varies in large limits: $b/a = 0.99, 0.97, 0.95, 0.90$ and 0.80. The frequency band is rather wide: $0 \leqslant x \leqslant 100$. In the case of a thin-walled shell ($h = 1/100$), two families of resonances were measured and identified, namely of the A and S_0 waves, and in the case of a thick-walled shell ($h = 1/5$) four families of resonances (of the S_0, A_0, S_1 and A_1 waves) were measured. In the case of a three-layered shell (two aluminum layers separated by a water layer; the relative thickness of each layer is $0.02a$), four families of resonances were identified. The results of measurements are presented in the form of the Regge trajectories and total form functions. It is significant that in the case of a thick-walled shell ($h = 1/5$) the resonances with $n \sim 50$ were measured, and in the case of a three-layered shell the resonances with $n \sim 100$.

The author is grateful to Professors M. de Billy, G. Quentin, J. Ripoche and H. Überall for their kindness in sending reprints of papers concerning the pulsed resonance identification method.

7 Peripheral Waves in the Scattering by Elastic Cylinders and Spheres

In this chapter, the steady-state problem of an incident plane acoustic wave scattered by an infinitely long elastic cylinder and by a sphere immersed in a liquid are considered. An analysis of the influence of the longitudinal (c_l) and transverse (c_t) velocities in the material of the scatterer on the form function and on the properties of the peripheral waves is presented. The form function was calculated in a series form. For its interpretation, the resonance scattering theory as the most suitable approach was used. The computation was accomplished in the narrow frequency band $0 \leqslant x \leqslant 50$ for the case of aluminum cylinder and sphere immersed in water. The incident plane wave generates the Rayleigh wave and transverse Whispering Gallery waves in the cylinder and the sphere. It is found that these waves depend weakly on c_l, while strongly depending on c_t.

7.1 Scattering from an Elastic Cylinder

Let us use the procedure presented in Chap. 2 for the case of a solid elastic cylinder. We shall consider the two-dimensional problem of a normally incident wave. In the far field the pressure scattered by the cylinder is given by (2.15), the partial form function is defined by (2.14) and the total form function by (2.13). We shall employ a rigid background (see (2.47) and (2.48)) and for the sake of simplicity, we shall introduce the following notations

$$p = |f|, \qquad \zeta_n = |\psi_n^{(r)}| . \tag{7.1}$$

We shall call $p(x)$ the form function and $\zeta_n(x)$ the modal resonance component.

As the results of computations and measurements have shown, the specularly reflected wave and the diffracted (creeping) waves are rather well modeled in scattering by an acoustically rigid cylinder and can be described as waves of corresponding type. Thus, the meaning of the difference introduced in (2.47) is the isolation of the shock waves in a "pure form", as caused by the peripheral waves revolving around the cylinder. In (2.48) the second term corresponds to the reflection and creeping wave component (the background) and the sum to the resonant component. The bigger the x is, the better the modal resonance component is defined.

The form function and the modal resonance components were calculated in the case of an aluminum cylinder immersed in water. The following parameters are used

aluminum: $\rho_1 = 2.79 \times 10^3 \,\text{kg/m}^3$, $c_1 = 6380 \,\text{m/s}$, $c_t = 3100 \,\text{m/s}$,

water: $\rho = 1 \times 10^3 \,\text{kg/m}^3$, $c = 1470 \,\text{m/s}$.   (7.2)

These parameters were chosen in the experimental investigation of the same problem in [7.1], which allows our results to be compared with experiment.

In order to investigate the influence of c_1 and c_t on the properties of peripheral waves, computations of $p(x)$ and $\zeta_n(x)$ were carried out for cylinders of hypothetical materials immersed in water. In one series of computations, c_t was held constant and c_1 was taken of the form

$$c_1' = qc_1 ,$$
               (7.3)

where for variant numbers II, III, IV and V $q = 0.90, 0.95, 1.05$ and 1.10, respectively.

In the other series of computations, c_1 was held constant and c_t was taken of the form

$$c_t' = qc_t ,$$
               (7.4)

where for variant numbers VI, VII, VIII and IX $q = 0.90, 0.95, 1.05$ and 1.10, respectively.

The basic variant, for which $c_1' = c_1$, $c_t' = c_t$, shall be called the first.

The observation point was situated in the far field at backscattering. The computation was carried out with step size $l_x = 10/256$ in the domain $0 \leqslant x \leqslant 150$ for the basic variant and in the domain $0 \leqslant x \leqslant 50$ for all other variants.

In Fig. 7.1 the form function is given. The calculation was accomplished for the basic variant. The total number of the extrema of the curve in Fig. 7.1 exceeds one hundred. The extrema differ in position, magnitude and pattern. The form function plot is rather complicated. As is well known, this plot reflects the result of the interference of peripheral waves and, as a rule, represents a small difference of large quantities. In the domain considered, $0 \leqslant x \leqslant 50$, the incident plane wave generates ten families of peripheral waves in the cylinder, each of which becomes apparent in the form of a sequence of resonances (see Figs. 7.2–10). The peripheral waves are characterized by magnitude, phase velocity, Q factor and phase, each depending on x. Therefore, it is no wonder that the result of interference of the waves has such a complicated, or more accurately, discontinuous form. Even for fixed velocities c_1 and c_t the positions of the extrema in the form function cannot always be connected with the resonance of a particular peripheral wave. In the case considered, the total number of resonances of the peripheral waves equals ninety, but the number of identified extrema of the form function curve, corresponding to these resonances, is only twenty eight. The identified resonances are labeled in Fig. 7.1. A similar situation

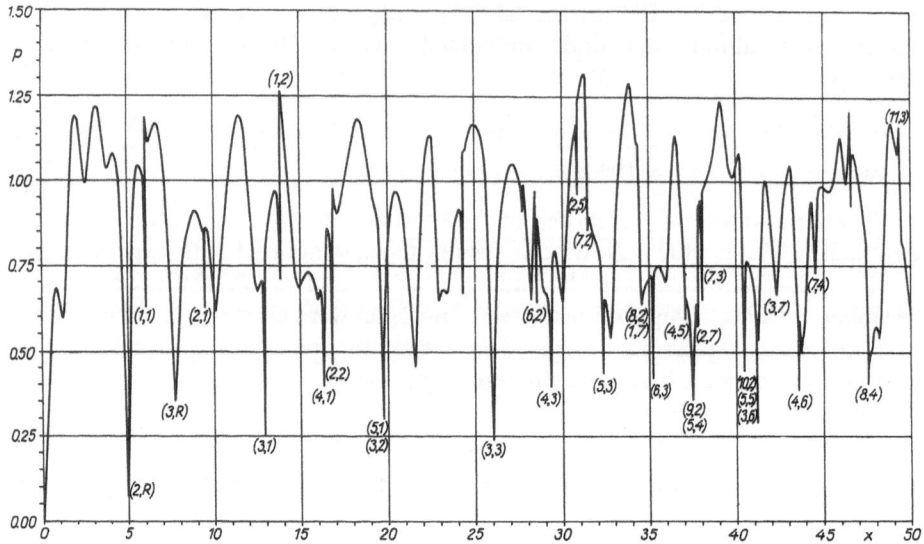

Fig. 7.1. The form function of the acoustic pressure scattered by a solid aluminum cylinder immersed in water as a function of $x = ka$ for the parameters defined in (7.2). Two symbols (n, l) label the partial mode resonance position

arises for all the computed variants of the problem. We cannot indicate a priori how the extrema of the form function curve will shift when c_l and c_t are changed.

The resonances are labeled by two symbols (n, l). The first defines the ordinal number of the resonance and the second indicates the type of the peripheral wave (the family of resonances); $l = R$ corresponds to the Rayleigh wave and $l = 1, 2, 3, \ldots$ to the transverse Whispering Gallery waves. Here the notations from [7.2] are used, although it should be noted that in [7.1] the Rayleigh wave is not distinguished and the following notations are used: $l = 1$ corresponds to the Rayleigh wave and $l = 2, 3, 4, \ldots$ to the transverse Whispering Gallery waves. From the dispersion curves computed for the first (basic) variant, it is obvious that for every peripheral wave with x increasing, the dispersion curve tends to c_t. This means that in the considered domain, $0 \leqslant x \leqslant 50$, any longitudinal Whispering Gallery waves are not generated.

The acoustic spectrogram computed for the basic variant is presented in Fig. 7.2. It is concordant with the one given in [7.3] (see Fig. 5.11). It should be noted that the extrema of the form function curve do not always explicitly correspond to the resonance frequencies (spectral lines in Fig. 4.2). As a result of the interference of the waves, the extrema positions on the form function curve are slightly shifted with respect to the spectral lines. Even in the case of very narrow, high quality resonances, other resonance(s) (n', l) can influence the extrema on the form function curve which are mainly connected with the (n, l) resonance. As a rule, the resonances of the peripheral wave with a high Q factor or magnitude are well observed on the form function curve. When the resonance

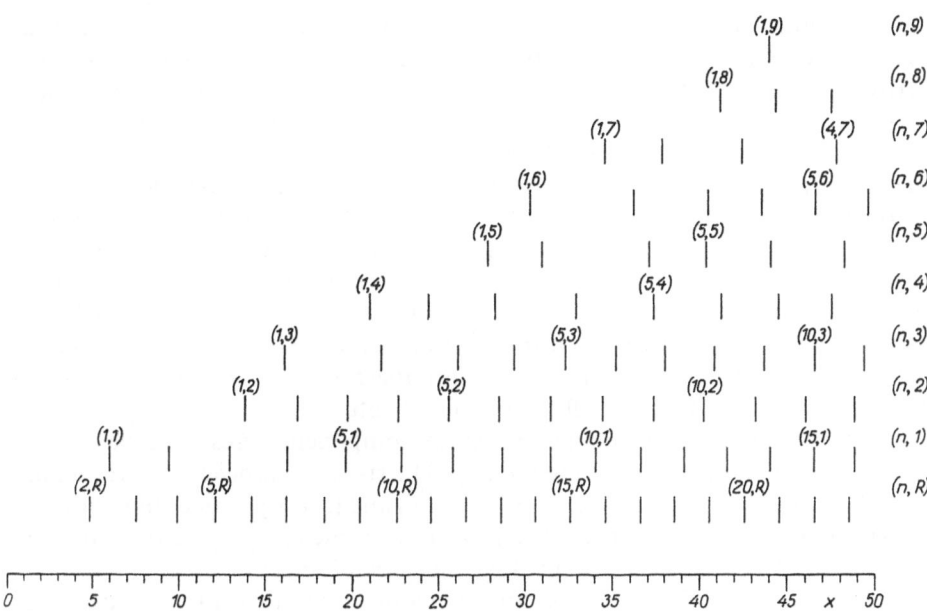

Fig. 7.2. The acoustic spectrogram corresponding to Fig. 7.1

has both of these properties, its chance of being distinguished on the form function curve is enhanced. One can distinctly see on the curve the extrema caused by the narrow (high Q factor) resonances. On many occasions one can clearly observe extrema corresponding to the constructive interference of two or more waves. Thus, for example, the extremum on the curve situated at $x = 40.39$ is connected with resonances (10, 2), (5, 5) and (3, 6) (see Fig. 7.1).

It should be mentioned that both resonances and antiresonances of the peripheral waves may produce extrema. This is especially true for waves with a large magnitude, but not very high quality. As is well known, such a phenomenon also takes place in the case of scattering by an elastic shell (see, for example, the plots of the form function and the contribution of the Rayleigh wave, Figs. 4.1 and 4.4). It is typical that sometimes one can see on the form function curve extrema which have no particular distinction; they have neither high Q factor nor magnitude and correspond to rather large indices. This is due to the fact that other waves do not significantly influence the form function curve in the region of the x domain where they are situated.

Only one domain of the form function curve is distinguished for all the nine calculated variants. On the x axis it is situated between 0 and ~ 5. This is the so-called quasi-rigid region [7.4]. In terms of the peripheral waves it is restricted from above by a value corresponding to the second resonance of the Rayleigh wave. As is well known, the first resonance of this wave is generated on neither a sphere nor a cylinder. For seven of the nine computed variants of the form function, the position on the x axis and magnitude of the first four maxima and three minima of the form function curve do not change for different variants. This is possible if, and only if, the properties of the waves generating these extrema depend neither on c_l nor on c_t. This is indeed the case, since these extrema are caused by the slip waves, re-radiated by the creeping waves revolving around the cylinder. These waves are often called the Franz waves. They propagate mainly in the liquid on the surface of the cylinder and their properties are defined not by the elastic cylinder but by the liquid. In the sixth and seventh variants, the second resonances of the Rayleigh wave is situated so low on the x axis that it "shades" the resonance of the creeping wave which are small in magnitude. In these two variants the third minimum and the fourth maximum are not seen on the form function curve, while they clearly manifest themselves in the basic variant. Thus, for these two variants the quasi-rigid region is narrower than for all the other variants.

On the plots of the modal resonance components presented below, the resonances of creeping waves are missing. The standard procedure of resonance isolation used here, in the form given by (2.47) and (2.48), just does not allow the separation of these resonances. The properties of the creeping waves generated on a solid elastic cylinder are investigated in detail in [7.2].

The analysis of the computed form function shows that a variation of c_l or c_t to the extent of 5%–10% affects the form function by changing the position, magnitude and number of the form function extrema, but so far it is not possible to correlate the variation of the parameters of the elastic cylinder with changing the properties of the form function.

The plots of modal resonance components of the Rayleigh wave and the transverse Whispering Gallery waves are shown in Figs. 7.3–10. They are computed for the first variant. The plots for other variants have a similar form. The results computed for variants II–V show that the 5%–10% variation of c_l only weakly affects the position of modal resonance components. In Table 7.1 the resonance positions of the Rayleigh wave for these four variants are compared. It is evident that at small values of n, increasing c'_l has almost no effect on

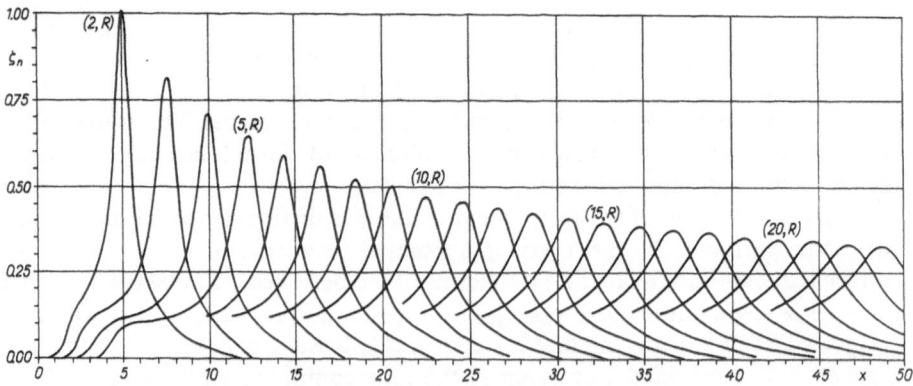

Fig. 7.3. The isolated modal resonances of the Rayleigh wave

the resonance positions. With n increasing, the influence of the c_1' becomes apparent, but even for the largest chosen value, $n = 23$, it is rather small. The 20% variation of c_1', from 0.90 c_1 to 1.10 c_1, shifts the 23rd resonance to the right only by 0.66 on the x axis. At fixed n, the resonance magnitudes also change little. They decrease by 0.003 when c_1' increases from 0.90 c_1 to 1.10 c_1. The resonance magnitudes calculated for the first variant are presented in Table 7.2.

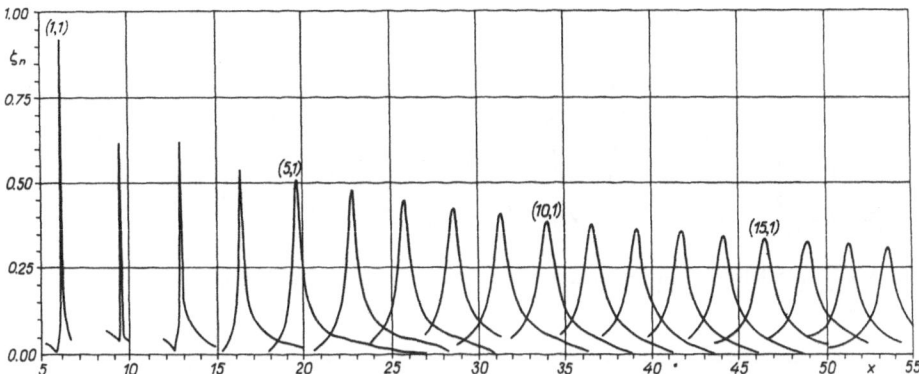

Fig. 7.4. The isolated modal resonances of the first ($l = 1$) transverse Whispering Gallery wave

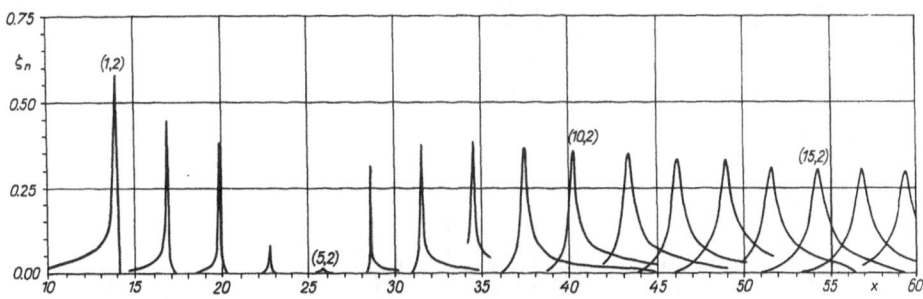

Fig. 7.5. The isolated modal resonances of the second ($l = 2$) transverse Whispering Gallery wave

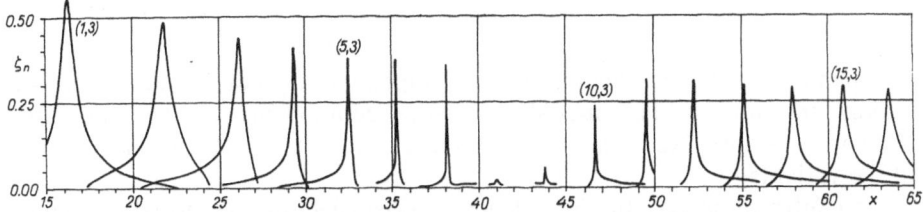

Fig. 7.6. The isolated modal resonances of the third ($l = 3$) transverse Whispering Gallery wave

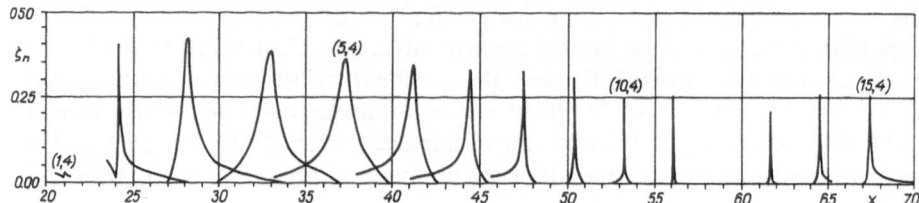

Fig. 7.7. The isolated modal resonances of the fourth ($l = 4$) transverse Whispering Gallery wave

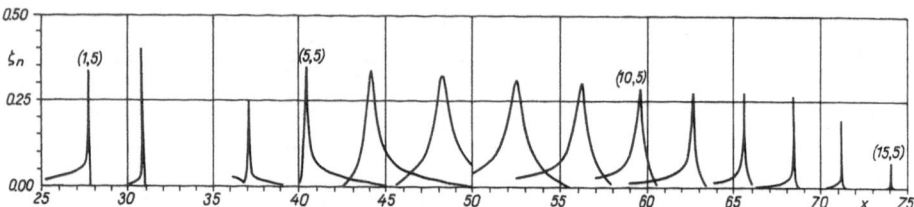

Fig. 7.8. The isolated modal resonances of the fifth ($l = 5$) transverse Whispering Gallery wave

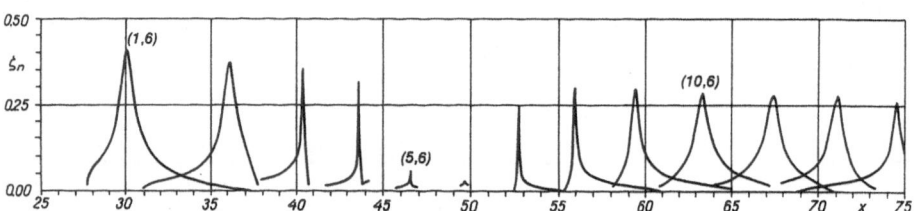

Fig. 7.9. The isolated modal resonances of the sixth ($l = 6$) transverse Whispering Gallery wave

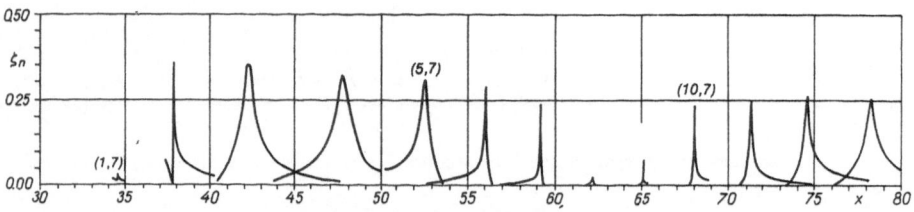

Fig. 7.10. The isolated modal resonances of the seventh ($l = 7$) transverse Whispering Gallery wave

The qualities of the modal resonance components do not change from one variant to another. The modal resonance components of the transverse Whispering Gallery waves are also not much affected by variations in c_1 for any of the variants.

In contrast, the variation of c_t has a marked effect on the modal resonance components connected with both the Rayleigh wave and the Whispering

Table 7.1. The positions of the modal resonances of the Rayleigh wave for different values of c_i'

n	Variant			
	II	III	IV	V
2	4.88	4.88	4.88	4.88
3	7.58	7.58	7.58	7.58
4	9.88	9.92	9.96	9.96
5	12.07	12.11	12.15	12.19
6	14.18	14.26	14.30	14.34
7	16.25	16.33	16.41	16.45
8	18.32	18.40	18.48	18.52
9	20.35	20.43	20.55	20.59
10	22.34	22.46	22.58	22.62
11	24.34	24.45	24.61	24.69
12	26.37	26.48	26.64	26.72
13	28.36	28.48	28.67	28.71
14	30.31	30.47	30.66	30.74
15	32.30	32.46	32.70	32.77
16	34.30	34.45	34.63	34.77
17	36.29	36.45	36.72	36.80
18	38.24	38.44	38.71	38.79
19	40.23	40.43	40.70	40.82
20	42.19	42.42	42.70	42.81
21	44.18	44.38	44.69	44.80
22	46.13	46.37	46.68	46.80
23	48.13	48.36	48.67	48.79

Gallery waves. This can be clearly seen from the data presented in Tables 7.2–9 where, for variants VI–IX and the first (basic) variant, the position on the x axis (upper row) and the magnitude (lower row) of the modal resonance components are given.

It can be seen from the data presented in the tables that, at fixed n and with increasing c_t', the resonance position shifts to the right on the x axis and its magnitude diminishes slightly. The larger n, the larger the effect of c_t' on both the resonance position and its magnitude. This is particularly clearly seen on the low quality (slanting) resonances of the partial modes. In principle, it should also be observed for narrow (high Q factor) resonances, but with the (rather large) computation step size chosen, $l_x = 10/256$, they are occasionally lost.

From the data presented in the tables, it can be seen that the 20% variation of c_t' (from $0.90\,c_t$ to $1.10\,c_t$) strongly shifts the position of partial mode resonances to the right; for example, resonance $(21, R)$ by 6.75, resonance $(15, 1)$ by 3.95, resonance $(12, 2)$ by 3.40, and resonance $(10, 3)$ by 2.03.

It should be noted that for every peripheral wave, the first resonance of the partial mode is practically fixed on the x axis. The wave is just too long to "feel" the c_t variation. This fact evokes the following speculation. Although each of the

Table 7.2. The positions (upper row) and the magnitudes (lower row) of the modal resonances of the Rayleigh wave for different values of c_1'

n	Variant				
	VI	VII	I	VIII	IX
2	4.88	4.88	4.88	4.88	4.88
	1.02	1.02	1.02	1.02	1.02
3	7.58	7.58	7.58	7.58	7.62
	0.820	0.819	0.819	0.818	0.818
4	9.88	9.92	9.92	9.96	10.00
	0.718	0.716	0.716	0.715	0.714
5	12.03	12.11	12.15	12.19	12.23
	0.650	0.649	0.647	0.646	0.645
6	14.10.	14.18	14.26	14.38	14.45
	0.601	0.599	0.597	0.595	0.594
7	16.09	16.25	16.37	16.48	16.64
	0.562	0.560	0.558	0.555	0.553
8	18.05	18.24	18.44	18.63	18.83
	0.531	0.528	0.525	0.523	0.521
9	19.92	20.20	20.47	20.74	21.02
	0.505	0.502	0.498	0.495	0.492
10	21.80	22.15	22.54	22.85	23.20
	0.483	0.479	0.475	0.472	0.468
11	23.59	24.10	24.53	25.00	25.39
	0.464	0.460	0.455	0.451	0.448
12	25.39	26.02	26.56	27.11	27.62
	0.448	0.442	0.438	0.433	0.429
13	27.19	27.89	28.59	29.22	29.84
	0.433	0.427	0.422	0.417	0.413
14	28.95.	29.80	30.59	31.37	32.07
	0.419	0.413	0.408	0.403	0.398.
15	30.70	31.68	32.58	33.48	34.34
	0.407	0.401	0.395	0.390	0.385
16	32.42	33.55	34.61	35.59	36.56
	0.396	0.390	0.384	0.378	0.373
17	34.18	35.39	36.60	37.73	38.79
	0.386	0.379	0.373	0.367	0.362
18	35.90	37.27	38.59	39.84	41.05
	0.377	0.370	0.363	0.357	0.352
19	37.62.	39.14	40.58	41.95	43.28
	0.368	0.361	0.354	0.348	0.343
20	39.34	40.98	42.58	44.10	45.55
	0.360	0.352	0.346	0.340	0.334
21	40.98	42.85	44.57	46.21	47.73
	0.352	0.345	0.338	0.332	0.326

Table 7.2. Continued

n	Variant				
	VI	VII	I	VIII	IX
22	42.70	44.73	46.56	48.32	
	0.345	0.337	0.331	0.325	
23	44.41	45.56	48.55		
	0.338	0.331	0.323		
24	46.17	48.44			
	0.332	0.324			

partial mode resonances describes the elastic scatterer, observations of some of them give better information about the object. In order for the influence of the elastic properties of the scatterer to become prominent, it is advantageous to observe some resonance of higher order of n, if the latter possesses sufficient magnitude and high Q factor. The resonances with $n \sim 10$–20, as a rule, fulfill this condition.

In contrast to the curve representing the form function (see Fig. 7.1), the curves of the modal resonance components permit a relation to be established between the variation of the elastic parameters of the scatterer, for example, its velocities c_1 and c_t, and the positions of the resonance lines. Thus the information about the elastic scatterer in the "raw" form cannot be effectively used. Decomposition of the total form function into the specularly reflected, creeping and peripheral waves and successive analysis of the latter permit us to obtain some more information about the scatterer. A recently developed method of form function filtering [7.5] permits the isolation of the families of resonances and identification of the ordinal number of each resonance.

The resonance position of a standing wave (partial mode) corresponds to the resonance of a travelling (peripheral) wave. Using the known position of the partial mode resonance, one can find the phase velocity of the peripheral wave

$$c_l^{\text{ph}}(x) = c \, \frac{x(n, l)}{n} \,, \tag{7.5}$$

where $c_l^{\text{ph}}(x)$ is the phase velocity of the peripheral wave of the lth family, and $x(n, l)$ is the position on the x axis of the nth resonance of the partial mode of the same family.

The relative phase velocity is suitable for purposes of comparison

$$\frac{c_l^{\text{ph}}}{c_t} = \left(\frac{c}{c_t} \right) \frac{x(n, l)}{n} \,. \tag{7.6}$$

The dispersion curves of relative velocities of the Rayleigh wave and the first three transverse Whispering Gallery waves are shown in Figs. 7.11–14. The dispersion curves of successive waves (at $l > 3$) have a similar character.

Table 7.3. The positions (upper row) and the magnitudes (lower row) of the modal resonances of the first ($l = 1$) transverse Whispering Gallery wave for different values of c_t'

n	Variant				
	VI	VII	I	VIII	IX
1	5.94	5.98	6.05	6.13	6.17
	0.670	0.621	0.917	0.618	0.908
2	9.14	9.30	9.45	9.61	
	0.305	0.508	0.623	0.0691	
3	12.38	12.66	12.93	13.16	13.40
	0.635	0.615	0.620	0.622	0.547
4	15.59	15.98	16.33	16.88	16.99
	0.572	0.564	0.547	0.552	0.546
5	18.67	19.18	19.65	20.08	20.43
	0.522	0.515	0.509	0.502	0.496
6	21.64	22.23	22.81	23.28	23.71
	0.485	0.477	0.471	0.467	0.463
7	24.45	25.16	25.78	26.33	26.80
	0.456	0.450	0.443	0.439	0.436.
8	27.19	27.97	28.63	29.22	29.73
	0.432	0.427	0.,421	0.416	0.414
9	29.76	30.66	31.41	32.03	32.54
	0.413	0.407	0.402	0.398	0.395
10	32.30	33.24	34.02	34.69	35.23
	0.397	0.391	0.386	0.383	0.380
11	34.77	35.74	36.60	37.30	37.89
	0.383	0.377	0.373	0.369	0.366
12	37.19	38.24	39.14	39.88	40.51
	0.370	0.365	0.361	0.357	0.354
13	39.53	40.66	41.60	42.42	43.05
	0.359	0.354	0.349	0.346	0.343
14	41.88	43.05	44.06	44.92	45.59
	0.349	0.344	0.340	0.336	0.333
15	44.18	45.39	46.48	47.38	48.13
	0.339	0.335	0.331	0.328	0.325
16	46.44	47.73	48.87	49.80	
	0.331	0.327	0.323	0.319	
17	48.71				
	0.323				

It is obvious from the plots presented that, for every peripheral wave, the family of the dispersion curves for different c_t' is similar to a divergent spray. At first (at small n and x) the spray diverges quickly; then its width gradually becomes established. The larger the value of c_t', the higher the dispersion curve in

Table 7.4. The positions (upper row) and the magnitudes (lower row) of the modal resonances of the second ($l = 2$) transverse Whispering Gallery wave for different values of c_t'

n	Variant				
	VI	VII	I	VIII	IX
1	13.75	13.83	13.87	13.91	13.94
	0.601	0.605	0.588	0.571	0.575
2	16.64	16.76	16.88	16.95	16.99
	0.460	0.531	0.459	0.483	0.519
3	19.95	19.61	19.77	19.88	19.96
	0.367	0.405	0.388	0.431	0.403
4	22.23	22.46	22.62	22.81	22.97
	0.0993	0.0924	0.0878	0.296	0.301
5			25.55	25.78	
			0.0175	0.0207	
6	27.77	28.13	28.48	28.79	29.14
	0.415	0.425	0.316	0.249	0.250
7	30.55	30.98	31.45	31.88	32.30
	0.404	0.355	0.380	0.400	0.389
8	33.32	33.87	34.41	34.96	35.51
	0.375	0.387	0.385	0.380	0.374
9	36.13	36.76	37.38	38.05	38.71
	0.370	0.370	0.368	0.366	0.362
10	38.91	39.61	40.35	41.09	41.84
	0.359	0.358	0.353	0.351	0.348
11	41.64	42.42	43.14	44.06	44.84
	0.350	0.346	0.343	0.339	0.337
12	44.37	45.23	46.05	46.91	47.77
	0.339	0.335	0.331	0.329	0.326
13	47.07	47.97	48.87	49.73	
	0.329	0.326	0.323	0.319	
14	49.73				
	0.320				

this family is situated. In the $x - (c^{\mathrm{ph}}/c_t)$ plane, the resonances of the same n value are situated on straight lines. The larger n, the smaller is the angle between this corresponding line and the x axis, and the more the phase velocities of the peripheral wave generated on the cylinder differ for different c_t'. This fact can be used to extract information about the properties of an elastic cylinder from the observed dependences. In addition to the acoustic spectrogram, dispersion curves of the phase velocities of the Rayleigh wave and some low order ($l = 1, 2, 3$) transverse Whispering Gallery waves can be used to obtain information on the nature of the scatterer.

Table 7.5. The positions (upper row) and the magnitudes (lower row) of the modal resonances of the third ($l = 3$) transverse Whispering Gallery wave for different values of c_t'

n	Variant				
	VI	VII	I	VIII	IX
1	16.17	16.17	16.17	16.17	16.17
	0.561	0.561	0.561	0.561	0.561
2	21.68	21.68	21.68	21.64	21.64
	0.485	0.485	0.484	0.485	0.485
3	25.94	25.98	26.05	26.05	26.05
	0.441	0.440	0.441	0.442	0.441
4	28.98	29.18	29.30	29.41	29.49
	0.412	0.417	0.406	0.414	0.414
5	31.84	32.07	32.27	32.46	32.58
	0.400	0.385	0.373	0.379	0.395
6	34.61	34.92	35.16	35.39	35.55
	0.372	0.232	0.373	0.326	0.370
7	37.34	37.70	38.01	38.28	38.52
	0.335	0.309	0.352	0.362	0.348
8			48.86	41.17	41.48
			0.0182	0.0238	0.143
9	42.77	43.24	43.71	44.10	44.49
	0.0599	0.0932	0.0550	0.243	0.0284
10	45.47	46.02	46.56	47.03	47.50
	0.169	0.286	0.236	0.279	0.325
11	48.13	48.79	49.41		
	0.299	0.311	0.320		

Table 7.6. The positions (upper row) and the magnitudes (lower row) of the modal resonances of the fourth ($l = 4$) transverse Whispering Gallery wave for different values of c_t'

n	Variant				
	VI	VII	I	VIII	IX
1	20.70	20.90	21.05		
	0.0306	0.0291	0.0305		
2	23.83	24.10	24.34	24.53	24.69
	0.449	0.458	0.406	0.354	0.365
3	27.81	28.01	28.20	28.40	28.55
	0.428	0.425	0.423	0.423	0.422
4	32.62	32.73	32.85	32.93	33.01
	0.395	0.394	0.394	0.393	0.393
5	37.27	37.30	37.34	37.34	37.38
	0.369	0.369	0.369	0.369	0.369
6	41.05	41.17	41.21	41.25	41.29
	0.351	0.352	0.351	0.351	0.351
7	44.10	44.34	44.49	44.61	44.69
	0.333	0.338	0.337	0.336	0.337
8	46.91	47.23	47.50	47.70	47.85
	0.302	0.320	0.327	0.325	0.326
9	49.65				
	0.310				

Table 7.7. The positions (upper row) and the magnitudes (lower row) of the modal resonances of the fifth ($l = 5$) transverse Whispering Gallery wave for different values of c_t'

n	Variant				
	VI	VII	I	VIII	IX
1	27.19	27.46	27.73	27.93	28.13
	0.181	0.183	0.334	0.223	0.320
2	30.23	30.55	30.86	31.13	31.33
	0.0531	0.0423	0.0406	0.0641	0.0896
3					
4	36.25	36.64	37.03	37.38	37.73
	0.173	0.204	0.255	0.180	0.253
5	39.38	39.88	40.31	40.74	41.09
	0.356	0.351	0.351	0.352	0.344
6	43.24	43.67	44.06	44.45	44.80
	0.342	0.341	0.339	0.338	0.337
7	47.73	48.01	48.24	48.52	48.75
	0.327	0.326	0.325	0.324	0.323

Table 7.8. The positions (upper row) and the magnitudes (lower row) of the modal resonances of the sixth ($l = 6$) transverse Whispering Gallery wave for different values of c'_t

n	Variant				
	VI	VII	I	VIII	IX
1	30.20	30.16	30.16	30.16	30.16
	0.410	0.410	0.411	0.411	0.411
2	36.13	36.13	36.13	36.13	36.09
	0.375	0.375	0.375	0.375	0.376
3	39.84	40.16	40.43	40.63	40.82
	0.297	0.356	0.349	0.349	0.353
4	42.85	43.20	43.55	43.87	44.18
	0.244	0.250	0.315	0.337	0.280
5	45.78	46.17	46.56	46.95	47.30
	0.175	0.0435	0.0560	0.176	0.173

Table 7.9. The positions (upper row) and the magnitudes (lower row) of the modal resonances of the seventh ($l = 7$) transverse Whispering Gallery wave for different values of c'_t

n	Variant				
	VI	VII	I	VIII	IX
1			34.49	34.80	35.08
			0.0363	0.0293	0.0415
2	37.07	37.42	37.77	38.13	38.44
	0.352	0.350	0.361	0.329	0.316
3	42.23	42.27	42.34	42.46	42.62
	0.347	0.347	0.343	0.346	0.345
4	47.70	47.70	47.73	47.77	47.77
	0.327	0.326	0.327	0.326	0.326

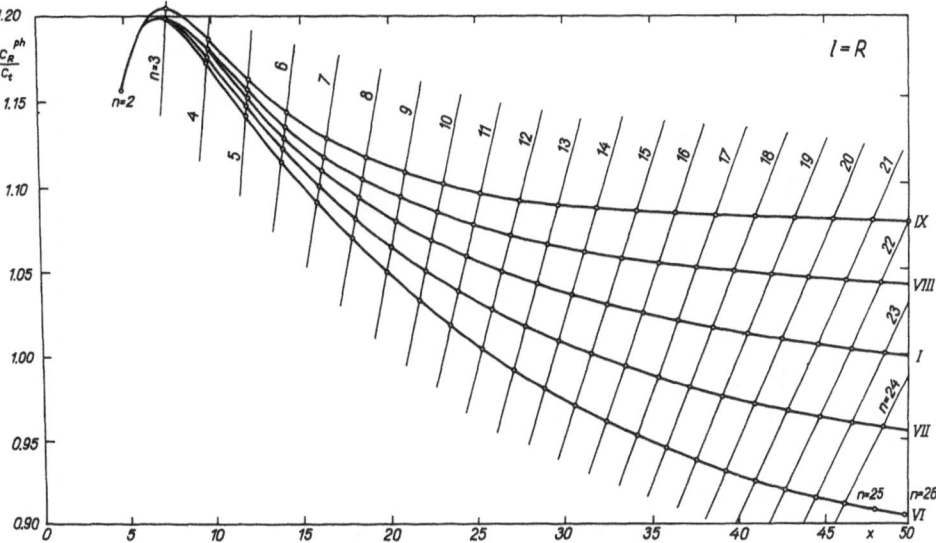

Fig. 7.11. The dispersion curves of the relative phase velocity of the Rayleigh wave ($l = R$) for different values of c'_t

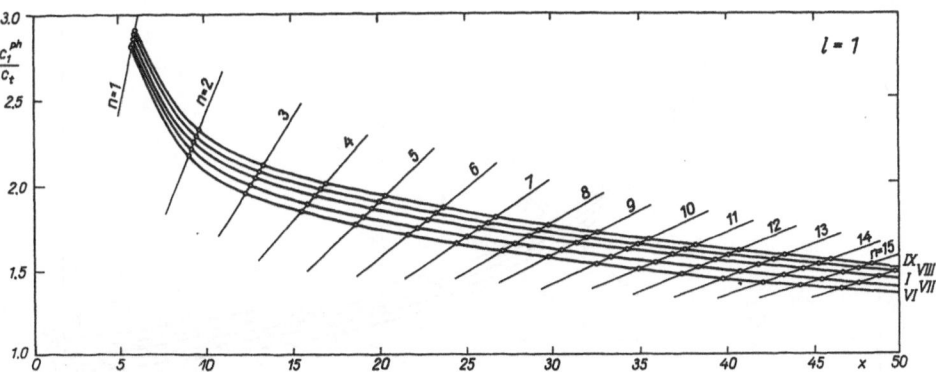

Fig. 7.12. The dispersion curves of the relative phase velocity of the transverse Whispering Gallery wave ($l = 1$) for different values of c'_t

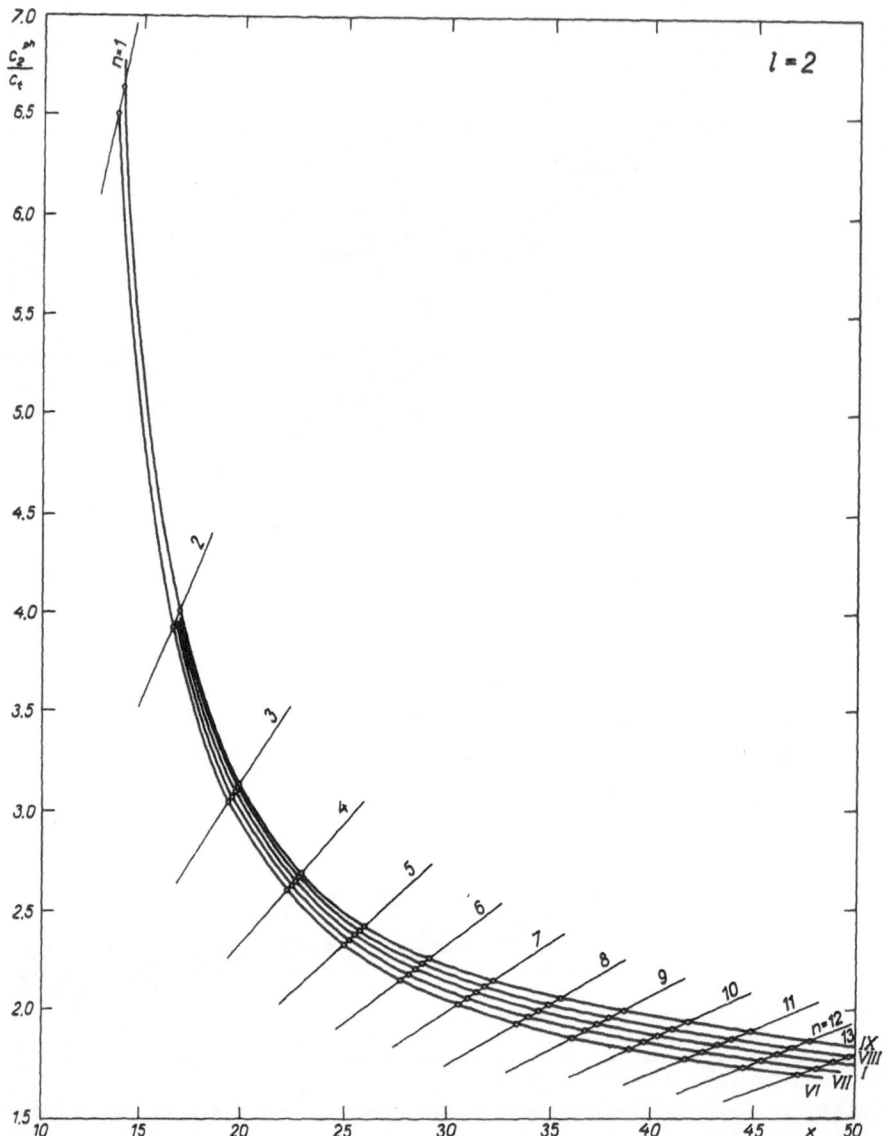

Fig. 7.13. The dispersion curves of the relative phase velocity of the transverse Whispering Gallery wave ($l = 2$) for different values of c'_t

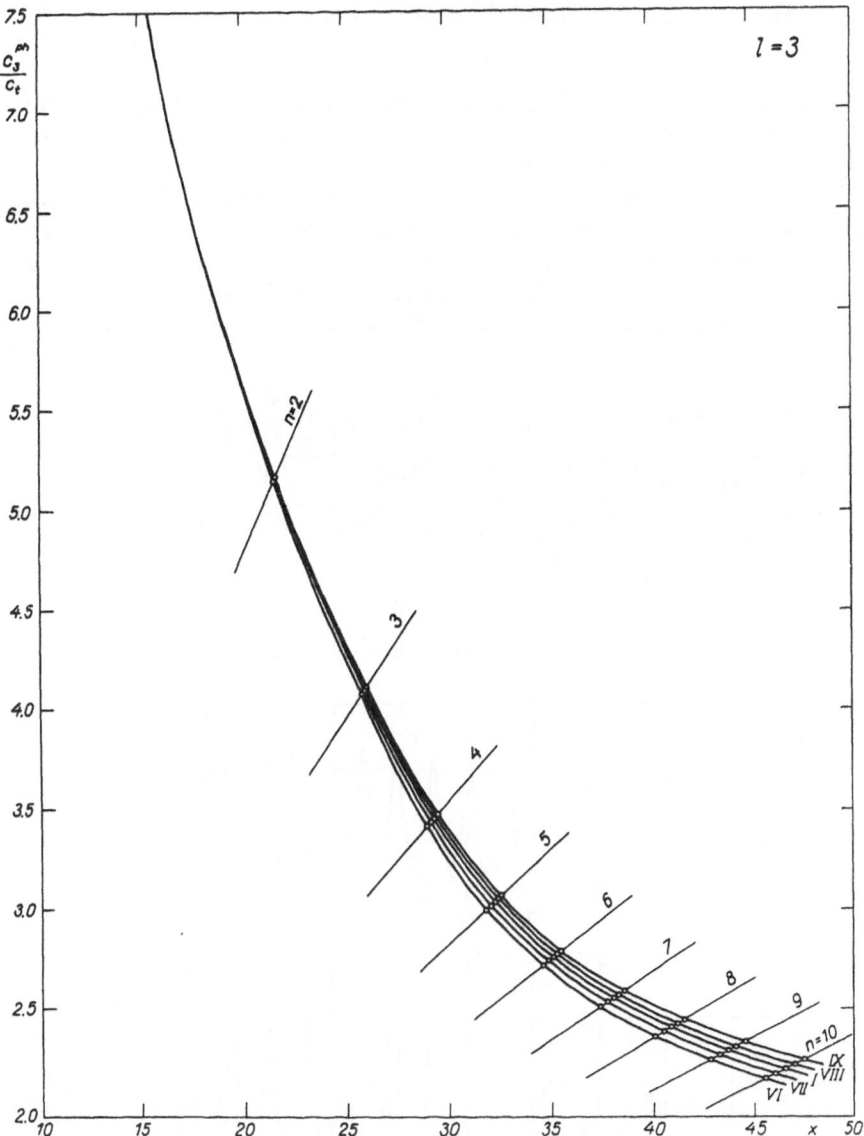

Fig. 7.14. The dispersion curves of the relative phase velocity of the transverse Whispering Gallery wave ($l = 3$) for different values of c'_t

7.2   Scattering from a Solid Elastic Sphere

Similarly to the above-described analysis, an analysis of the peripheral waves generated by the incident plane acoustic wave on an elastic sphere has been carried out. We shall present some of the results obtained.

The form function was calculated according to (3.8) and (3.9). The observation point was situated in the far field at backscattering. The physical parameters are defined by (7.2). The computation was carried out in the domain $0 \leqslant x \leqslant 200$ with the computation step size $l_x = 10/256$. The resonance components of the partial modes were computed for the form function analysis. The

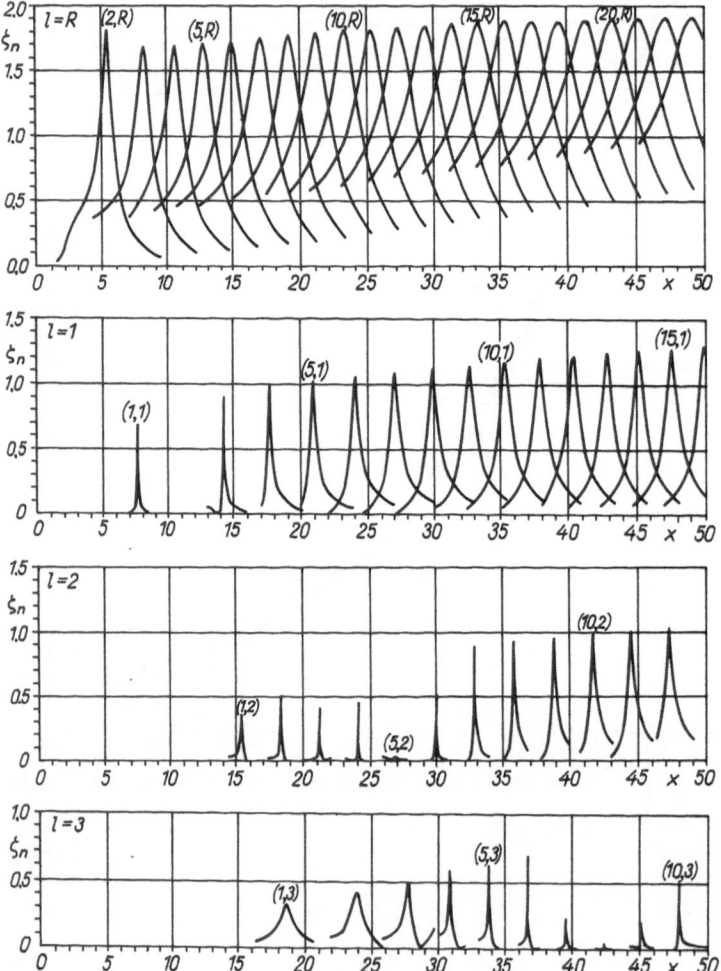

Fig. 7.15. The isolated modal resonances of the Rayleigh wave ($l = R$) and the three first transverse Whispering Gallery waves ($l = 1, 2, 3$) for the case of an aluminum sphere immersed in water

rigid background was used (see (3.24–26)) and single-type resonances were joined into families.

In order to investigate the influence of the velocities c_l and c_t on the properties of the peripheral waves, two series of computations were carried out. The scattering by spheres from hypothetical materials immersed in water was investigated. In one series of computations we set $c'_l = qc_l$ (with $c'_t = c_t$), and in the other $c'_t = qc_t$ (with $c'_l = c_l$) where $q = 0.9, 1.1$.

In Fig. 7.15, for the basic variant ($c'_l = c_l$, $c'_t = c_t$), the contributions of the resonance components of the partial modes for the peripheral waves with $l = R, 1, 2, 3$ are presented.

Aluminum, in comparison with tungsten carbide (see Figs. 4.1, 4.4 and 4.5), is a rather soft material. At $x \leqslant 100$, the form function, calculated for the case of scattering on a tungsten carbide sphere, shows a contribution of the Rayleigh wave noticeably bigger than that given by the Whispering Gallery waves. On the contrary, as the results of computations have shown, for scattering on an aluminum sphere, the contribution of the shock waves caused by the Whispering Gallery waves to the total form function is essential and commensurable with that of the Rayleigh wave. The Q factor of the Whispering Gallery resonances is bigger than that of the Rayleigh wave and therefore the resonances of the Whispering Gallery waves can be clearly observed on the form function curve.

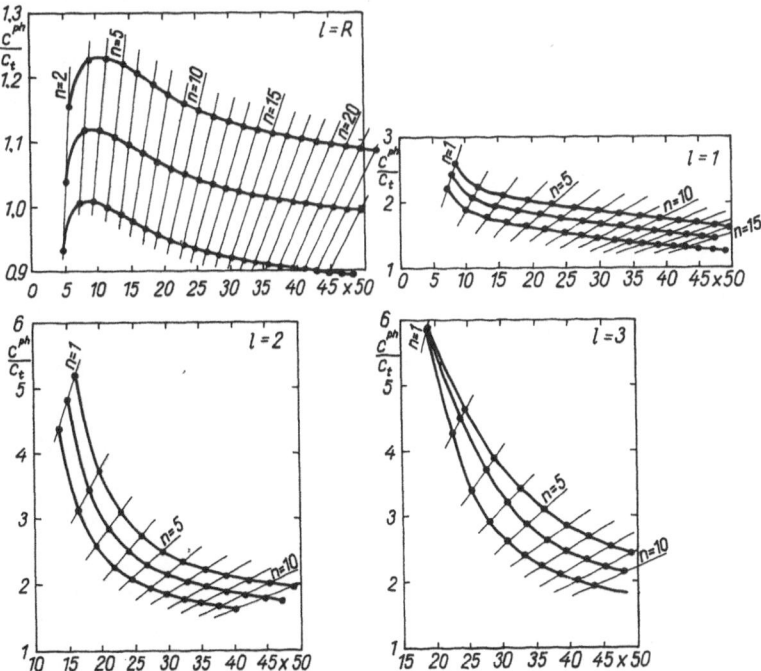

Fig. 7.16. The dispersion curves of the relative phase velocities of the Rayleigh wave ($l = R$) and the three first transverse Whispering Gallery waves ($l = 1, 2, 3$) for different values of c'_t

In Fig. 7.16, the dispersion curves of the relative phase velocities of the peripheral waves for $l = R, 1, 2, 3$ are shown for the case of different c'_t values but with $c'_l = c_l$. The upper curve corresponds to $q = 1.1$, the middle one to $q = 1.0$ (the basic variant), and the lower one to $q = 0.9$. These curves look similar to those described above for the case of scattering by solid cylinders.

The relative phase velocity of the peripheral wave can be found by using the resonance position $x(n, l)$, according to the formula

$$\frac{c_l^{ph}}{c_t} = \left(\frac{c}{c_t}\right) \frac{x(n, l)}{n + 1/2} \quad (n = 1, 2, 3 \ldots; l = R, 1, 2, 3 \ldots). \tag{7.7}$$

For the cases of $l = R$ and $l = 1$, the dispersion curves of the phase velocities are almost equidistant and proportional to q. In the cases of $l = 2, 3, \ldots$ the difference in the phase velocities is biggest in the vicinity of the cutoff frequency. With x increasing, the difference in the velocities (at fixed x values) diminishes. The resonances of the same order n are situated on straight lines. The larger the n, the smaller is the angle of inclination of this line to the x axis, and the greater is the difference (at fixed n value) in the phase velocities for different values of q.

The results presented above were published in [7.6, 7].

8 Analysis of Peripheral Waves via Backscattering by Circular–Cylindrical Shells

The results of applying resonance scattering theory to the problem of a plane acoustic wave scattering from a thick-walled circular–cylindrical shell are presented. The computation of the form function and isolated modal resonances have been carried out in a broad frequency band. Attention is focused on the analysis of the Lamb-type peripheral waves S_j and A_j ($j = 0, 1, 2, 3, 4$).

8.1 Introductory Remarks

Let us consider the two-dimensional problem of a plane pressure wave scattered by a circular–cylindrical shell. The direction of the initial-wave propagation is perpendicular to the longitudinal axis of the shell. We shall confine ourself to an investigation of the scattered pressure only at a point in the farfield for backscattering. The material of the shell is assumed to be homogeneous and isotropic, and its cross section circular. The shell is evacuated inside.

In the well-known solution of the problem as obtained in [8.1], the relative thickness of the shell $h = 1 - b/a$ enters as one of the parameters, and practically does not affect the computational algorithm. From the physical point of view, this is a very important parameter and the solution is strongly influenced by it. Below, we shall analyse the pressure scattered by a rather thick cylindrical shell (tube) with $h = 1/10$. This tube is in an intermediate position between the solid elastic cylinder (with $b/a = 0$) and the thin shell (when $h \ll 1$).

In the first limiting case the secondary (scattered) acoustic pressure is provided by the shock waves re-radiated by the peripheral waves: the Rayleigh-type wave and Whispering Gallery waves (see Chap. 7). In the second limiting case, the scattered pressure is provided by the shock waves re-radiated by the peripheral waves of another type: by normal (Lamb-type) waves propagating in a thin curved layer. The thinner the shell, the smaller the number of types of these waves in the restricted x range from $x = 0$ onward.

As in these limiting cases, the pressure scattered by a tube is made up from contributions furnished by the reflected wave (and wave refracted in the curved layer) and by the slip waves re-radiated by the creeping (Franz-type) waves. The contribution of creeping waves is investigated in detail in [8.2–5] and we shall not discuss it here.

Below, the contribution of the Lamb-type waves to the scattered pressure will be considered. In contrast to Chap. 2, in which the main attention was paid to just the formalism of the resonance scattering theory, we shall use this theory here as the main working tool. We shall analyse the scattered pressure by understanding the composition of waves forming the scattered pressure, and shall describe each of these waves. Attention will be paid not only to their spectral characteristics (resonance frequencies of partial modes), but also to their magnitudes and damping coefficients.

The computation of the form function $f(x)$ is carried out in a broad x range $(0 \leqslant x \leqslant 400)$ with a computation step size $l_x = 5/256$; the isolated modal resonances $\zeta_n(x) = |\psi_n^{(r)}(\varphi = \pi, x)|$, see (2.58), are computed with double step size $l_x = 10/256$. This choice is stipulated by the rapid alteration of both the functions $f(x)$ and $\zeta_n(x)$ and allows us to find the resonance positions with sufficient accuracy.

The physical parameters of the shell material and the liquid surrounding the shell are chosen as

aluminum: $\rho_1 = 2.79 \times 10^3$ kg/m^3, $c_1 = 6380$ m/s, $c_t = 3100$ m/s ,

water: $\rho = 1 \times 10^3$ kg/m^3, $c = 1470$ m/s .   (8.1)

Here, c_1 and c_t are the velocities of the longitudinal and transverse waves in the shell material.

At fixed $h = 1/10$, the high-x domain considered in the computations is needed only for generating in the shell the symmetric and antisymmetric Lamb-type waves of high order. The zero-order symmetric (S_0) and antisymmetric (A_0) wave, and the A and A_1 waves, have been studied earlier in [8.6–8]. The behavior of the Lamb-type peripheral waves of higher order has not yet been considered. The computed form function is presented in Fig. 8.1.

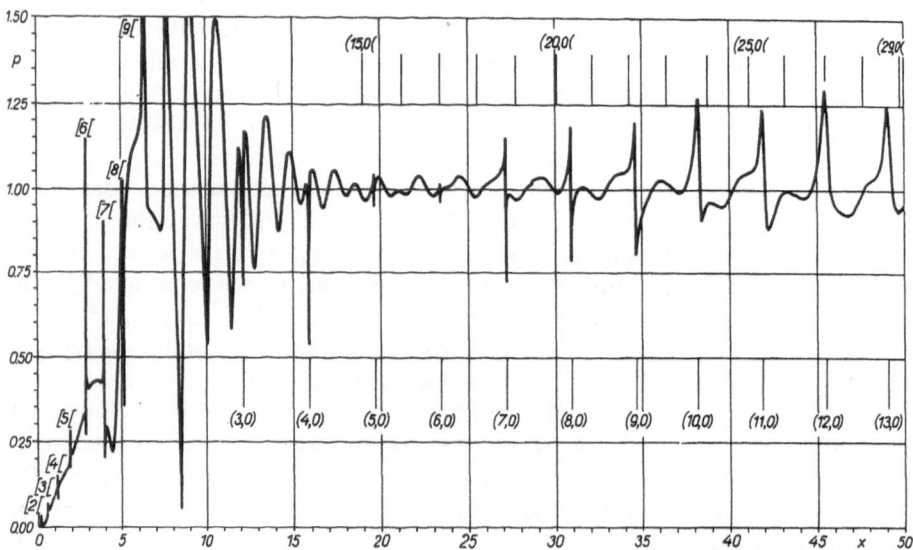

Fig. 8.1.1

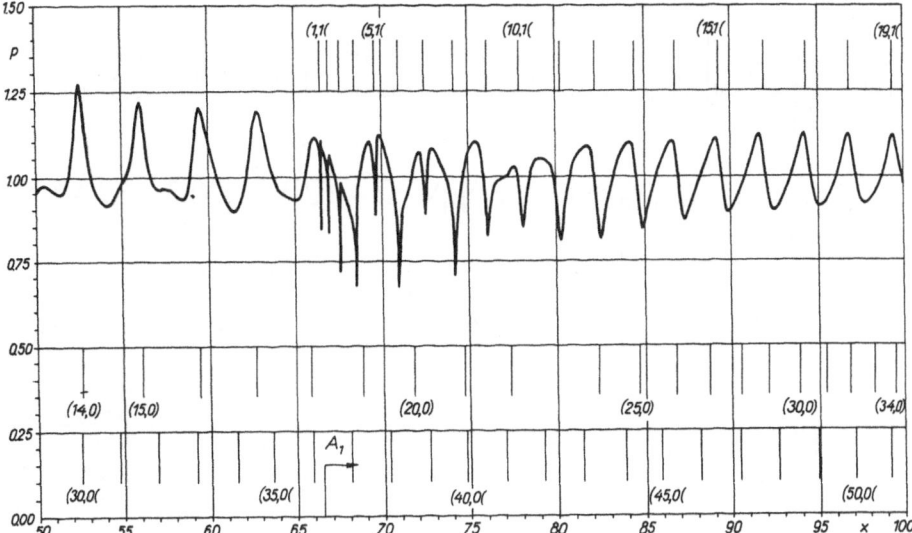

Fig. 8.1.2

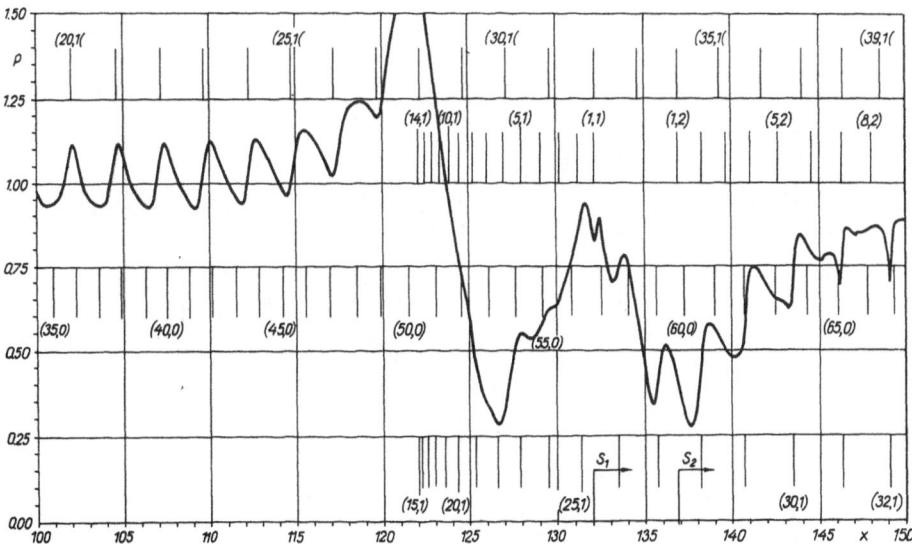

Fig. 8.1.3

Fig. 8.1.1–8.1.8 (p. 132–136). The form function for backscattering from an aluminum shell with $h = 1/10$ in water, in the range $0 \leqslant x \leqslant 400$

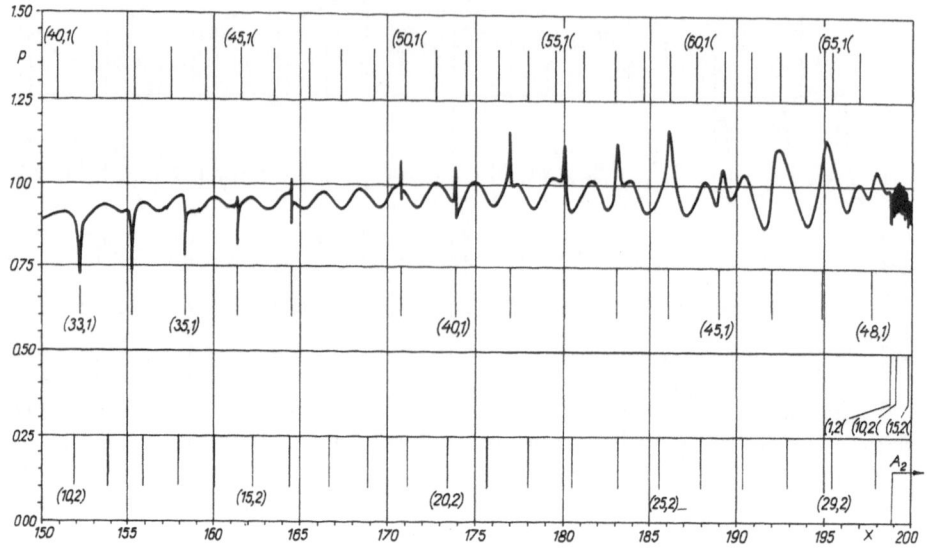

Fig. 8.1.4

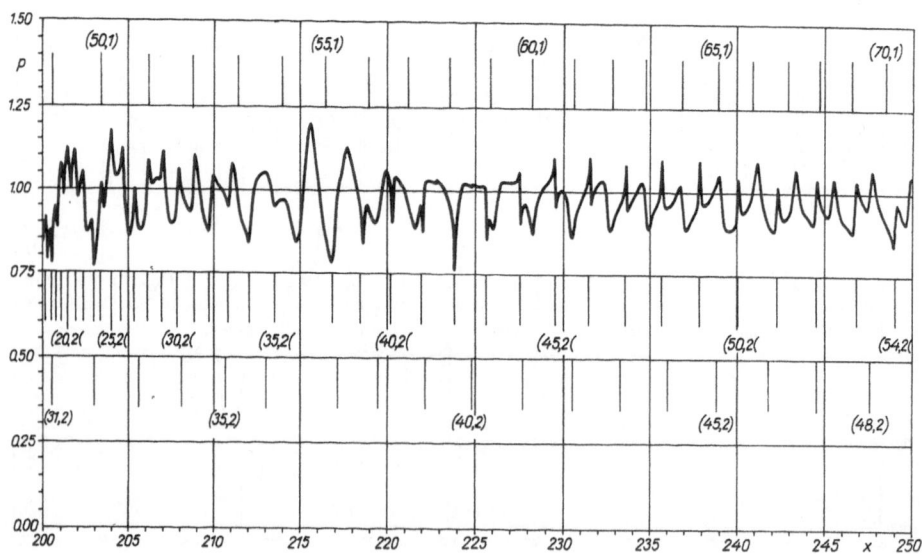

Fig. 8.1.5

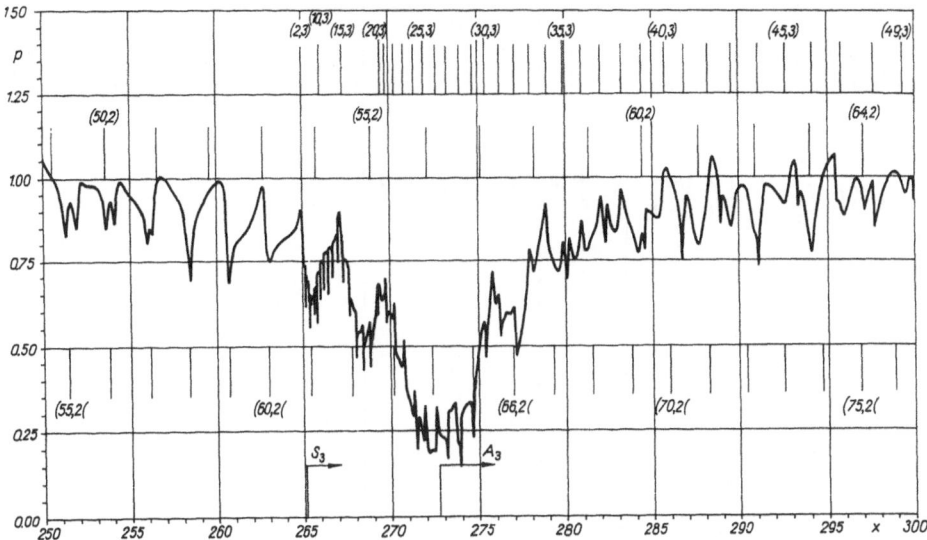

Fig. 8.1.6

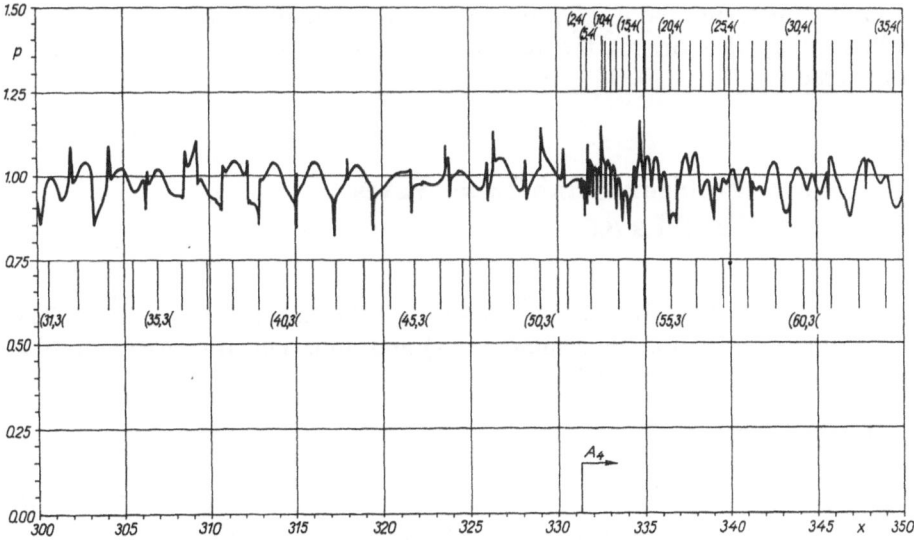

Fig. 8.1.7

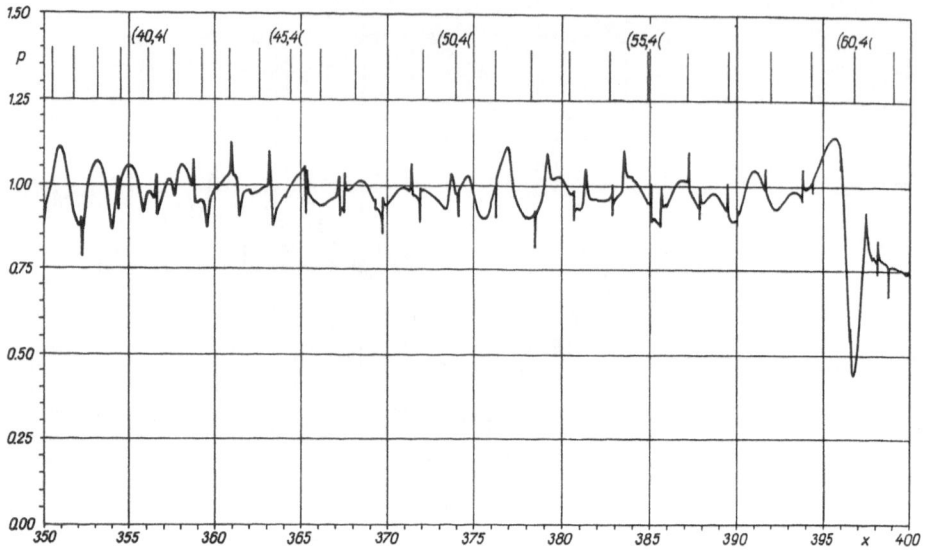

Fig. 8.1.8

Before passing to the form function analysis, we shall examine the model problem concerning the wave propagation in a thin elastic layer and present results on applications of the resonance scattering theory to the problem considered.

8.2 Model Problem: Lamb-Type Waves in a Plane "Dry" Layer

Let us consider an equivalent (in material and thickness) plane dry layer. The dispersion equation for the symmetric (S) and antisymmetric (A) Lamb-type waves has the form (see [8.9])

$$E = 0 \quad \text{and} \quad F = 0 \tag{8.2}$$

with

$$E = \varphi_1 \coth p_l - \varphi_2 \coth p_t, \quad F = \varphi_1 \tanh p_l - \varphi_2 \tanh p_t,$$

$$\varphi_1 = \frac{(2 - y^2)^2}{4}, \quad \varphi_2 = \frac{4\sqrt{(1 - \gamma_0^2 y^2)(1 - y^2)}}{y^4},$$

$$p_l = \frac{z}{y}\sqrt{1 - \gamma_0^2 y^2}, \quad p_t = \frac{z}{y}\sqrt{1 - y^2}, \quad \gamma_0 = \frac{c_t}{c_l}, \quad y = \frac{c^{\text{ph}}}{c_t},$$

$$z = k_t d, \quad k_t = \omega/c_t, \quad 2d = a - b.$$

$$\tag{8.3}$$

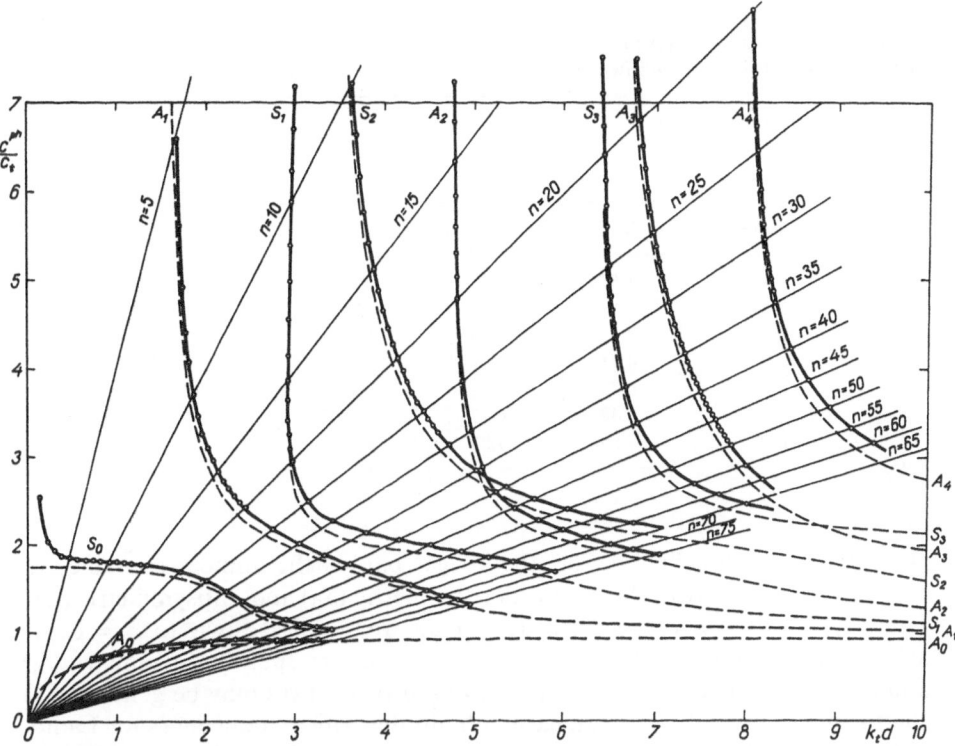

Fig. 8.2. The dispersion curves for the relative phase velocity of the Lamb-type peripheral waves revolving around the shell (solid lines, the resonance positions being labeled by circles) and the dispersion curves for the corresponding Lamb-type waves propagating on a plane dry layer (dashed lines)

From these equations the dependence of relative (divided by c_t) phase velocities y on the variable z is found.

The roots of (8.2) depend on the parameter γ_0 which is a function of Poisson's ratio v only

$$\gamma_0 = \sqrt{\frac{1-2v}{2(1-v)}}. \tag{8.4}$$

With the chosen parameters of (8.1), we obtain $\gamma_0 = 0.4859$.

In Fig. 8.2, the dispersion curves of the relative phase velocities of the Lamb-type waves on a plane dry layer are shown as the dashed lines. The computation was carried out with step size $l_z = 0.01$.

The full number of symmetric (Q_S) and antisymmetric (Q_A) waves which can appear on a plane layer of a given thickness $2d$ at the frequency ω is defined in [8.9–11]

$$Q_S = 1 + [q] + [0.5 + \gamma_0 q], \quad q = z/\pi,$$
$$Q_A = 1 + [\gamma_0 q] + [0.5 + q]. \tag{8.5}$$

Table 8.1. The critical frequencies for the Lamb-type wave generation in a plane dry aluminum layer and in the evacuated cylindrical aluminum shell immersed in a liquid (water)

Wave type	Plane dry layer $z = k_t d$		Empty shell	
	Formula	Value	$z = k_t d$	$x = ka$
A_1	$\pi/2$	1.57	1.58	66.52
S_1	π	3.14	3.13	132.11
S_2	$\pi/2\gamma_0$	3.23	3.25	136.88
A_2	$3\pi/2$	4.71	4.71	198.83
S_3	2π	6.28	6.28	265.08
A_3	π/γ_0	6.47	6.47	272.73
A_4	$5\pi/2$	7.85	7.85	331.33

Here the square brackets denote the integer part of the number contained in them. As is well known, the zero-order symmetric (S_0) and antisymmetric (A_0) Lamb-type waves may be generated at all frequencies down to $\omega = 0$. As ω increases past the so-called critical values of ωd, corresponding symmetric (S_1) and antisymmetric (A_1), $l = 1, 2, 3, \ldots$, Lamb-type waves may be generated in succession. Using (8.5), one can easily obtain the critical frequencies for Lamb-type waves with $l \geqslant 1$. In the case of an aluminum layer they are given in Table 8.1. The critical frequencies of the corresponding peripheral waves generated in the shell by an incident plane acoustic wave are also presented in Table 8.1. In the case of the shell, the critical frequencies are found at the first resonances of the peripheral waves. Even for a very thick shell (with $h = 1/10$), the actual values of the critical frequencies given in the variable $z \equiv k_t d$ are rather close to the corresponding frequencies of the Lamb-type waves in a plane dry layer defined by this evidently simple formula. The curvature of the cylindrical layer and the contact with the liquid affect the values of the critical frequencies only weakly. Therefore, for a rough, preliminary judgement on the critical frequencies of the peripheral waves generated in the shell, one can use the values corresponding to the thin dry plane layer.

8.3   Application of the Resonance Scattering Theory

Although some properties of the form function, obtained by the direct summation of the Rayleigh series, can be explained by an application of the heuristic considerations taken from the solution of the model problem of the Lamb-type waves in a plane layer, some special procedure should be used for its more accurate interpretation. We have chosen the resonance scattering theory as the simplest one to realize.

The resonance scattering theory allows us to isolate the resonance components of the partial modes. The resonances of the same kind are joined into families. The dispersion curves can be used as one of the indications upon which the resonances are united into families. The similarity of the dispersion curve obtained from the resonances of partial modes with the standard one – related to the Lamb-type wave in a plane dry layer – can be used as the classification test. A change of the resonance properties can be observed when n increases.

As is generally accepted, the resonance of the partial mode is denoted by two numbers n, l. The first of these defines the ordinal number of the resonance and the second indicates the family. The resonances of the symmetric wave would be labeled as (n, l) and those of the antisymmetric one as $(n, l($. The resonances of the fluid-borne bending (A) wave are labeled just by $[n[$. As will be shown below, the properties of the waves generated in the shell (circular–cylindrical layer), which is in contact with the liquid on the outer surface and free ("dry") on its inner surface, are similar to the properties of the normal (Lamb-type) waves in a plane dry layer. Therefore, for these waves we use standard names: the symmetric waves are denoted by S_l and the antisymmetric waves by A_l.

As one would expect, the waves S_0, A_0 and A do not possess any critical frequency. The resonances of these waves can be observed from zero on the x axis. A refinement is necessary here. Although the resonances of the A wave may appear beginning from $x = 0$, actually the resonances of noticeable magnitude can be observed only in a rather narrow range of x which we denote the "strong-bending domain". The location in x, or in $z \equiv k_t d$, of this range depend on the relative thickness of the shell h and the material parameters of the shell. The smaller is h, the higher in x the strong-bending domain is situated. For a thin-walled shell at small values of n, the magnitudes of the resonances are small. They begin to be noticeable only from a certain n value upward. In the example considered of a thick-walled shell, the A wave resonances can already be observed from $n = 3$ on.

As can be seen from Fig. 8.3, the resonance magnitude of the partial modes of the S_0 wave diminishes with growing n. The Q factor, which can be gleaned from the angle of opening of the resonance curve at the resonance frequency, at first increases (the angle diminishes) up to $n \sim 8$ and then steadily decreases. The procedure of the resonance scattering theory allows us to isolate the partial modes of the S_0 wave up to $n \sim 80$, while in the form function curve the resonances are distinguishable only up to $n \sim 17$. Since the resonance magnitudes of the partial modes for $n = 15$–65 change negligibly (Fig. 8.3), there can be two explanations for this. First, the drastic decrease of the Q factor of the S_0 wave at $n = 15$–65 and an almost in-antiphase superposition of the successive resonance components of the S_0 wave, or, secondly, the excitation at $x = 66.52$ of the A_1 wave with resonances of a high Q factor for $n \leqslant 10$. The magnitude of these resonances is commensurable with the magnitude of the S_0 wave resonances.

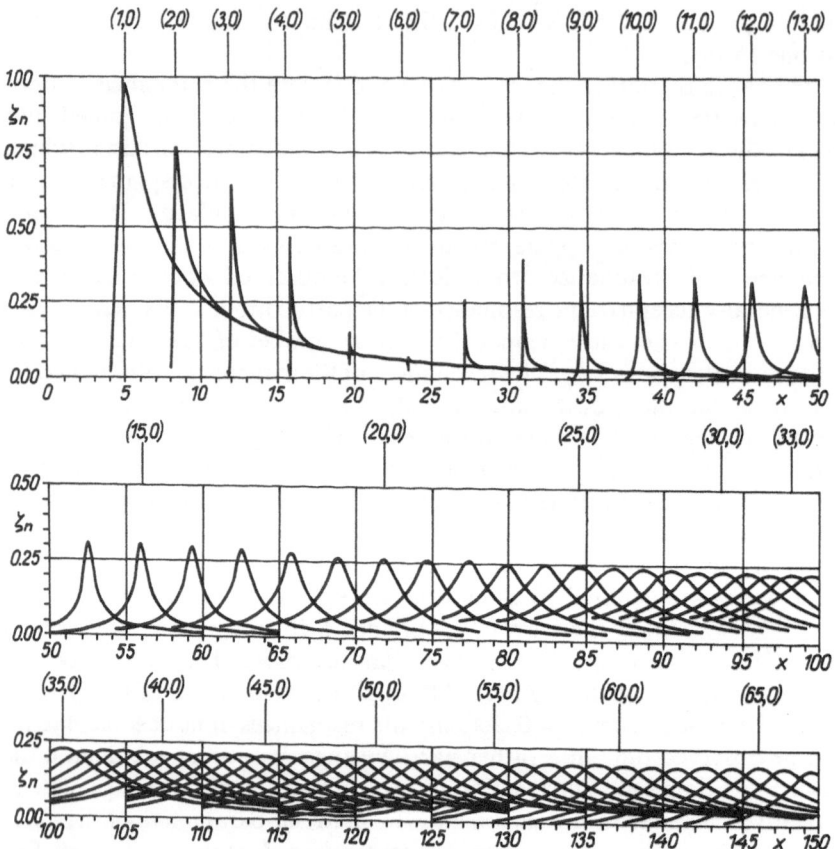

Fig. 8.3. The isolated modal resonances $(n, 0)$ of the S_0 wave calculated for the aluminum shell with $h = 1/10$ in water, obtained by suitable (rigid) background subtraction, and displayed versus $x = ka$

It is evident from Fig. 8.4 that the resonance magnitudes of the partial modes of the A_0 wave decrease monotonically with n. The Q factor of the wave changes in the same way. It is difficult to distinguish the resonances of this wave on the form function curve.

As in the limiting problem (of Lamb-type waves in a plane dry layer), with increasing frequency the S_0 and A_0 waves become more similar to each other. For example, this is obvious from the $\zeta_n(x)$ curves of these waves in the range $x = 80$–100.

An evident formula describing the peripheral wave can be obtained using the data of the resonances of partial modes. Such formulae are presented in [8.12] for the Rayleigh and Whispering Gallery waves excited in a solid elastic sphere by a plane acoustic wave.

It can be seen from Fig. 8.5 that the resonances of the S_1 wave generated at the critical frequency $x = 132.11$ move on the x axis first from the right to the left

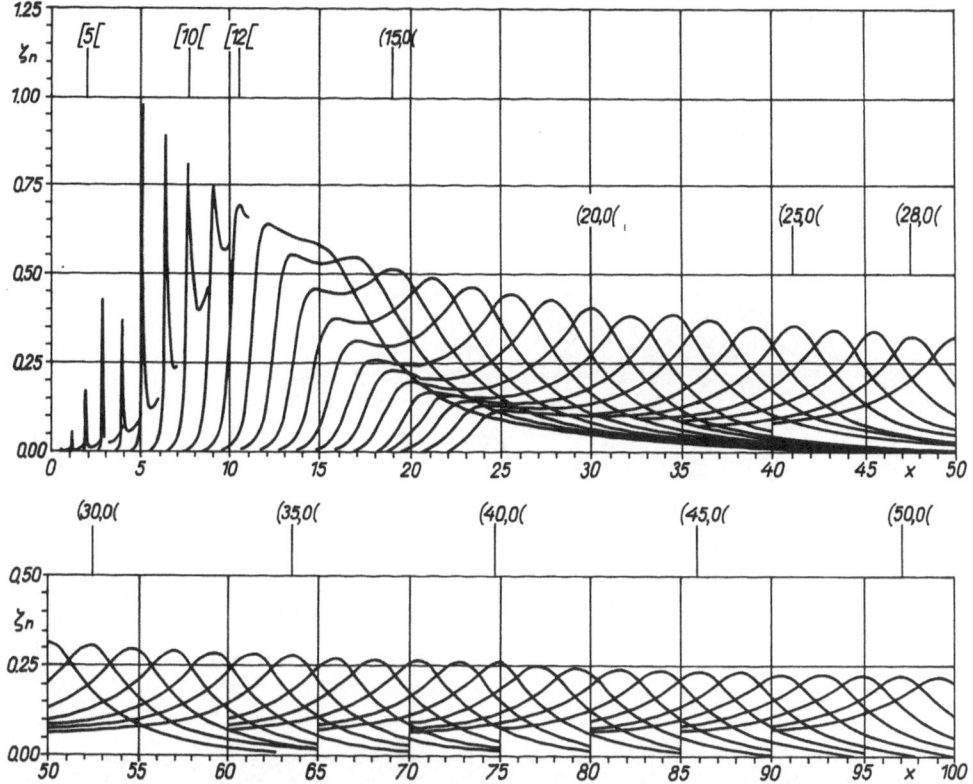

Fig. 8.4. The isolated modal resonances [n[of the A wave, and (n, 0 (of the A_0 wave

(for $n = 1$–14) and then, as for all the other waves, from the left to the right. On the dispersion curve of the phase velocity this phenomenon is reflected in the "backbending" of the curve. Such a fact is well known for the Lamb-type wave S_1 in a plane dry layer, which once again indicates the resemblance of the waves in the two problems, namely of the scattering of an acoustic wave by an elastic shell, and of wave propagation in a plane dry layer. With n increasing, the resonance magnitude of the S_1 wave slowly diminishes. The Q factor of the wave first slowly increases and afterwards, at $n \gtrsim 35$, rapidly diminishes.

Beginning from the critical frequency $x = 66.52$, the A_1 wave will be generated (Fig. 8.6). With n increasing, the resonance magnitude of the partial mode slowly diminishes. The Q factor of the A_1 wave up to $n \sim 30$ is rather high in comparison with other waves. Thus, the resonances of this wave can be well distinguished in the total form function.

The S_2 wave possesses a small Q factor as it is evident from the $\zeta_n(x)$ curves in Fig. 8.7. With n increasing, the resonance magnitude of the partial mode of the S_2 wave slowly diminishes and the Q factor increases. The resonances of this wave are almost indistinguishable on the total form function curve.

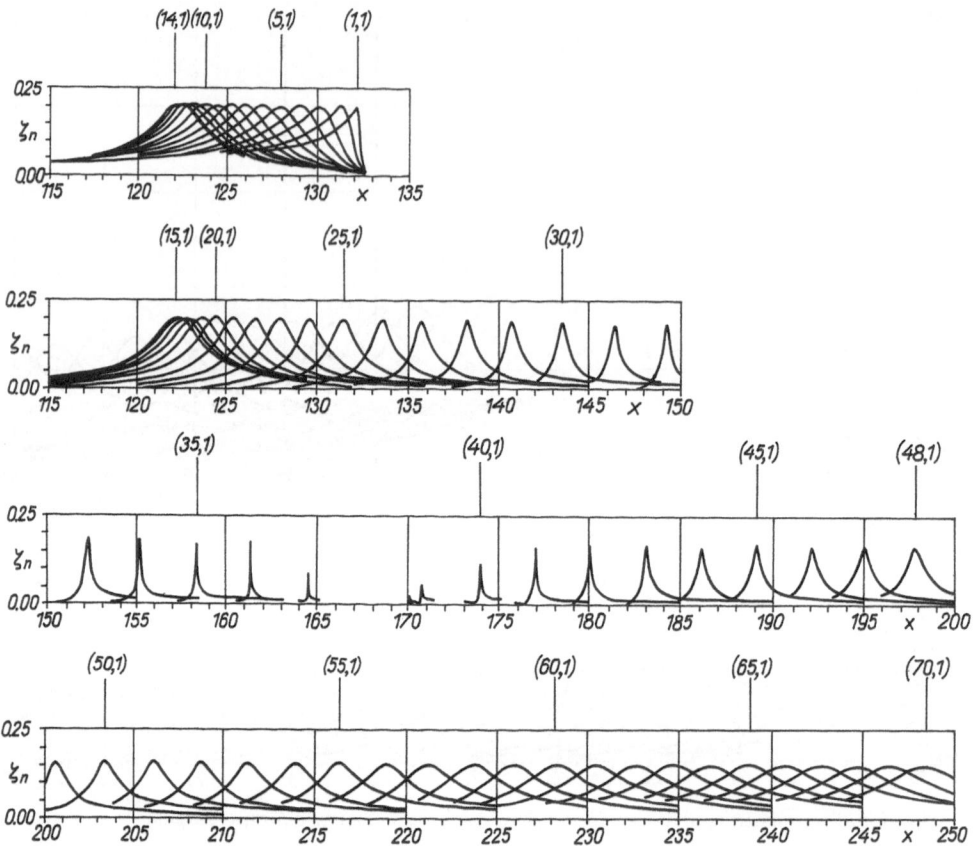

Fig. 8.5. The isolated modal resonances $(n, 1)$ of the S_1 wave

The first thirteen resonances of the partial modes of the A_2 wave follow each other very closely. With n increasing (after $n \sim 15$), the interval of the successive resonances of this wave gradually increases while the magnitude and the Q factor diminish. The resonances of this wave become apparent very clearly on the form function curve up to $n \sim 50$. In the vicinity of the intersection of two dispersion curves, namely those corresponding to the S_2 and the A_2 waves which takes place at $x \sim 213$, it is rather difficult to separate the modal resonance component. Therefore, for $n = 37$, such a component is not shown in Fig. 8.8, and for $n = 36$ and $n = 37$ the resonance components of the S_2 wave are shown by dashed lines in Fig. 8.7.

The first twenty resonances of the partial modes of the S_3 wave follow each other on the x axis at very small intervals (Fig. 8.9). The magnitude and the Q factor of the S_3 wave resonances first increase with n increasing (up to $n \sim 25$) and then diminish. Before the A_3 wave is generated in the shell, the resonances of the S_3 wave are well distinguished on the total form function curve.

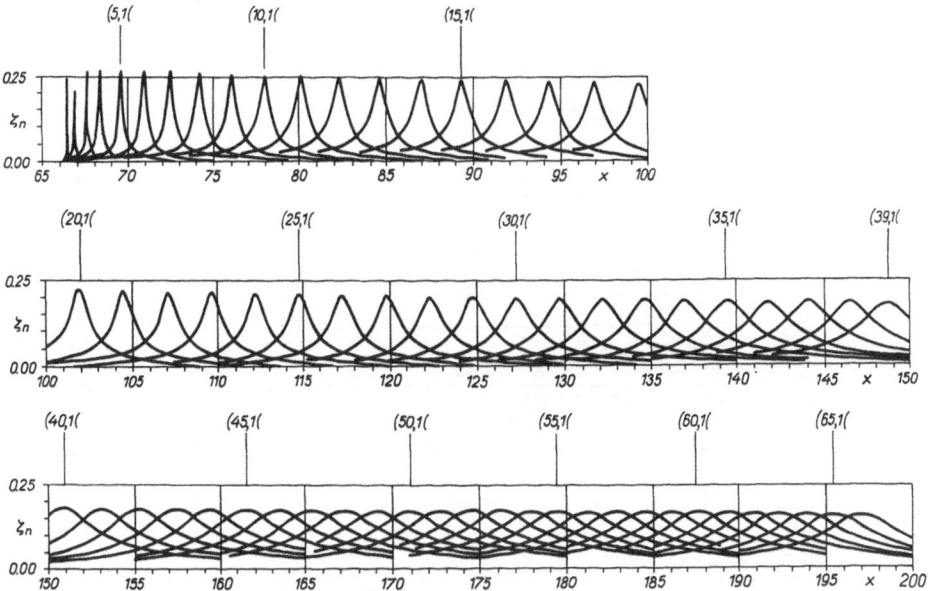

Fig. 8.6. The isolated modal resonances $(n, 1($ of the A_1 wave

Above the critical frequency $x = 272.73$, the A_3 wave is generated (Fig. 8.10). The resonance magnitudes of the partial modes of this wave almost do not change with n increasing but the Q factor quickly increases, especially after $n \sim 55$. In spite of the fact that the intervals between the successive resonance frequencies of this wave are rather big, the resonance curves are situated very close to each other and therefore they are practically indistinguishable at the chosen scale of the plot. Therefore, up to $n = 15$, the resonance curves are shown only for $n = 1, 5, 10$ and 15. For n values, different from the mentioned ones, the resonance curves change in the same way. Because of a small Q factor and an almost antiphase superposition of the successive resonances, the resonances of the A_3 wave are not distinguishable on the form function curve.

The isolated modal resonances of the A_4 wave are shown in Fig. 8.11.

It is typical that the maximal magnitudes of the resonance components of the S_0 and A wave are of order of unity, while for the Lamb-type waves with $l > 1$ they do not exceed 0.25.

The resonances of the waves with small Q factor, as a rule, are not distinguishable on the form function curve. In such a case, at a fixed x value both some successive partial modes with low Q factor and the partial modes of other waves contribute to the form function. Thus, the contribution of the peripheral wave to the form function can be both clear and hidden. The question of the so-called hidden resonances is discussed in Chap. 10 for the case of scattering by an elastic spherical shell.

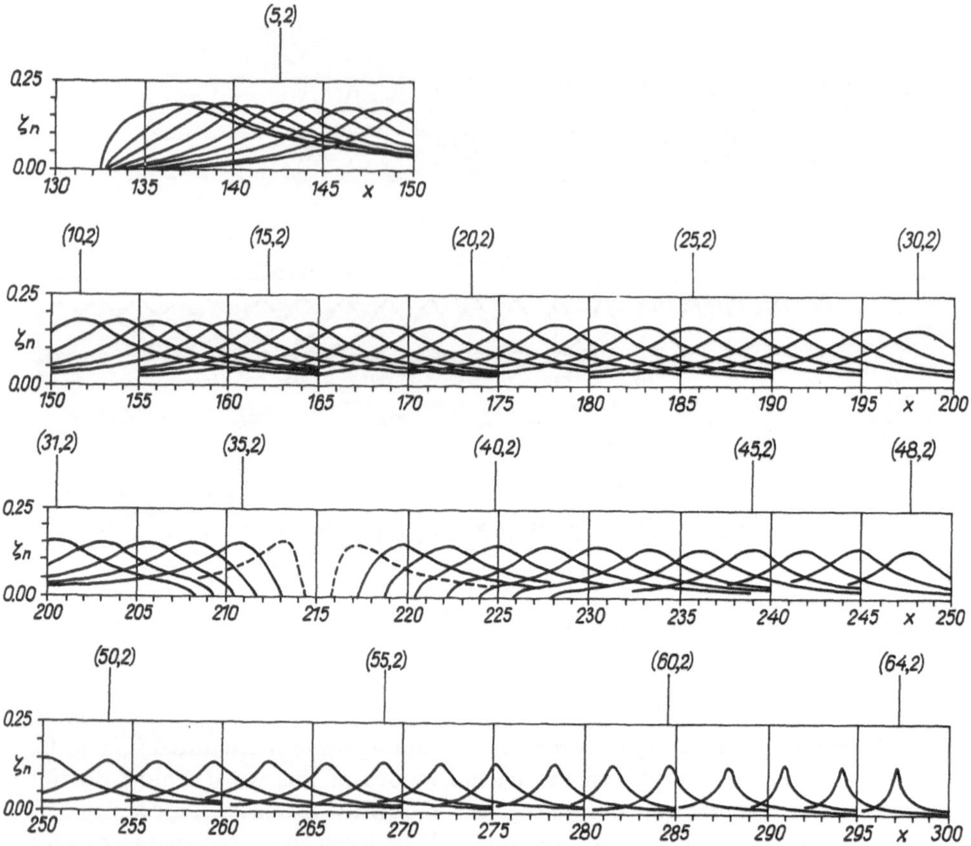

Fig. 8.7. The isolated modal resonances $(n, 2)$ of the S_2 wave

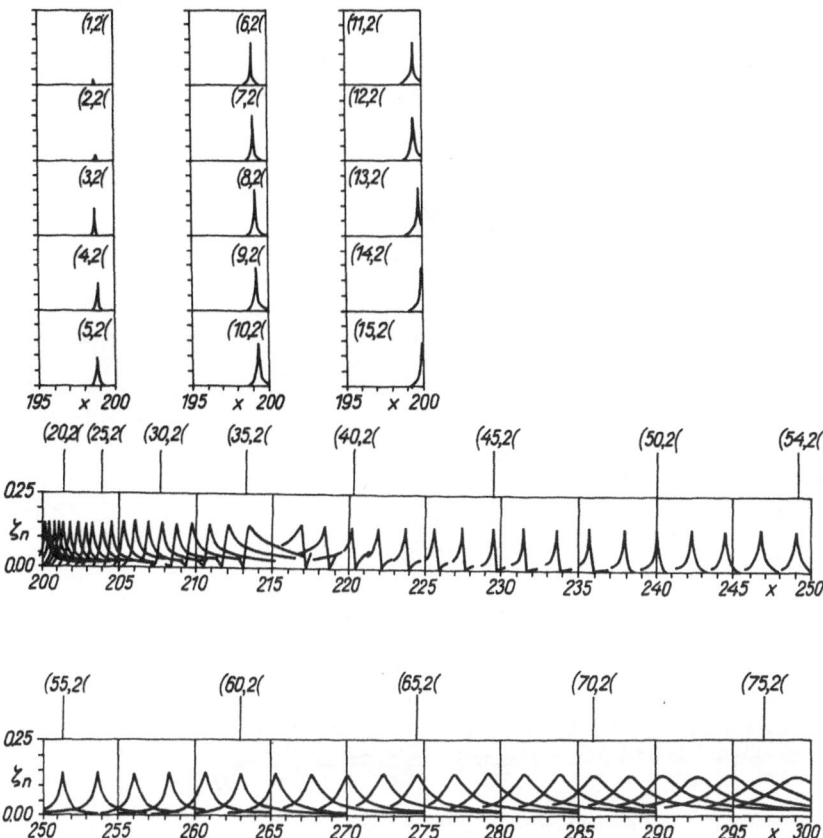

Fig. 8.8. The isolated modal resonances $(n, 2($ of the A_2 wave

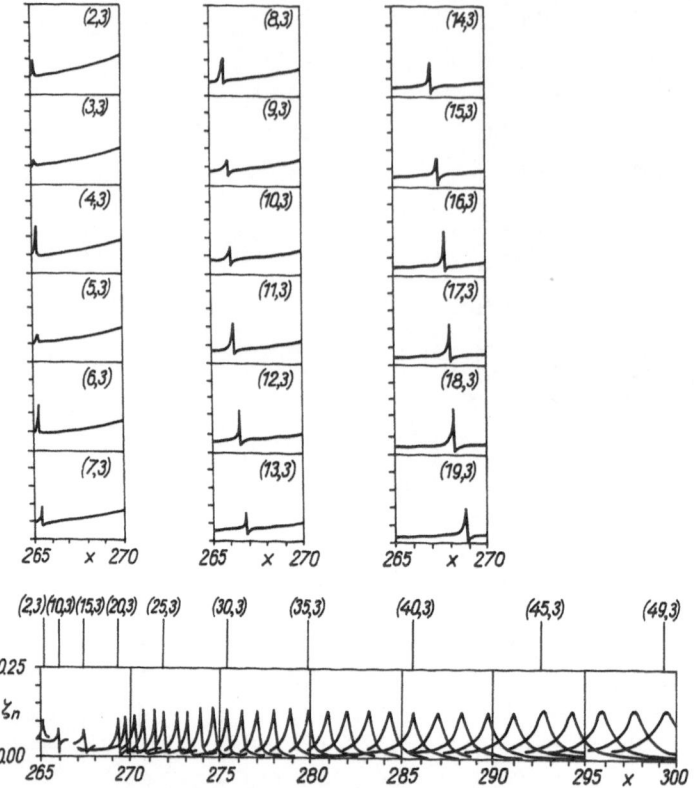

Fig. 8.9. The isolated modal resonances $(n, 3)$ of the S_3 wave

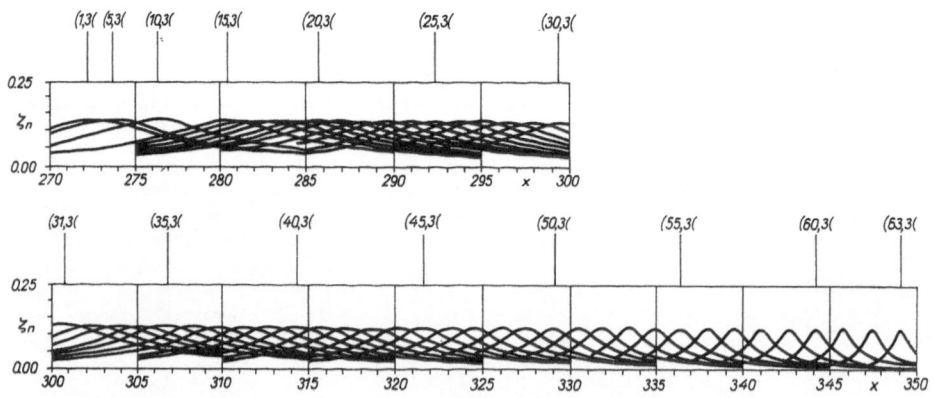

Fig. 8.10. The isolated modal resonances $(n, 3($ of the A_3 wave

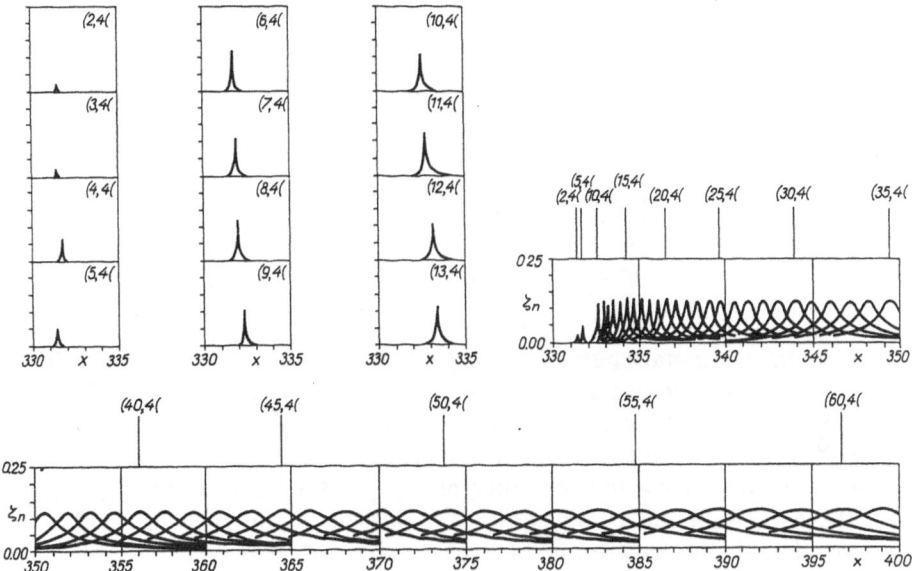

Fig. 8.11. The isolated modal resonances $(n, 4($ of the A_4 wave

It should be noted that the presented curves of $\zeta_n(x)$ do not exhaust all the excited resonances. For waves of different families the upper number n of the $\zeta_n(x)$ curve is different. This number has been chosen from the condition of representability of the family. We stop showing the resonances when it is clear how the resonance curve should behave with n increasing.

8.4 Form Function Analysis

The resonance frequencies of the partial modes are labeled in Fig. 8.1. As is well known, these frequencies correspond to the resonance frequencies of the peripheral waves revolving around the shell. The resonance of the peripheral wave occurs when exactly n $(n = 1, 2, 3, \dots)$ wavelengths fit over the circumference of the shell

$$n\lambda_{nl} = 2\pi a, \quad \lambda_{nl} = \frac{2\pi a}{x_{nl}} v_{nl}^{\mathrm{ph}}, \quad v_{nl}^{\mathrm{ph}} = \frac{c_{nl}^{\mathrm{ph}}}{c}$$

$$(n = 1, 2, 3, \dots; l = 0, 1, 2, \dots),$$

(8.6)

where λ_{nl} is the wavelength of the peripheral wave, c_{nl}^{ph} is its phase velocity at the resonance frequency x_{nl}, and v_{nl}^{ph} is the relative phase velocity. The resonance frequencies coincide with the eigenfrequencies of the liquid-loaded elastic shell.

Table 8.2. The positions and magnitudes of the modal resonances generated in the shell in the vicinity of $x = 126$

Wave type	n	x	$\zeta_n(x)$
S_0	53	125.82	0.201
A_0	64	127.15	0.200
S_1	7	126.02	0.201
S_1	22	126.56	0.201
A_1	30	127.34	0.200

From the computational results (Fig. 8.1) it is obvious that in the vicinity of the critical frequencies, as a rule, some essential variation of the form function curve occurs. In particular, the form function curve becomes more jagged. In those form function domains where the critical frequencies of two peripheral waves are situated close to each other, as, e.g., for the waves S_1 and S_2; S_3 and A_3, the form function curve shows a well-defined dip or a peak. This is connected with the generation of a new wave which may be of significant magnitude and whose phase differs from (at a dip) or is close to (at a peak) the phase of the superposition of the peripheral waves observed at the frequencies just below the critical frequency.

In the example considered, in the form function curve in the range $120 \leqslant x \leqslant 150$ a deep and broad dip appears. At two points of the form function curve the magnitude is rather small: $p(x) = 0.284$ at $x = 126.66$ and $p(x) = 0.278$ at $x = 137.58$. Such a shape of the form function curve can be explained by the generation of the S_1 and S_2 waves. In the range $120 \leqslant x \leqslant 150$ the main contribution to the form function is furnished by the Lamb-type waves S_0, A_0, S_1 and A_1, and by the specularly reflected wave with magnitude of the order of unity.

The positions and magnitudes of the partial mode resonances generated in the shell in the vicinity of $x \sim 126$ are presented in Table 8.2. If all these resonances occur at some $x = x_0$ and the corresponding Lamb-type waves have the same phase at $x = x_0$, then the magnitude of their superposition will exceed unity. Assuming that the phase of each of these waves is opposite to the phase of the specularly reflected wave, the magnitude of the form function at $x = x_0$ will be equal to zero. Actually the resonances of the S_0, A_0, S_1 and A_1 waves are not observed at the same frequency and are somewhat out of phase. Therefore, the minimal magnitude of the form function is about 0.3. In this way the contributions of the waves with resonances nearest to $x = x_0$ affect the form function. On account of the small Q factor of the Lamb-type waves in the range $120 \leqslant x \leqslant 130$, at $x = x_0$ also the neighboring resonances affect the form function: from 49 to 58 of the S_1 wave, from 60 to 69 of the A_0 wave, from 1 to 25 of the S_1 wave, and from 27 to 33 of the A_1 wave.

The deep and broad dips are typical for the form function curve. Their position on the x axis depends on the relative thickness of the shell h. For the

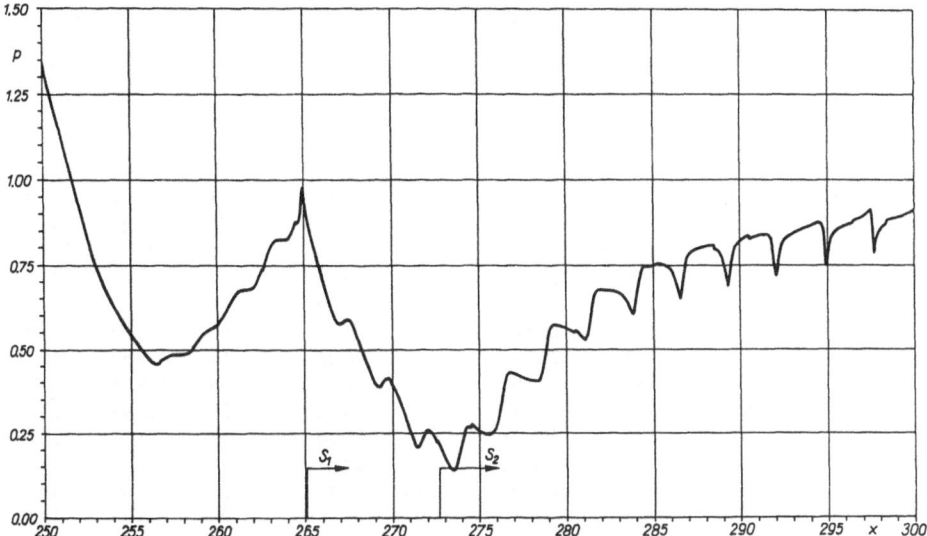

Fig. 8.12. The form function for backscattering from an aluminum shell with $h = 1/20$ in water

same material parameters (see (8.1)) but with the shell being thinner (with $h = 1/20$), such a dip occurs at $250 \leqslant x \leqslant 280$ (Fig. 8.12). The minimal magnitude of the form function in this range is $p(x) = 0.139$ and occurs at $x = 273.48$. Just in the vicinity of this dip the generation of two peripheral waves begins: the S_1 at $x = 265.00$ and the S_2 wave at $x = 272.70$.

The described dip in the form function curve is caused by the generation of two symmetric waves. An analogous dip can also occur when a symmetric and an antisymmetric waves are generated. In the case of the shell with $h = 1/10$, such a dip can be observed at $260 \leqslant x \leqslant 285$ and is caused by the S_3 and A_3 waves.

8.5 Phase Velocities

The relative phase velocity of the peripheral wave can be calculated when the position of the nth resonance of the lth family is known

$$v^{\mathrm{ph}} = \frac{c^{\mathrm{ph}}}{c}, \qquad v^{\mathrm{ph}}[x(n, l)] = \frac{x(n, l)}{n}, \qquad v^{\mathrm{ph}}[x(n, l(\,] = \frac{x(n, l(}{n}, \tag{8.7}$$

where $x(n, l)$ and $x(n, l(\,$ are the positions of the resonances of the symmetric and antisymmetric Lamb-type waves, respectively.

Using $z = k_t d$ as an independent variable, the velocities of the peripheral waves generated in shells with different relative thicknesses can be compared.

Table 8.3. The positions on the x axis of the modal resonances of the Lamb-type peripheral waves revolving around the shell

n	S_0	S_1	S_2	S_3	A_0	A_1	A_2	A_3	A_4
1	5.35	132.11	136.88			66.52	198.83	272.73	331.33
2	8.63	131.17	138.01	265.08		66.91	198.83	272.85	331.33
3	12.23	130.08	139.45	265.12		67.58	198.83	273.01	331.41
4	15.90	128.98	141.02	265.20		68.48	198.87	273.28	331.48
5	19.65	127.93	142.66	265.27		69.61	198.91	273.59	331.60
6	23.44	126.95	144.38	265.39		70.98	198.94	273.98	331.72
7	27.15	126.02	146.13	265.51		72.50	198.98	274.45	331.88
8	30.90	125.20	147.96	265.66		74.18	199.06	275.00	332.07
9	34.61	124.45	149.88	265.82		76.05	199.14	275.59	332.27
10	38.28	123.79	151.80	266.02		78.01	199.26	276.25	332.50
11	41.95	123.24	153.79	266.25		80.12	199.38	276.95	332.77
12	45.55	122.81	155.82	266.48		82.30	199.49	277.73	333.05
13	49.10	122.50	157.89	266.72		84.61	199.65	278.59	333.40
14	52.62	122.30	160.00	267.03	16.76	86.95	199.80	279.49	333.71
15	56.05	122.27	162.15	267.30	19.02	89.38	200.04	280.83	334.10
16	59.38	122.34	164.34	267.66	21.21	91.84	200.27	281.45	334.49
17	62.66	122.58	166.52	268.01	23.40	94.34	200.51	282.50	334.96
18	65.82	123.01	168.79	268.40	25.55	96.88	200.82	283.59	335.40
19	68.91	123.59	171.05	268.79	27.73	99.45	201.13	284.73	335.90
20	71.84	124.38	173.40	269.26	29.96	102.03	201.52	285.90	336.46
21	74.65	125.35	175.74	269.73	32.11	104.61	201.91	287.11	336.99
22	77.34	126.56	178.09	270.20	34.34	107.19	202.38	288.36	337.62
23	79.88	127.97	180.51	270.74	36.60	109.77	202.85	289.65	338.24
24	82.27	129.61	182.93	271.29	38.83	112.30	203.40	290.98	338.95
25	84.49	131.48	185.35	271.88	41.09	114.84	204.02	292.34	339.65
26	86.60	133.56	187.81	272.50	43.32	117.38	204.65	293.71	340.49
27	88.55	135.82	190.31	273.16	45.59	119.88	205.35	295.12	341.21
28	90.39	138.24	192.81	273.87	47.81	122.38	206.13	296.52	342.03
29	92.15	140.82	195.35	274.61	49.96	124.84	206.95	297.97	342.92
30	93.75	143.52	197.89	275.39	52.30	127.34	207.85	299.41	343.87
31	95.31	146.33	200.47	276.21	54.57	129.80	208.83	300.86	344.84
32	96.76	149.22	203.05	277.03	56.84	132.23	209.88	302.34	345.90
33	98.16	152.23	205.59	277.93	59.10	134.69	211.02	303.83	346.99
34	99.57	155.23	208.16	278.91	61.32	137.07	212.23	305.31	348.13
35	100.90	158.32	210.70	279.88	63.59	139.49	213.59	306.80	349.30
36	102.23	161.41	213.09	280.90	65.82	141.84		308.28	350.55
37	103.52	164.53	217.19	281.99	68.09	144.18	215.16	309.27	351.88
38	104.80	167.66	219.61	283.13	70.31	146.48	216.81	311.29	353.20
39	106.13	170.78	222.19	284.34	72.58	148.75	218.59	312.77	354.65
40	107.42	173.87	224.88	285.59	74.80	150.98	220.27	314.26	356.09
41	108.71	176.99	227.62	286.88	77.03	153.16	221.99	315.74	357.66
42	110.04	180.08	230.35	288.24	79.26	155.31	223.79	317.23	359.22
43	111.37	183.13	233.16	289.65	81.48	157.38	225.63	318.67	360.90
44	112.73	186.13	236.02	291.13	83.67	159.45	227.54	320.16	362.58
45	114.10	189.10	238.87	292.66	85.90	161.45	229.49	321.60	364.34
46	115.51	192.07	241.76	294.26	88.13	163.44	231.52	323.09	366.17
47	116.95	194.96	244.65	295.90	90.31	165.35	233.59	324.53	368.05
48	118.36	197.81	247.58	297.66	92.54	167.27	235.70	326.02	369.96

Table 8.3. Continued

n	S_0	S_1	S_2	S_3	A_0	A_1	A_2	A_3	A_4
49	119.84	200.63	250.55	299.41	94.72	169.10	237.85	327.46	371.95
50	121.29	203.36	253.55	301.29	96.91	170.90	240.04	328.95	373.98
51	122.77	206.09	256.60	303.20	99.10	172.70	242.27	330.39	376.05
52	124.34	208.75	259.65	305.20	101.29	174.45	244.49	331.88	378.20
53	125.82	211.37	262.70	307.27	103.48	176.17	246.76	333.36	380.35
54	127.46	213.91	265.78	309.38	105.66	177.85	249.06	334.84	382.58
55	129.02	216.41	268.91	311.56	107.81	179.53	251.37	336.33	384.84
56	130.63	218.87	272.03	313.83	110.00	181.21	253.67	337.86	387.11
57	132.27	221.25	275.16	316.17	112.15	182.81	256.02	339.38	389.45
58	133.91	223.59	278.32	318.55	114.30	184.45	258.32	340.94	391.84
59	135.59	225.90	281.48	321.02	116.45	186.05	260.66	342.50	394.22
60	137.27	228.16	284.61	323.55	118.59	187.66	262.97	344.10	396.68

For this purpose, the dependence of the relative phase velocity y on z is often used. The corresponding values are connected by the relations

$$y = \left(\frac{c}{c_t}\right) v^{ph}, \qquad z = \frac{1}{2}\left(\frac{c}{c_t}\right) hx . \tag{8.8}$$

The dispersion curves of the Lamb-type peripheral wave phase velocities are shown in Fig. 8.2 as a solid line. The positions of the partial mode resonances are labeled on them. It is obvious from this figure that at fixed z, the phase velocity of a peripheral wave revolving around the shell is somewhat bigger than the corresponding one in the plane dry layer. The difference in velocity is partly connected with the curvature of the cylindrical layer, and partly with the liquid loading. If we chose a problem of a plane layer on which one side is liquid-loaded and the other free ("dry") as a model problem, then the dispersion curves will be situated somewhat beyond the dashed curves in Fig. 8.2, and the difference in velocity will be smaller. On the other hand, if the relative thickness of the shell h is smaller, say $h = 1/100$, then the dispersion curves corresponding to the shell (solid lines in Fig. 8.2) will lower towards the dashed lines and the difference will again be smaller. The positions of the resonances of partial modes for the Lamb-type waves are presented in Table 8.3 and for the fluid-borne A wave in Table 8.4.

According to the definition (8.7), we obtain in the variables (8.8)

$$\frac{y[x(n, l)]}{z[x(n, l)]} = \frac{y[x(n, m)]}{z[x(n, m)]} = \frac{y[x(n, l(]}{z[x(n, l(]} = \frac{y[x(n, m(]}{z[x(n, m(]} = \frac{2}{nh} . \tag{8.9}$$

This means that in the y–z plane the resonances of the same order n but of different families (l and m), both for symmetric and antisymmetric Lamb-type peripheral waves, are situated on straight lines passing through the origin with the slope coefficient equal to $2/nh$. For n values divisible by five the resonances

Table 8.4. The positions on the x axis of the modal resonances of the fluid-borne A wave revolving around the shell

n	x	n	x
3	0.60	8	5.20
4	1.21	9	6.45
5	1.99	10	7.73
6	2.93	11	9.10
7	4.02	12	10.51

are shown in Fig. 8.2. Relation (8.9) can be used to check the accuracy of computation.

Using the procedure of the resonance scattering theory, we have obtained the dispersion curves of the phase velocity of the peripheral waves revolving around the shell. Usually the calculation of such curves is a topic of separate analysis. Often they are found from the solution of the problem of the eigenfrequencies of a fluid-loaded elastic body by equating the corresponding determinant to zero. Although the resulting dispersion curves are suitable for arbitrary loading, actually they do not contain, and cannot contain, any information on the magnitude of the vibrations. Therefore, in one part of the dispersion curve calculated in such a way, depending on the loading, considerable magnitudes of partial modes may be present and in another part they may be negligible. The magnitudes of the fluid-borne A wave are a typical example of such a possibility: inside the strong-bending domain they are very large and outside they are negligibly small.

8.6   Asymptotic Formula for Resonance Frequencies

Having considered a model problem on high-frequency long-wave eigenvibrations of a thin circular elastic ring, which is a cross section of the cylindrical shell, Kaplunov has obtained the asymptotic formula for the position of the resonance frequencies [8.13]

$$z = z_q - \eta^2(n^2 p_q + v_q) \quad (q = 1, 2) , \tag{8.10}$$

where

$$p_1 = (2z_1 \gamma_0^2)^{-1} + 4z_1^{-2} \cot z_1 ,$$

$$p_2 = (2z_2)^{-1} + 4z_2^{-2} \cot(\gamma_0 z_2) ,$$

$$v_1 = -2[1 - 1/(16\gamma_0^2)]z_1^{-1} + \varepsilon c/4\eta c_t z_1 ,$$

$$v_2 = -15/(8z_2), \qquad z_1 = \tfrac{1}{2}\pi j \gamma_0^{-1} , \qquad z_2 = \tfrac{1}{2}\pi j \quad (j = 1, 2, 3, \ldots) ,$$

$$\gamma_0 = c_t/c_1 = [(1 - 2v)/2(1 - v)]^{1/2}, \qquad \varepsilon = \rho c/\rho_1 c_t ,$$

$$\eta = d/R_0, \qquad R_0 = \tfrac{1}{2}(a + b) . \tag{8.11}$$

The index $q = 1$ (with $j = 1$ and $j = 2$ in the expressions for z_1) corresponds to the S_2 and A_3 waves; the index $q = 2$ (with $j = 1, 2, 3, 4$ in the expressions for z_2) corresponds to the waves A_1, S_1, S_3 and A_4, respectively.

The term $v_q (q = 1, 2)$ takes into account the influence of the curvature of the shell. The second component in the expression for v_1 depends on the relative impedance. At $v_q = 0$ the formula (8.10) gets transformed into an approximation following from the Rayleigh–Lamb equations (8.2). From (8.10) it follows that, at $n \sim 1$, the contribution of the v_q term could be commensurable with the contribution of the first term in the brackets, but at $1 \leqslant n \leqslant \eta^{-1}$, as one would expect, the curvature of the shell influences only weakly the position of the resonance frequencies. The error of the asymptotic formula is of the order $0[\eta^2 n^2(\eta^2 n^2 + \varepsilon)]$. Thus, the formula is suitable for the estimation of the resonance frequencies at $n \ll \eta^{-1}$. It is obvious that the smaller η, the more resonance frequencies are correctly described by the formula. In practical situations (when $\varepsilon \sim 10^{-2}$), the acoustic medium affects the position of the resonance frequencies weakly.

The positions of resonances have been computed according to (8.10) in the case of an aluminum shell with $h = 1/10$ immersed in water. The appropriate parameters are given by (8.1).

In Table 8.5, the number of resonances n_* is given for which the positions of the resonance frequencies calculated according to the formula (8.10) do not differ from the exact one by more than one computation step size. This number n_* depends on the type (S_l or A_l) and the family ($l = 1, 2, 3, \ldots$) of the wave.

The influence of the contribution furnished by the term which corresponds to the curvature of the shell on the accuracy of the asymptotic formula (8.10) can be estimated by comparing the resonance frequencies calculated according to (8.10), and those calculated by the same formula, but with $v_q = 0$. In Table 8.6, the results of such calculations are presented for the resonance frequencies of the A_1 and A_3 waves. The first column in this table gives the exact values of the resonance frequencies, computed according to the procedure of the resonance scattering theory (see Table 8.3), the second column corresponds to the calculation according to (8.10), and the third one to the calculation according to

Table 8.5. The number of resonances n_* correctly predicted by the asymptotic formula (8.10) for the Lamb-type symmetric S_l and antisymmetric A_l waves [8.13, Table 1]

Type and family of the wave	S_l			A_l			
	S_1	S_2	S_3	A_1	A_2	A_3	A_4
n_*	1	1	13	4	8	8	12

Table 8.6. Resonance frequencies of the A_1 and A_3 waves. The first column corresponds to the exact value, the second column corresponds to the calculation according to (8.10), and the third column corresponds to the calculation according to (8.10), but with $v_q = 0$ [8.13, Table 3]

n	A_1			A_3		
	1	2	3	1	2	3
1	66.52	66.52	66.38	272.73	272.73	272.73
2	66.91	66.92	66.78	272.85	272.86	272.84
3	67.58	67.59	67.45	273.01	273.01	273.02
4	68.48	68.52	68.38	273.28	273.30	273.27
5	69.61	69.72	69.58	273.59	273.63	273.60
6	70.98	71.19	71.05	273.98	274.02	274.00
7	72.50	72.93	72.79	274.45	274.49	274.47
8	74.18	74.92	74.79	275.00	275.04	275.01
9	76.05	77.19	77.06	275.59	275.65	275.63
10	78.01	79.73	79.59	276.25	276.34	276.31

(8.10), but with $v_q = 0$. In the case of the A_1 wave, one can see the difference in the results, and, consequently, here the curvature strongly affects the results. With increasing frequency, the influence of the curvature diminishes, as can be clearly observed from the results related to the A_3 wave.

Using the resonance frequencies calculated according to (8.10), one can obtain the dispersion curves of the phase velocities. In the y–z plane (see Fig. 8.2), the number of correctly described resonance frequencies is defined by the slope coefficient of the ray, $\tan \gamma = 2/n_* h$, passing through the origin. The resonances situated above this ray are described exactly. The calculation of the position of the resonance frequencies in the case of an aluminum shell with different relative thicknesses

$$h = 2^{-(2+p)} \quad (p = 1, 2, 3) \tag{8.12}$$

has shown that the dispersion curves for different h values practically coincide, and the position on the y axis of similar resonances differ only negligibly. The resonance frequencies of the A_1 wave are presented in Table 8.7. Therefore, if the asymptotic formula properly describes the first four resonance positions at $h = 1/8$, then it describes exactly the first eight resonance positions at $h = 1/16$, and the first sixteen resonance positions at $h = 1/32$. The calculation for the case of a shell with $h = 1/32$ has confirmed these considerations. In Table 8.8, for the A_1 wave, the exact (first column) and asymptotic (second column) values of the resonance frequencies are given.

For the chosen material parameters [see (8.1)], and the case of $h = 1/10$, only the first resonance position is found exactly for the S_1 and S_2 waves. This is caused by the proximity of the cutoff frequencies of the S_1 and S_2 waves for the

Table 8.7. Resonance frequencies of the A_1 wave for shells with different relative thicknesses [8.13, Table 4]

$h = 1/8$	$h = 1/16$	$h = 1/32$
$y_2 = 1.5965$	—	$y_8 = 1.5892$
$y_4 = 1.6555$	$y_8 = 1.6451$	$y_{16} = 1.6419$
$y_8 = 1.8604$	$y_{16} = 1.8361$	$y_{32} = 1.8271$

Table 8.8. The exact (first column) and asymptotic (second column) values of the resonance frequencies of the A_1 wave for the case of scattering by an aluminum shell with $h = 1/32$ [8.13, Table 5]

n	1	2	n	1	2
1	212.07	212.04	14	219.41	219.38
2	212.19	212.16	15	220.47	220.43
3	212.38	212.35	16	221.60	221.54
4	212.66	212.62	17	222.77	222.72
5	213.01	212.97	18	224.02	223.96
6	213.44	213.39	19	225.31	225.25
7	213.95	213.89	20	226.64	226.60
8	214.49	214.46	21	228.05	228.01
9	215.16	215.11	22	229.53	229.46
10	215.86	215.82	23	231.02	230.97
11	216.64	216.61	24	232.58	232.52
12	217.50	217.46	25	234.18	234.13
13	218.44	218.39	26	235.86	235.78

case of aluminum (with Poisson's ratio $v = 0.3455$). For materials with $v = 1/3$ these cutoff frequencies coincide altogether. At v values substantially distinct from $1/3$, the exactness of the presented formula for the resonance positions of the S_1 and S_2 waves does not lead to accurate predictions for all other waves. This can be proved by the calculation of the resonance frequencies of the S_1 and S_2 waves for the case of an empty zinc shell with $h = 1/10$ immersed in water. The physical parameters for zinc are the following

$$\rho_1 = 7.1 \times 10^3 \text{ kg/m}^3, \qquad c_1 = 4170 \text{ m/s}, \qquad c_t = 2440 \text{ m/s},$$
$$v = 0.24 . \tag{8.13}$$

In the latter case the asymptotic formula correctly describes the position of the three first resonance frequencies of the S_1 and S_2 waves. The exact values of the resonance frequencies on the x axis are presented in Table 8.9.

For v values close to $1/3$, one may use instead of formula (8.10), as a most rough approximation neglecting the curvature of the shell and some other

Table 8.9. The exact values of the three first resonance positions of the S_1 and S_2 waves computed for a zinc shell with $h = 1/10$ [8.13, Table 2]

n	S_1	S_2
1	89.10	104.45
2	88.94	104.73
3	88.71	105.16

Table 8.10. Resonance frequencies of the S_1 and S_2 waves. The first column corresponds to the exact value, the second column corresponds to the calculation according to (8.14), and the third column corresponds to the calculation according to formula (8.10) [8.13, Table 6]

n	S_1			S_2		
	1	2	3	1	2	3
1	132.11	132.04	132.07	136.88	136.81	136.96
2	131.17	131.00	130.00	138.01	137.84	138.60
3	130.08	129.77	128.07	139.49	139.08	141.49
4	128.98	128.45	124.57	141.02	140.39	145.45
5	127.93	127.10	120.07	142.66	141.75	150.54
6	126.95	125.73	114.57	144.38	143.12	156.77
7	126.02	124.34	108.08	146.13	144.50	164.17
8	125.20	122.96	100.58	147.96	145.89	172.61
9	124.45	121.56	92.08	149.88	147.29	182.25
10	123.79	120.16	82.58	151.80	148.69	192.98

factors, the following formula

$$z_n = \pi\left[\left(\frac{1}{2} + \frac{1}{4\gamma_0}\right) \mp \left\{\frac{1}{4}\left(1 - \frac{1}{2\gamma_0}\right)^2 + \left(\frac{2\eta n}{\pi^2}\right)^2\right\}^{1/2}\right]. \qquad (8.14)$$

Here the minus sign corresponds to the S_1 wave and the plus sign corresponds to the S_2 wave. In Table 8.10, for the aluminum shell with $h = 1/10$, the positions of resonance frequencies of the S_1 and S_2 waves are presented. Here the exact solution (first column), the calculational data according to formula (8.14) (second column), and the calculational data according to formula (8.10) (third column) are given. It can be seen that, at $v \sim 1/3$, formula (8.14) describes the resonance frequencies better than formula (8.10).

Some of the results presented above were first published in [8.13].

9 Analysis and Synthesis of the Backscattered Acoustic Pressure from a Circular–Cylindrical Shell

The problem of acoustic scattering by a circular shell attracted the attention of investigators long ago. First it was examined in a simplified formulation in [9.1–3], where the motion of the elastic body was described by the equations of thin shell theory, and later in a more general form, using the equations of linear elasticity theory. The formal solution of the problem in series form was given in [9.4]. Since then, hundreds of papers have been published concerning this problem. Reviews of previous work can be found, for example, in [9.5–27].

It was once thought that the classical solution, in spite of being exact, absolute, and uniformly convergent, was unsuitable for computation, especially at high frequencies, because of slow convergence. Now we know that the series begins to converge at $n \sim x$ (where n is the summation index and $x \equiv ka$, k being the wave number and a the cylinder radius), and converges for practical purposes at $n \sim 2x$. In order to accelerate the convergence of the series, the latter is often transformed into integral form using the Sommerfeld–Watson transformation.

The integral has singularities – poles, saddle points, etc. By a suitable deformation of the integration contour, the integral evaluation is reduced to a calculation of the residues of the poles and an integration along the curve passing through the saddle point. At high x values, the results obtained in this way describe the reflected, diffracted and re-radiated waves. However, only the main term (the zero term) can usually be found in the asymptotic formulas. The calculation of the successive terms (the corrections) of the asymptotic expansion is rather difficult.

The integral Sommerfeld–Watson transformation, combined with asymptotic (short-wavelength) methods of complex variable theory, represents a strict mathematical instrument following the physical nature of the scattering process and therefore permitting us to describe all the essential features of this process. Using this transformation, the dependence on x of the pole families was calculated in the complex plane for shells with different relative thicknesses. It is well known that the real part of the pole is related to the phase velocity of the peripheral wave and the imaginary part is related to the damping factor. Also, the critical entrance (and re-radiation) angles were found and the resonant nature of the scattering process was explained. By evaluating the residues of the poles, the dependence on x of the peripheral wave magnitudes were found. As far as we know, until now the total form function of the pressure scattered by the shell has not been constructed using this method.

Recently, the Sommerfeld–Watson transformation was used with the partial-wave series for the scattering of a plane wave by a fluid-loaded elastic sphere [9.28]. Expressions for the scattering due to each class of elastic surface waves were interpreted in terms of contributions from repeated circumnavigations. These expressions were summed up in closed form as in the analysis of Fabry–Perot resonators. The form function was synthesized by adding this sum to the specular reflection [9.29]. Later in [9.30], the geometrical theory of diffraction was extended to describe the contributions of the surface elastic waves to the backscattering from spheres and cylinders in water at high frequencies. The synthesized form function consists of the reflected wave and of two re-radiated waves caused by the Rayleigh and Whispering Gallery waves. In the range $30 < x < 100$, the synthesis is in good agreement with the solution from the partial-wave series and the experimental results [9.28, 31]. This approach is described in Chap. 4. In contrast to the investigation presented in [9.29], our attention will be directed towards the contributions furnished by two waves: the Lamb-type wave S_0 and the Stoneley-type wave A. The latter is called the fluid-borne A wave in [9.32, 33].

The resonance scattering theory (see Chaps. 2 and 3) is a different approach for a qualitative description of the physical mechanism of the scattering process. This method is commonly used in quantum theory and nuclear physics. Recently the resonance scattering theory was used to describe each series term. According to this approach, each term is represented as a sum of two summands, one of which is nonresonant (the background) and the other resonant by nature. It is assumed that, in the case of thick-walled shells, the "potential term" equals the corresponding term in the problem of scattering by a rigid cylinder. With decreasing shell thickness, the potential term would turn into the corresponding term in the problem of scattering by a soft cylinder. The combination of the resonant terms into families (the Regge poles) allows one to find the resonance position of the lth peripheral wave. Using the Sommerfeld–Watson transformation and calculating the residues of the poles, the amplitude and width of the nth resonance of the lth peripheral wave can be found. The numerical results obtained according to the resonance scattering theory concern the phase and group velocities of the different types of peripheral waves (in particular, the Rayleigh wave, the S_0 and A_1 Lamb-type waves and the fluid-borne A wave) and the position of successive resonances (index n) of different (index l) types of peripheral waves. These results are in excellent agreement with the experimental data [9.34, 35] as compared to the prediction of the resonance scattering theory [9.36, 37]. However, we shall note that the synthesis of the form function based on the data obtained by the resonance scattering theory has not yet been carried out.

The rapid development of computer techniques has shown that the slow convergence of the series of eigenmodes in fact does not impede the computations. So, for example, by 1968 the form function of the pressure backscattered by a spherical shell in the range $355 \leqslant x \leqslant 395$ had already been computed [9.38], and by 1975 the form function of the pressure scattered by an elastic sphere and cylinder in the range $900 \leqslant x \leqslant 950$ had been computed [9.39] by the direct summation of the series.

The numerical results, the asymptotic solution obtained by means of the Sommerfeld–Watson transformation, and the results of the resonance scattering theory are in good agreement, certainly in those domains of x where they are valid, but they are mutually complementary when one of them fails for some reason. The theoretical results are in agreement with the data produced in laboratory experiments carried out in various research centers over the last 30 years [9.5, 28, 31, 34, 35, 40–62].

There are now many qualitative methods for describing the scattering process, but none of them can present with sufficient accuracy the individual components of the scattered field and the result of their interaction. In particular, one of the drawbacks of the asymptotic methods used is that they can be applied only at rather large x values, although the form function must often be described down to $x = 0$.

The coordinates of the pole corresponding to the peripheral waves depend on x and are found using the complicated secular equation in a broad x range, and the calculation must be carried out in small step sizes. Although this may be accomplished analytically, in order to achieve sufficient accuracy computers are often used.

Regardless of the method by which the backscattered form function was found, once the form function has been obtained, inevitably the question of its interpretation and utilization will emerge. In particular, it is interesting to determine how the form function is physically put together, which components it is composed of, and how these components interact. It is fascinating to try to understand how the initial physical parameters of the problem affect the different aspects of the scattering process and the whole form function. Usually, such a question arises when considering two complementary problems, one concerning extraction of the information about an elastic body from its form function, and the other of changing the physical parameters of the body, even as far as replacement of its structure, in order to hide, if possible, the form function identity of the elastic scatterer.

The aim of this chapter is to carry out an analysis of the form function, to separate its basic elements, and to describe them adequately in the whole representative x range, generally speaking beginning from $x = 0$, to combine from those components the form function, and to compare it with the calculated one. It was found that such a program could be realized, and the designed form function is well concordant with the computed one. The case of a very thin shell was investigated earlier in [9.63]; in that case, the re-radiation was caused only by the S_0 wave. Here, an approximate analytical description of the A wave is given and the result of its interaction with the S_0 wave is described. For the latter, our prior description [9.63] is modified for the case of a shell with moderate thickness.

In our reasoning, we start from the calculated backscattered form function itself, the resonant nature of which was detected in the pioneering theoretical

[9.2, 3] and experimental [9.41] work, and try to use all the extensive qualitative results of our predecessors concerning this problem.

9.1   Analysis of the Acoustic Pressure Scattered by the Shell

Let us consider a thin empty cylindrical shell submerged in an unbounded acoustic medium. The plane harmonic pressure wave (2.1) strikes the shell and is scattered by it. The direction of propagation is perpendicular to the longitudinal axis of the shell. Therefore, the problem considered is two dimensional. Using the method of separation of variables, one can obtain the exact solution of the steady-state problem. At a fixed observation point, the solution for the pressure scattered by the shell is given by (2.54). In the far field, this takes the form

$$p_s = p_* \left(\frac{a}{2r} \right)^{1/2} \exp\left[i(kr - \omega t) \right] \sum_{n=0}^{\infty} 2(i\pi x)^{-1/2} \varepsilon_n R_n \cos n\varphi \; . \tag{9.1}$$

If, according to [9.36], the partial scattering function and the form function are introduced (see Chap. 2), then (9.1) may be presented in the form (2.47), (2.48) in which $\psi_n^{(r)}(\varphi)$ is defined as in (2.58).

At fixed $\varphi = \varphi_0$ we shall introduce the following notations

$$p = |f|, \qquad \zeta_n = |\psi_n^{(r)}| \; . \tag{9.2}$$

Here, $p(x)$ is the modulus of the far field form function and $\zeta_n(x)$ is the modulus of the nth partial mode scattering amplitude. Here, a rigid background is chosen. As will be shown below, even for a shell with moderate thickness (with $h = 1/64$), such a choice of background turns out to be effective and, in particular, allows us to separate the resonance component.

The calculation was carried out for the observation point situated in the far field in the source direction ($r/a = 10^4$, $\varphi = \pi$) for an Armco iron shell (with $h = 1/64$) immersed in water

Armco iron: $\rho_1 = 7.7 \times 10^3$ kg/m³, $c_1 = 5960$ m/s, $c_t = 3240$ m/s ,

water: $\rho = 1.0 \times 10^3$ kg/m³, $c = 1493$ m/s . $\tag{9.3}$

Here c_1 and c_t are, respectively, the longitudinal and shear velocities of Armco iron. The computation step size l_x was chosen as

$$l_x = 10/256 \; . \tag{9.4}$$

The resonances of the peripheral wave are usually labeled by two numbers: n, l. The first defines the ordinal number of resonance and the second defines its family. The resonances of the Lamb-type zero order symmetric wave S_0 will be designated by $(n, 0)$ and those of the A wave by $[n[$. In the latter case the second number is dropped for simplicity. The resonances of the S_0 and A waves are indicated in Fig. 9.1. The plots of $\zeta_n(x)$ of the S_0 and A waves are shown in

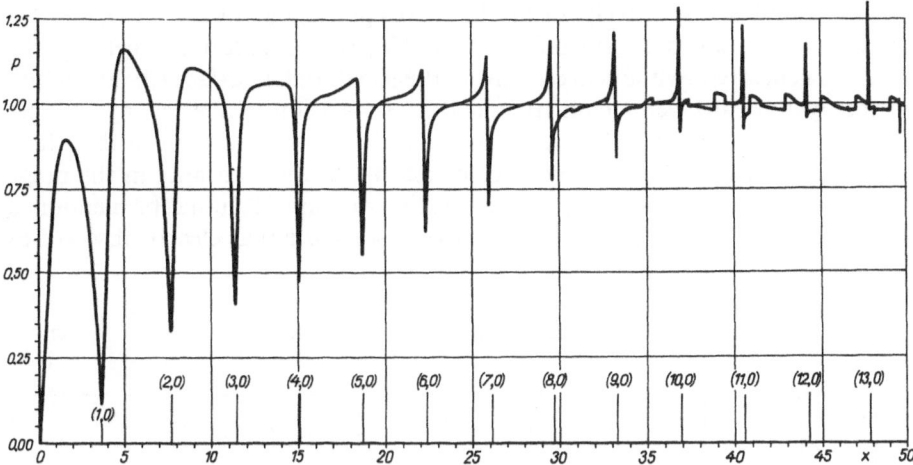

Fig. 9.1.1

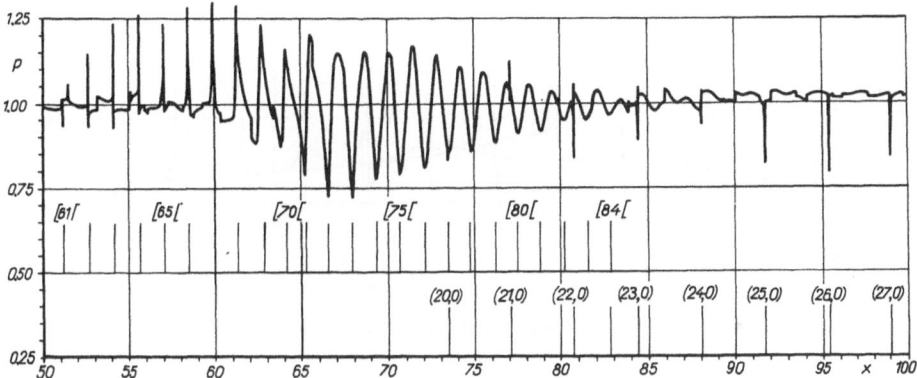

Fig. 9.1.2

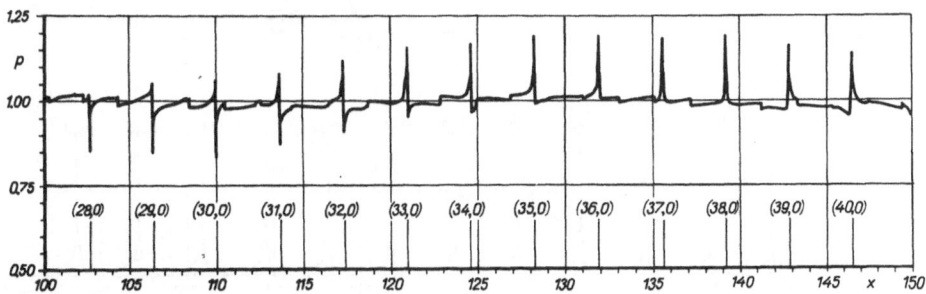

Fig. 9.1.3

Figs. 9.1.1–9.1.3. The form function modulus for backscattering from an Armco iron shell with $h = 1/64$ in water. This is the exact solution from the partial-wave series when the motion of the shell is described by the equations of linear elasticity theory (ET). The successive resonances of the S_0 wave are designated by $(n, 0)$ and those of the A wave by $[n[$ [9.65, Fig. 1]

Figs. 9.2 and 9.3, respectively. From the curves presented in Figs. 9.1–3, one can see that in the range considered, $0 \leqslant x \leqslant 150$, the backscattered field on the whole is generated by the reflection from the shell's outer surface (the refraction in the cylindrical layer included), and the re-radiation is affected by two peripheral waves – S_0 and A. The influence of the S_0 wave is seen in the whole x range considered, but the influence of the A wave is seen only in the range $40 \leqslant x \leqslant 90$, which we call the strong-bending domain. It should be mentioned that in this x range, the contribution of the S_0 wave is not as clearly observed as

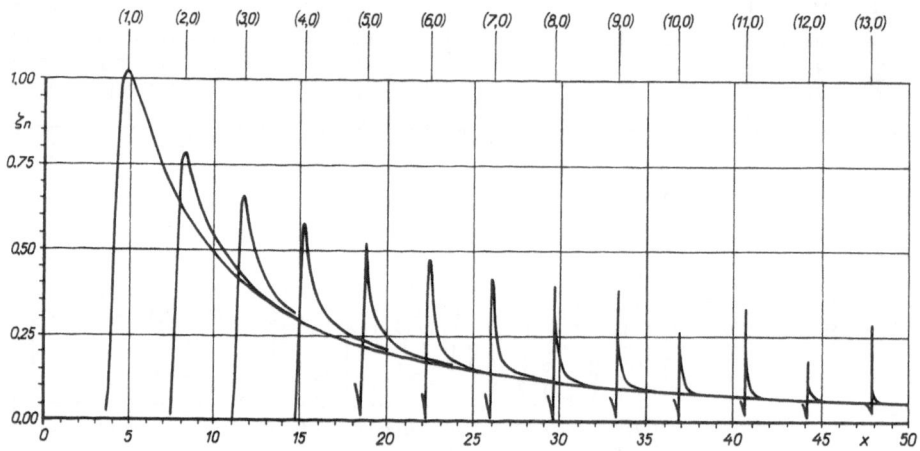

Fig. 9.2.1

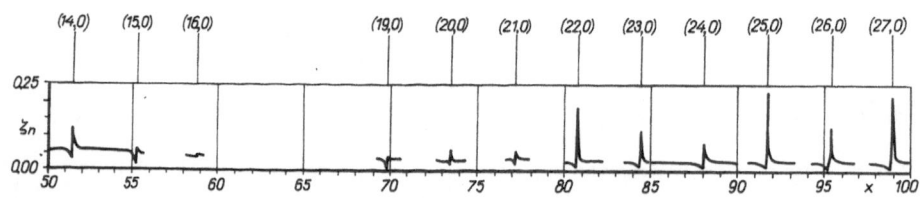

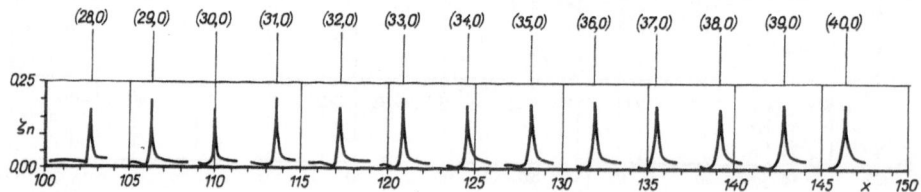

Fig. 9.2.2

Figs. 9.2.1 and 9.2.2. The isolated modal resonances $(n, 0)$ of the S_0 wave of the Armco iron shell in water, obtained by suitable (rigid) background subtraction, and displayed versus $x \equiv ka$ [9.65, Fig. 2]

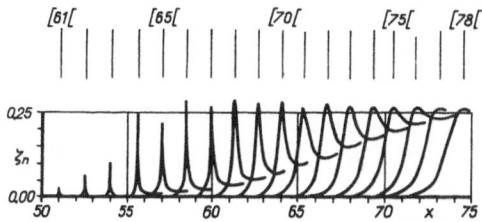

Fig. 9.3. The isolated modal resonances $[n[$ of the A wave of the Armco iron shell in water [9.65, Fig. 3]

it is outside this range. For the chosen thickness $h = 1/64$, the strong-bending domain caused by the A wave is situated comparatively low on the x axis. Let $(x_{n,0})$ and $[x_n[$ define the resonance positions of the S_0 and A waves, respectively. For the sake of brevity, we shall introduce the notations

$$x_n = (x_{n,0}), \qquad y_n = [x_n[\, . \tag{9.5}$$

The nondimensional phase $[v_j^{\mathrm{ph}}(x)]$ and group $[v_j^{\mathrm{gr}}(x)]$ velocity as well as the critical angle of excitation (and re-radiation) $\varphi_j(x)$ of the S_0 and A waves are determined, respectively, by

$$v_1^{\mathrm{ph}}(x_n) = x_n/n, \qquad v_2^{\mathrm{ph}}(y_n) = y_n/n \, ,$$

$$v_1^{\mathrm{gr}}(x_n) \cong x_n - x_{n-1}, \qquad v_2^{\mathrm{gr}}(y_n) \cong y_n - y_{n-1} \quad (n = 1, 2, 3, \ldots) \, , \tag{9.6}$$

$$\varphi_j(x) = \arcsin[v_j^{\mathrm{ph}}(x)]^{-1} \quad (j = 1, 2) \, ,$$

where the index 1 is used to label the functions related to the S_0 wave, and the index 2 to label the functions related to the A wave.

In Table 9.1, the positions and the values of the extrema generated by the S_0 wave are presented; and in Table 9.2, those generated by the A wave are presented. In Tables 9.1 and 9.2 the resonances are listed in the $0 \leqslant x \leqslant 100$ range, which is narrower than the range used in Fig. 9.1. This stems from our intention to restrict the synthesis of the form function to this range only, for two reasons. Firstly, from studying Figs. 9.1 and 9.2, one can see that in the range $100 \leqslant x \leqslant 150$, only one peripheral wave, namely the S_0 wave, contributes to the scattered pressure field. Therefore, the description of the far field form function in this range is much simpler than in the range $0 \leqslant x \leqslant 100$. Secondly, at $x > 100$, one can use either the approach outlined below for the range $0 \leqslant x \leqslant 100$, or take advantage of a simpler approach applied in [9.63] for the case of a very thin shell.

The computation was carried out with step size $l_x = 10/256$, and therefore the true positions and values of both the functions $|f(x)|$ and $\arg f(x)$ at resonance frequencies may differ from the computed ones. The use of a large computation step size apparently caused the deviation of $|f(x)|$ from the interpolated values (for example, in the vicinity of $n = 10$ and $n = 12$ for the resonances of the S_0 wave), and it also caused small differences between the positions of the extremas of $|f(x)|$ and the zeroes of $\arg f(x)$.

Table 9.1. The coordinates and the values of the successive resonances caused by the S_0 wave with the description of the motion of the shell by the equation of ET, TT and BT and their approximation [9.65, Table I]

n	LET x_n	LET $|f(x_n)|$	LET $\arg f(x_n)$ max	LET $\arg f(x_n)$ min	RST $\tilde x_n$	RST $\zeta_n(\tilde x_n)$	RST $v_1^{ph}(x_n)$	TT $\tilde x_n$	TT $|\bar f(\tilde x_n)|$	BT $\tilde x_n$	BT $|\tilde f(\tilde x_n)|$	BT $\arg \tilde f(\tilde x_n)$ max	BT $\arg \tilde f(\tilde x_n)$ min	App $\tilde{\tilde x}_n$	App $|P_1(\tilde x)_n|$	App $\hat x_n$	App $|f_n(\hat x_n)|$
1	3.67	0.121	3.71	3.67	4.80	1.022	3.67	3.67	0.121	3.67	0.121	3.71	3.67	3.87	0.977	3.67	0.196
2	7.66	0.331	7.42	7.85	8.13	0.791	3.83	7.66	0.336	7.66	0.337	7.38	7.85	7.70	0.714	7.64	0.314
3	11.41	0.407	11.25	11.56	11.60	0.622	3.80	11.41	0.418	11.37	0.417	11.21	11.56	11.41	0.603	11.41	0.397
4	15.08	0.478	14.97	15.20	15.12	0.580	3.77	15.04	0.487	15.04	0.487	14.92	15.20	15.08	0.530	15.08	0.470
5	18.71	0.547	18.63	18.83	18.71	0.521	3.74	18.71	0.555	18.71	0.555	18.63	18.87	18.67	0.481	18.71	0.533
6	22.34	0.633	22.30	22.46	22.34	0.471	3.72	22.34	0.630	22.34	0.630	22.27		22.30	0.446	22.34	0.607
7	25.98	0.704	25.94	26.13	25.98	0.424	3.71	25.98	0.716	26.02	0.717	25.94		25.94	0.402	25.98	0.673
8	29.61	0.794		29.57	29.61	0.395	3.70	29.65	0.775	29.65	0.774	29.57		29.61	0.343	29.61	0.690
9	33.20	1.219		33.24	33.24	0.390	3.69	33.20	1.223	33.20	1.225	33.24		33.24	0.396	33.20	1.142
10	36.88	1.291	36.91	37.10	36.88	0.273	3.69	36.88	1.241	36.88	1.249	36.91		36.88	0.384	36.88	1.266
11	40.51	1.090	40.55	40.51	40.55	0.337	3.68	40.51	1.236	40.55	1.235	40.59		40.55	0.373	40.53	1.301
12	44.18	1.105	44.22	44.18	44.22	0.170	3.68	44.18	1.258	44.18	1.244	44.22	44.14	44.18	0.361	44.18	1.273
13	47.85	1.255	47.89	47.85	47.85	0.287	3.68	47.85	1.287	47.85	1.267	47.89	47.81	47.85	0.351	47.85	1.318
14					51.52	0.118		51.52	1.308	51.52	1.293	51.56	51.48				
15					55.20	0.061		55.20	1.276	55.20	1.295	55.23	55.16				
16					58.83	0.046		58.87	1.204	58.87	1.249	58.91	58.82				
17					62.50	0.039		62.77	1.182	62.54	1.187	62.58	62.50				
18					66.17	0.036		66.21	1.116	66.21	1.139	66.25	66.17				
19					69.84	0.035		69.88	1.142	69.88	1.104	69.96	69.84				
20					73.48	0.058		73.20	1.179	73.55	1.078		73.51				
21					77.48	0.050		77.19	1.270	77.23	1.057	77.11	77.19				
22	80.78	0.849		80.78	80.78	0.184	3.67	80.82	0.809	80.82	0.944	80.82	80.86	80.82	0.259	80.82	0.892
23	84.41	0.976		84.45	84.45	0.110	3.67	84.49	0.914	84.49	0.951	84.49	84.53	84.49	0.249	84.47	0.907
24	88.09	0.918	88.09	88.13	88.09	0.084	3.67	88.00	1.044	88.16	0.962	88.16	88.20	88.13	0.240	88.14	0.883
25	91.76	0.823	91.72	91.76	91.76	0.227	3.67	91.91	1.028	91.88	0.974	91.83	91.88	91.80	0.230	91.82	0.820
26	95.43	0.963	95.39	95.43	95.43	0.122	3.67	95.74	1.012	95.55	0.966	95.51	95.55	95.47	0.220	95.49	0.789
27	99.06	0.827	99.06	99.10	99.06	0.215	3.67	99.53	1.006	99.22	0.960	99.18	99.82	99.14	0.211	99.16	0.874

Column groups: Linear elasticity theory (x_n, $|f(x_n)|$, $\arg f(x_n)$ max/min); RST ($\tilde x_n$, $\zeta_n(\tilde x_n)$, $v_1^{ph}(x_n)$); Timoshenko-type shell theory ($\tilde x_n$, $|\bar f(\tilde x_n)|$); Bending-free shell theory ($\tilde x_n$, $|\tilde f(\tilde x_n)|$, $\arg \tilde f(\tilde x_n)$ max/min); Approximation ($\tilde{\tilde x}_n$, $|P_1(\tilde x)_n|$, $\hat x_n$, $|f_n(\hat x_n)|$).

Table 9.2. The coordinates and the values of the successive resonances caused by the A wave with the description of the motion of the shell by the equations of ET, TT and their approximation [9.65, Table II]

n	Linear elasticity theory				RST			Timoshenko-type shell theory		Approximation							
	y_n	$|f(y_n)|$	$\arg f(y_n)$ max	$\arg f(y_n)$ min	$\tilde{y}_n$	$\zeta_n(\tilde{y}_n)$	$v_2^{ph}(y_n)$	$\bar{x}_n$	$|\bar{f}(\bar{x}_n)|$	y_n^*	$|f_{oo}(y_n^*)|$	$\tilde{y}_n$	$|f_*(\tilde{y}_n)|$	$\tilde{y}_n$	$|P_2(\tilde{y}_n)|$	$\hat{y}_n$	$|f_a(\hat{y}_n)|$
60	49.69	0.919					0.8282										
61	51.19	0.934	51.19	51.17			0.8392			51.19	0.102					51.21	1.037
62	52.68	1.146	52.70	52.66	52.70	0.064	0.8497			52.68	0.151	52.58	0.106	52.81	0.153	52.73	1.079
63	54.16	1.234	54.16	54.14	54.18	0.106	0.8597			54.16	0.268	54.18	0.161	54.20	0.183	54.18	1.138
64	55.63	1.258	55.64	55.61	55.63	0.241	0.8692			55.63	0.265	55.82	0.206	55.61	0.218	55.59	1.199
65	57.09	1.233	57.11	57.05	57.07	0.221	0.8783			57.09	0.276	57.11	0.185	56.99	0.249	56.99	1.243
66	58.52	1.284	58.55	58.48	58.52	0.287	0.8867			58.52	0.285	58.40	0.294	58.38	0.276	58.38	1.267
67	59.94	1.301	60.00	59.88	59.92	0.274	0.8946	59.96	1.026	59.96	0.306	60.00	0.269	59.77	0.297	59.79	1.275
68	61.35	1.289	61.48	61.27	61.33	0.287	0.9022	61.37	1.215	61.33	0.315	61.45	0.302	61.13	0.314	61.18	1.263
69	62.75	1.230		62.65	62.70	0.281	0.9094	62.77	1.182	62.70	0.303	62.54	0.323	62.50	0.322	62.50	1.389
70	64.16	1.161		64.02	64.06	0.282	0.9166	64.14	1.213	64.02	0.277	64.02	0.254	63.89	0.323	64.02	1.221
71	65.23	0.788		65.39	65.39	0.278	0.9187	65.43	0.867	65.39	0.285	65.39	0.289	65.25	0.316	65.10	0.758
72	66.60	0.728	66.37	66.76	66.72	0.276	0.9250	66.60	0.807	66.68	0.295	66.60	0.274	66.62	0.303	66.48	0.746
73	67.97	0.721	67.75		68.05	0.273	0.9311	68.16	0.765	67.95	0.280	67.93	0.257	67.97	0.284	67.87	0.737
74	69.34	0.777	69.12	69.59	69.34	0.271	0.9370	69.49	0.724	69.30	0.225	69.22	0.230	69.34	0.261	69.26	0.743
75	70.70	0.787	70.47		70.63	0.269	0.9427	70.86	0.754	70.63	0.222	70.43	0.223	70.68	0.235	70.64	0.768
76	72.17	0.811			71.91	0.266	0.9496	72.23	0.787	71.88	0.205	71.72	0.209	72.05	0.208	72.05	0.792
77	73.46	0.832	73.22	73.89	73.20	0.264	0.9540	73.48	0.839	73.46	0.174	73.55	0.234	73.40	0.180	73.48	0.703
78	74.86	0.854	74.59		74.53	0.261	0.9597	75.12	0.847	74.86	0.150			74.75	0.154	74.82	0.850
79	76.27	0.883	75.94				0.9654	76.60	0.835					76.09	0.129	76.21	0.874
80	77.58	0.909	77.26	77.79			0.9698	78.01	0.829					77.44	0.106	77.60	0.904
81	78.93	0.917	78.59	79.23			0.9744	79.38	0.830					78.77	0.086	78.98	0.921
82	80.25	0.950					0.9787	80.82	0.809					80.12	0.068	80.12	0.931
83	81.52	0.950					0.9822	82.07	0.873					81.45	0.053	81.54	0.949
84	82.85	0.965					0.9863	83.40	0.894					82.80	0.042	82.97	0.961

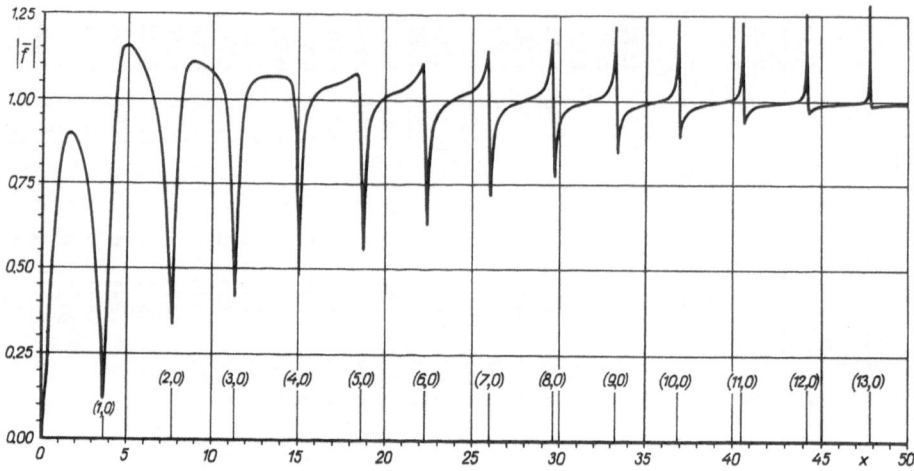

Fig. 9.4.1

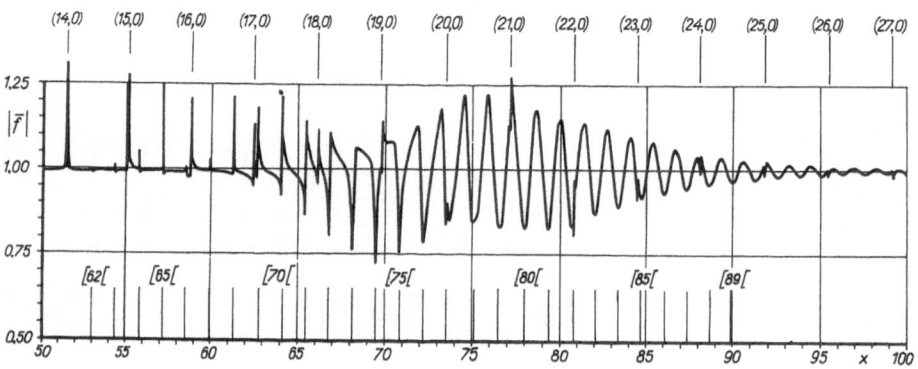

Fig. 9.4.2

Figs. 9.4.1 and 9.4.2. This figure is similar to Fig. 9.1 except that here the motion of the shell is described by the equations of the Timoshenko-type thin shell theory (TT) [9.65, Fig. 4]

The resonance positions $\check{x}_n$ and amplitudes $\zeta_n(\check{x}_n)$ of the S_0 wave which were computed according to the resonance scattering theory are given in Table 9.1. Table 9.2 presents the resonance positions $\check{y}_n$ and amplitudes $\zeta_n(\check{y}_n)$ of the A wave.

In Figs. 9.4 and 9.5, the moduli of the form function are presented for the case when the cylinder motion is described by the equations of the Timoshenko-type shell theory (TT) and by the equations of the bending-free (membrane) shell theory (BT), respectively. The bar indicates the values relating to the TT and the tilde indicates those relating to the BT.

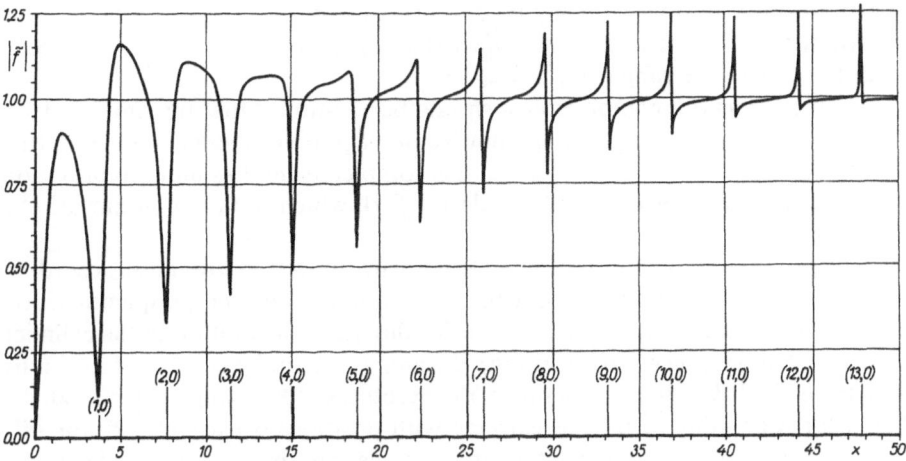

Fig. 9.5.1

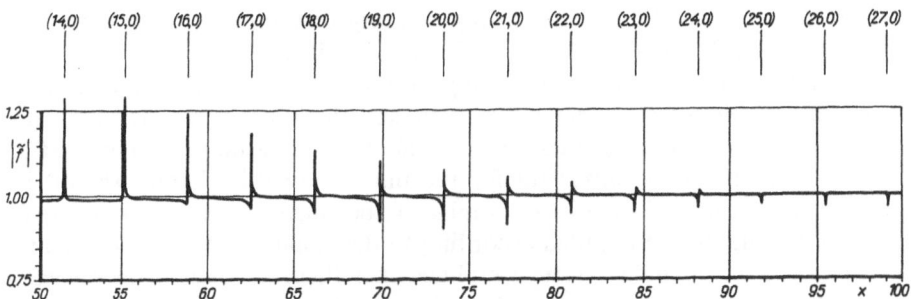

Fig. 9.5.2

Figs. 9.5.1 and 9.5.2. This figure is similar to Fig. 9.1 except that here the motion of the shell is described by the equations of the bending-free (membrane) thin shell theory (BT) [9.65, Fig. 5]

As can be seen from a comparison of Figs. 9.1 and 9.4 and the corresponding columns of Tables 9.1 and 9.2, the $p(x)$ curve, computed according to the TT, properly follows the curve calculated according to linear elasticity theory (ET). The position of resonances are defined accurately enough, but the values of $p(x)$ in the extremal points calculated according to the TT are somewhat smaller than those computed by the ET.

In the strong-bending domain, the curve calculated by TT essentially differs from the one corresponding to the ET: here there is an error both in the positions of the extrema (systematic shift) and in their values; above $x \geqslant 75$, the "oscillation" span of the curve in Fig. 9.4 is significantly larger than in Fig. 9.1.

In Fig. 9.4 above $x > 80$, the resonances of the S_0 wave are almost invisible, while they can be clearly observed in Fig. 9.1. If the TT is used, then, following the plot given in Fig. 9.4, one should assume that when $x > 50$, the S_0 wave contribution to $p(x)$ will quickly be diminished and, in the range $x > 100$, this contribution could be neglected. In fact this is incorrect, as can be clearly seen from Fig. 9.1 in the range $100 \leqslant x \leqslant 150$.

Summarizing this comparison, it should be mentioned that the simplified TT model is entirely justified for a qualitative description of the form function (the strong-bending domain included), but it is completely unfit as input information for obtaining an approximate formula for $f(x)$, which is the main aim of our investigation.

The graph presented in Fig. 9.5 is very easily understood because, according to the BT, in $f(x)$ the main contribution contains only one peripheral wave, namely, the S_0 wave. Up to $x < 30$, the BT describes the motion of the cylinder rather well and the $p(x)$ curves in Figs. 9.1 and 9.5 are quite consistent. It can be seen from Table 9.1 that in this range, the exactness of BT is high. At $x \geqslant 80$, the curve corresponding to BT has extrema with positions consistent with the ET, but their values are somewhat overstated, as can be seen from the curve corresponding to the ET, the dips of the form function modulus in the S_0 wave resonance condition are more profound. For us, the main interest is the description of the form function in the strong-bending domain, i.e., at $40 \leqslant x \leqslant 90$; here, the curve calculated according to the BT exhibits the contribution of the S_0 wave in $f(x)$, which is well "masked" in $p(x)$ in Fig. 9.1. In Fig. 9.5, the S_0 wave resonance positions are rather close to those calculated by the ET and, the higher is x, the more the resonance values differ from those computed by the ET.

One more property important for the purpose of this chapter should be mentioned. A comparison of the graphs of the form function modulus (see Figs. 9.1, 9.4 and 9.5), computed according to the equations of the ET, TT and BT, allows us to conclude that the simplified shell theories describe the reflected wave sufficiently well. Roughly speaking, in those figures at $x \gtrsim 10$, the modulus of reflected wave corresponds to the "carrying" line of the "vibrations". This is the line $p(x) = 1$.

9.2   Synthesis of the Approximate Formula for the Backscattering Form Function

Using the above-described procedure for the acoustic pressure generated by scattering from the shell and its wave composition, we shall now give obvious approximate formulas for the basic types of waves.

9.2.1   The Specularly Reflected Wave and the Waves Refracted in a Thin Layer

When the observation point is situated in the far field in the source direction ($r/a \gg 1$, $\varphi = \pi$), during reflection and refraction the longitudinal wave does not convert into a transverse wave (and vice versa). Thus, in this case, the specularly reflected wave and the waves refracted in the cylindrical layer may be described by the formula for the same process in a thin liquid layer of the same thickness.

For the plane liquid layer at normal incidence of the plane wave, we have

$$f_1(x) = v - (1 - v^2)\frac{\exp(ix\alpha_0)}{1 - v\exp(ix\alpha_0)}, \tag{9.7}$$

where

$$v = \frac{1 - \zeta}{1 + \zeta}, \qquad \zeta = \frac{\rho c}{\rho_1 c_t}, \qquad \alpha_0 = 2h\frac{c}{c_t}. \tag{9.8}$$

Here, $f_1(x)$ is the pressure component caused by the specularly reflected wave and the waves reflected in the layer.

In (9.7), the first term describes the contribution of the wave specularly reflected from the exterior surface, and the second term describes the waves refracted in the layer. The shell is considered to be thin and the divergence of the wave at refraction is neglected. From the interior shell surface the wave is reflected completely (with the reflection coefficient $v_0 = -1$ as for the acoustically soft body). Equation (9.7) is the exact expression for the case of a plane liquid layer. In the case of an elastic cylindrical shell, it still holds approximately; the thinner the layer and the bigger is x, the better it will be.

Our experience has shown the adequacy of (9.7) even at $x \sim 1$ in the case of an empty Armco iron shell with $h = 1/512$, immersed in water. However, an attempt to use it directly for the same material parameters of the shell and the liquid, the shell being thicker ($h = 1/64$), was discouraging. Namely, if from the exact solution in the series form $f(x)$ one subtracts $f_1(x)$, then the remainder should describe the contribution furnished by the peripheral waves,

$$f_0(x) = f(x) - f_1(x). \tag{9.9}$$

For the example considered, as shown before, at $x < 30$ the main contribution is given by the S_0 wave. Therefore, here the remainder $f_0(x)$ should contain this wave. In Fig. 9.6, $|f_0(x)|$ is shown. The positions of the maxima of the $|f_0(x)|$ curve hardly differ from the resonance positions of the curve $p(x)$ (see Fig. 9.1), but the minima of the $|f_0(x)|$ curve are not located at the antiresonance points of the $p(x)$ curve and are systematically shifted to the right. This shift diminishes with increasing x and eventually disappears. In order that the $|f_0(x)|$ minima be situated at the antiresonance points of the S_0 wave, the phase in (9.7) must be corrected and the formula itself presented in the form

$$f_{10}(x) = v - (1 - v^2)\frac{\exp(ix\alpha)}{1 - v\exp(ix\alpha)}, \qquad \alpha = \alpha_0\alpha(x). \tag{9.10}$$

At the chosen parameters of the shell and the liquid, the function $\alpha(x)$ may be described by the approximate formula

$$\alpha(x) = (1 + \alpha_1 x^{-1/2} + \alpha_2 x^{-3/2})^{-1}, \tag{9.11}$$

$$\alpha_1 = 0.357, \qquad \alpha_2 = 0.685.$$

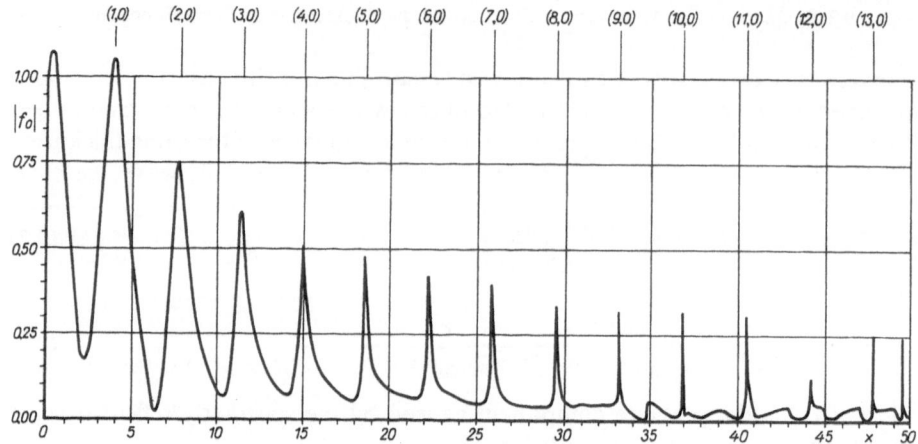

Fig. 9.6. The modulus of the remainder $|f_0(x)|$, see (9.9) [9.65, Fig. 6]

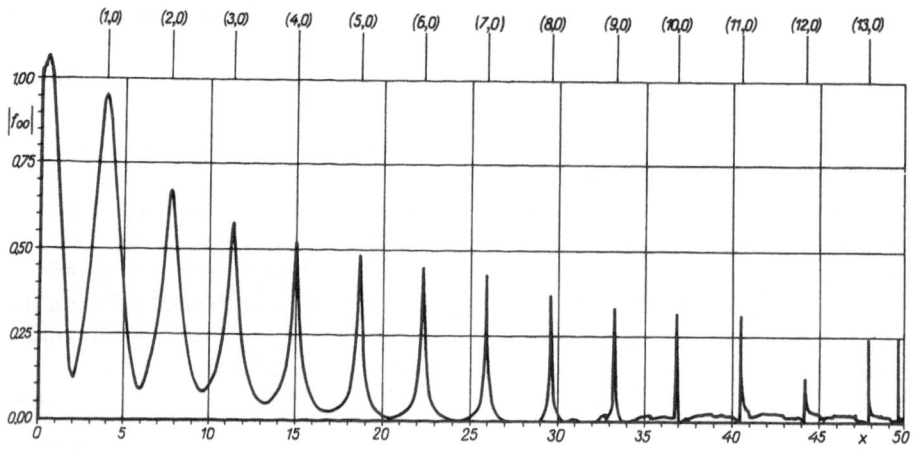

Fig. 9.7.1

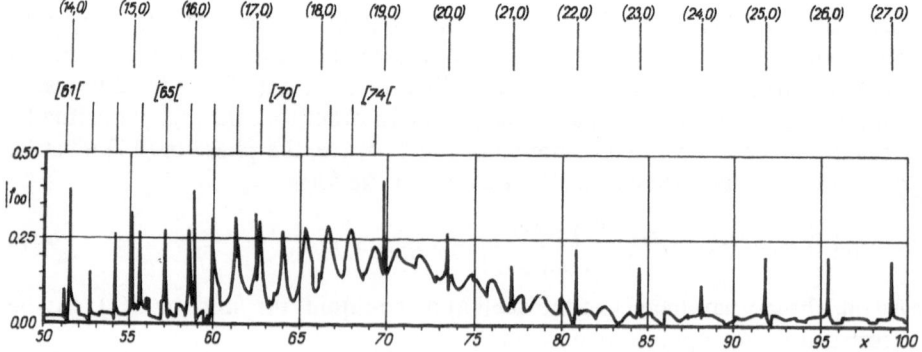

Fig. 9.7.2

Figs. 9.7.1 and 9.7.2. The modulus of the remainder $|f_{00}(x)|$, see (9.12) [9.65, Fig. 7]

Now, (9.9) takes the form

$$f_{00}(x) = f(x) - f_{10}(x) .$$ (9.12)

In Fig. 9.7, the graph of the remainder modulus $|f_{00}(x)|$ is presented. From this curve, one can see that at $x \leqslant 30$ its minima are situated at equal distances from the maxima, the way they should be. A comparison of Figs. 9.6 and 9.7 shows that the phase correction carried out by (9.10) and (9.11) changes $|f_0(x)|$ only at $x < 30$. At larger x, as can be seen from (9.11) and the graph in Fig. 9.7, the influence of the correction is negligible. This is exactly what we wanted to attain.

Let us define the maxima positions of $|f_{00}(x)|$ corresponding to the resonances of the A wave by y_n^*. The values of y_n^* and $|f_{00}(y_n^*)|$ are given in Table 9.2 and will be used below as the reference points in constructing the approximate formulas for the A wave.

Now, using our knowledge of $f_{00}(x)$, we shall move on to the construction of the approximate formulas describing the contributions of the S_0 and A waves. Starting with the locality considerations of the geometrical theory of diffraction and our experience in describing the S_0 wave for the case of a very thin shell, [9.63], we can write the formula for the contribution given by the S_0 wave (index 1) and the A wave (index 2) in the form

$$P_j(x) = F_j^{(0)}(x)\, \varepsilon_j^{(0)}(x) \exp[i\beta_j(x)] \frac{\exp[ix\gamma_j(x)]}{1 - \varepsilon_j(x) \exp[ix\delta_j(x)]} ,$$ (9.13)

where

$$\begin{aligned} \varepsilon_j(x) &= \exp[-2\pi\kappa_j(x)] , \\ \varepsilon_j^{(0)}(x) &= \exp[-2(\pi - \varphi_j)\kappa_j(x)] , \\ \delta_j(x) &= 2\pi/v_j^{\mathrm{ph}}(x) , \\ \gamma_j(x) &= 2\left[1 - \cos\varphi_j(x) + \frac{\pi - \varphi_j(x)}{v_j^{\mathrm{ph}}(x)}\right] , \end{aligned}$$ (9.14)

and $\varphi_j(x)$ and $v_j^{\mathrm{ph}}(x)$ are defined by (9.6).

In (9.13), $P_j(x)$ is the contribution to the pressure introduced by the peripheral wave, $F_j^{(0)}(x)$ is the "landing" amplitude of the peripheral wave, $F_j^{(0)}(x)\, \varepsilon_j^{(0)}(x)$ is the "launching" amplitude of the peripheral wave, $\beta_j(x)$ is the total phase jump at the wave excitation (and re-radiation) at the critical angle $\varphi_j(x)$, $\varepsilon_j(x)$ is the damping coefficient of the peripheral wave over one full turn, $\varepsilon_j^{(0)}(x)$ is the damping coefficient of the peripheral wave on the path from the critical entrance point to the critical re-radiation point, $\kappa_j(x)$ is the damping coefficient per unit length of propagation, $\delta_j(x)$ is the increase of the phase in one full turn, and $\gamma_j(x)$ is the increase of the phase for the wave propagation from the critical entrance point to the critical re-radiation point and double propagation from and to the reference plane. The function $F_j^{(0)}(x)$ represents the total contribution by the left- and right-hand peripheral waves. The internal losses during peripheral wave propagation are not taken into account. The damping is caused only by the radiation into the liquid.

For the S_0 and A wave, the points $x = x_n$, $x = y_n$ are called the resonance points, and $x = x_{n+1/2}$, $x = y_{n+1/2}$ are called, consequently, the antiresonance points. The positions of the resonance points are determined by the equations

$$x\delta_1(x) = 2\pi n, \qquad x\delta_2(x) = 2\pi n , \tag{9.15}$$

and the positions of the antiresonance points are determined by the equations

$$x\delta_1(x) = 2\pi(n + 1/2), \qquad x\delta_2(x) = 2\pi(n + 1/2) . \tag{9.16}$$

At the resonance points, $\exp[ix\delta_j(x)] = 1$ and the denominator in (9.13) attains the maximal value. Coincidence resonances of the function $f(x)$ occur when integer number of the peripheral wavelength fits the perimeter of the shell. In the case of a coincidence resonance, the peripheral wave circumnavigating the shell is superimposed on itself without any phase difference and therefore will be resonantly amplified. In the case of an antiresonance, the peripheral wave revolving around the shell is superimposed on itself in antiphase. During the propagation, the S_0 and A waves are damped slowly and may carry out quite a number of turns before dying out completely.

At the resonance frequency, the modulus of the re-radiated wave equals

$$|P_1(x_n)| = [F_1^{(0)}(x_n)\varepsilon_1^{(0)}(x_n)]/[1 - \varepsilon_n(x_n)] ,$$
$$|P_2(x_n)| = [F_2^{(0)}(y_n)\varepsilon_2^{(0)}(y_n)]/[1 - \varepsilon_2(y_n)] , \tag{9.17}$$

and at the antiresonance frequency, it is

$$|P_1(x_{n+1/2})| = \frac{F_1^{(0)}(x_{n+1/2})\varepsilon_1^{(0)}(x_{n+1/2})}{1 + \varepsilon_1(x_{n+1/2})} ,$$

$$|P_2(y_{n+1/2})| = \frac{F_2^{(0)}(y_{n+1/2})\varepsilon_2^{(0)}(y_{n+1/2})}{1 + \varepsilon_1(y_{n+1/2})} . \tag{9.18}$$

Let us draw the envelopes through the maxima and minima of $|P_j(x)|$ $(j = 1, 2)$. The envelopes passing through the maxima are determined by the expression $F_j(x)$ and those passing through the minima by the expression $\Phi_j(x)$. If $F_j(x)$ is given in obvious form, then the equation for $\Phi_j(x)$, according to (9.17) and (9.18), takes the following form:

$$\Phi_j(x) = E_j(x) F_j(x), \qquad E_j(x) = \frac{1 - \varepsilon_j(x)}{1 + \varepsilon_j(x)} \quad (j = 1, 2) . \tag{9.19}$$

Only the launching amplitude $F_j^*(x)$ is present in the formula for the scattered pressure

$$F_j^*(x) \equiv F_j^{(0)}(x)\varepsilon_j^{(0)}(x) = F_j(x)[1 - \varepsilon_j(x)] . \tag{9.20}$$

Using relation (9.20) in (9.13), we obtain

$$P_j(x) = F_j(x) \exp[i\beta_j(x)] \frac{\exp[i\gamma_j(x)][1 - \varepsilon_j(x)]}{1 - \varepsilon_j(x) \exp[ix\delta_j(x)]} . \tag{9.21}$$

The contribution to the pressure introduced by each peripheral wave $P_j(x)$, (9.21), would be completely determined if the functions $v_j^{ph}(x)$, $F_j(x)$, $\varepsilon_j(x)$ and $\beta_j(x)$ are known. The search for the approximate formulas of these functions shall be carried out separately for the S_0 and A waves.

9.2.2 The S_0 Wave

To describe $v_1^{ph}(x)$, we shall use the extrema positions of the curve $|f_{00}(x)|$ (see Fig. 9.7), corresponding to the resonances of the S_0 wave. We shall write $v_1^{ph}(x)$ in a form containing the free parameters A_0, A_1, A_2 and A_3

$$v_1^{ph}(x) = A_3 + (A_2 x^2 - A_1 x + A_0)^{-1} . \tag{9.22}$$

In the example considered, let us set

$$A_3 = 3.67 . \tag{9.23}$$

We shall find three other parameters from the linear system of algebraic equations

$$
\begin{aligned}
x_{(1)}^2 A_2 - x_{(1)} A_1 + A_0 &= [v_1^{ph}(x_{(1)}) - A_3]^{-1} , \\
x_{(2)}^2 A_2 - x_{(2)} A_1 + A_0 &= [v_1^{ph}(x_{(2)}) - A_3]^{-1} , \\
x_{(3)}^2 A_2 - x_{(3)} A_1 + A_0 &= [v_1^{ph}(x_{(3)}) - A_3]^{-1} ,
\end{aligned}
\tag{9.24}
$$

following from the condition that at the three reference points ($x_{(1)}$, $x_{(2)}$ and $x_{(3)}$), the formula (9.22) represents $v_1^{ph}(x)$ accurately. We shall place the reference points at where $v_1^{ph}(x)$ changes strongly. In the example considered, this is the domain $x < 50$. For the reference points, we use those corresponding to the 3, 8 and 13 resonances of the S_0 wave

$$
\begin{aligned}
x_{(1)} &= x_3 = 11.41, & x_{(2)} &= x_8 = 29.59, & x_{(3)} &= x_{13} = 47.85 , \\
v_1^{ph}(x_{(1)}) &= 3.803, & v_1^{ph}(x_{(2)}) &= 3.700, & v_2^{ph}(x_{(3)}) &= 3.681 .
\end{aligned}
\tag{9.25}
$$

Then the system of equations (9.24) has the solution

$$A_0 = 5.94, \quad A_1 = 0.3893, \quad A_2 = 0.04609 . \tag{9.26}$$

The validity of the description given by (9.22) can be easily checked, using it to locate the resonances x_n of the S_0 wave at

$$x_n = n v_1^{ph}(x) , \tag{9.27}$$

which are the maxima on the curve in Fig. 9.7, both below and above the strong-bending domain. The phase velocity $v_1^{\mathrm{ph}}(x)$ should be calculated rather accurately because the formula (9.27) increases the error n times.

Similarly, using the information about the $|f_{00}(x)|$ values at the resonance points x_n of the S_0 wave, we shall find free parameters B_0, B_1 and B_2 in the approximate formula of the amplitude $F_1(x)$

$$F_1(x) = [x^{1/2}(B_2 x^2 - B_1 x + B_0)]^{-1} . \tag{9.28}$$

It is appropriate to place the reference points both below and above the strong-bending domain. We shall use the 1, 6 and 27 resonances of the S_0 wave as the reference points

$$
\begin{aligned}
&x_{(1)} = x_1 = 3.95, \qquad x_{(2)} = x_6 = 2.31, \qquad x_{(3)} = x_{27} = 99.06 , \\
&|f_{00}(x_{(1)})| = 0.968, \quad |f_{00}(x_{(2)})| = 0.464, \quad |f_{00}(x_{(3)})| = 0.211 .
\end{aligned}
\tag{9.29}
$$

By solving the system of algebraic equations, we obtain

$$B_0 = 0.537, \qquad B_1 = 0.451 \times 10^{-2}, \qquad B_2 = 0.392 \times 10^{-4} . \tag{9.30}$$

The accuracy of the description (9.28) can be easily checked by computing the S_0 wave amplitude at the points of successive resonances and comparing the results with the ones corresponding to $|f_{00}(x_n)|$.

Let us now move on to the description of $\varepsilon_1(x)$. In those x ranges where $f_{00}(x)$ is caused mainly by the S_0 wave, we shall, as before, use the equality

$$|f_{00}(x)| = |P_1(x)| . \tag{9.31}$$

Then, according to (9.18–20), we obtain

$$E_1(x_{n+1/2}) = \frac{\Phi_1(x_{n+1/2})}{F_1(x_{n+1/2})} \equiv \frac{|f_{00}(x_{n+1/2})|}{F_1(x_{n+1/2})} , \tag{9.32}$$

and using (9.19) we find

$$\varepsilon_1(x_{n+1/2}) = [1 - E_1(x_{n+1/2})]/[1 + E_1(x_{n+1/2})] . \tag{9.33}$$

Inserting (9.32) into (9.33), we obtain

$$\varepsilon_1(x_{n+1/2}) = \frac{1 - |f_{00}(x_{n+1/2})|/F_1(x_{n+1/2})}{1 + |f_{00}(x_{n+1/2})|/F_1(x_{n+1/2})} . \tag{9.34}$$

For arbitrary x, the formula $\varepsilon_1(x)$ is assumed to be

$$\varepsilon_1(x) = (C_2 x^2 - C_1 x + C_0)^{-1} \qquad \text{at } x < x_0 , \tag{9.35}$$

$$\varepsilon_1(x) = 1 - (D_1/x) + (D_2/x^2) \qquad \text{at } x \geqslant x_0 , \tag{9.36}$$

where C_0, C_1, C_2, D_1 and D_2 are free parameters and x_0 is the parameter separating the ranges.

Here, it is appropriate to use the antiresonance points up to the strong-bending domain. In (9.35), we shall use the antiresonances

2, 4 and 6 of the S_0 wave, and in (9.36), the antiresonances 2 and 6

$$x_{(1)} = x_{2+1/2} = 9.57, \qquad x_{(2)} = x_{4+1/2} = 16.86, \qquad x_{(3)} = x_{6+1/2} = 24.12 ,$$

$$\varepsilon_1(x_{(1)}) = 0.7457, \qquad \varepsilon_1(x_{(2)}) = 0.8640, \qquad \varepsilon_1(x_{(3)}) = 0.9419 . \tag{9.37}$$

By solving the system of algebraic equations, we obtain

$$C_0 = 1.72, \qquad C_1 = 0.0473, \qquad C_2 = 0.835 \times 10^{-3} ,$$

$$D_1 = 0.720, \qquad D_2 = 16.4 . \tag{9.38}$$

According to (9.35), $\varepsilon_1(x)$ takes the maximal value at $x = x_*$

$$x_* = C_1/2C_2 . \tag{9.39}$$

In the example considered

$$x_* = 28.32, \qquad \varepsilon_1(x_*) = 0.955 . \tag{9.40}$$

We shall assume that the boundary $x = x_0$ coincides with x_*

$$x_0 = x_* . \tag{9.41}$$

In the strong-bending domain and in the ranges adjoining it (in the example considered at $40 \leqslant x \leqslant 90$, see Figs. 9.1 and 9.7), it is rather difficult to determine $\varepsilon_1(x)$ precisely, because here in the vicinity of the S_0 wave antiresonances, a strong influence of the A wave is observed. In this range it is not justified to use the formula of the type (9.35) because, after $x = x_*$ (9.40), the values of $\varepsilon_1(x)$ would diminish very rapidly and, at $x \sim 100$, they would be of order $\varepsilon_1 \sim 0.2$, which would be in clear contradiction with the input information.

As is well known, the coefficient $\varepsilon_1(x)$ determines the Q factor of the wave and could be interpreted geometrically as the opening angle of the $|f_{00}(x)|$ curve at the resonance points; the smaller this angle, the higher a Q factor the wave possesses. In the range $0 \leqslant x < 50$, it can be seen that, with x increasing, the angle decreases and, consequently, $\varepsilon_1(x)$ increases. In the range $90 \lesssim x \lesssim 100$, the opening angle of the $|f_{00}(x)|$ curve is roughly the same as in the range $x \sim 50$. Therefore, we have used the assumption that, after $x = x_*$, $\varepsilon_1(x)$ continues to increase and we have passed on to describing it by (9.36). In this way the combined formula has been generated. If we try to describe $\varepsilon_1(x)$ by a single formula in the whole range considered, $0 \leqslant x \leqslant 100$, then such a formula will be complicated and would contain many parameters. Since the requirements for the description of $\varepsilon_1(x)$ are not very great, the combined formula may be assumed to be justified. In the formula for $P_1(x)$, the coefficient $\varepsilon_1(x)$ enters as one among many others and, generally speaking, is weakly connected to them. Now, using the evident approximate formula for $\varepsilon_1(x)$, we can find $\kappa_1(x)$, $\varepsilon_1^{(0)}(x)$ and $F_1^{(0)}(x)$

$$\kappa_1(x) = (1/2\pi) \ln[1/\varepsilon_1(x)] ,$$

$$\varepsilon_1^{(0)}(x) = \exp(-\{[\pi - \varphi_1(x)]/\pi\} \ln[1/\varepsilon_1(x)]) , \tag{9.42}$$

$$F_1^{(0)}(x) = F_1(x)[1 - \varepsilon_1(x)][\exp(-\{[\pi - \varphi_1(x)]/\pi\} \ln[1/\varepsilon_1(x)])]^{-1} .$$

We shall now move on to the description of $\beta_1(x)$. Temporarily inserting $\beta_1(x) = 0$ into (9.13), we shall form the difference

$$\beta_{01} = \arg f_{00}(x) - \arg P_1(x) , \tag{9.43}$$

which we shall compute at the resonance points x_n of the S_0 wave. It is also profitable to use the combined formula for the function $\beta_1(x)$

$$\beta_1(x) = K_1[(x + K_2)/(x + K_3)] \qquad \text{at } x \leqslant x_{00} , \tag{9.44}$$

$$\beta_1(x) = K_4[(x^{1/2} - K_5)/(x^{1/2} - K_6)] \quad \text{at } x > x_{00} . \tag{9.45}$$

The free parameters $K_j (j = 1, 2, \ldots, 6)$ could be easily found from the condition that at fixed points the approximate formula for $\beta_1(x)$ exactly represents the desired dependence.

In the example considered, we shall choose some points corresponding to the resonances of the S_0 wave as the reference points: 2, 4 and 6 for (9.44) and 2, 5 and 9 for (9.45)

$$x_{(1)} = x_2 = 7.73, \qquad x_{(2)} = x_4 = 15.04, \qquad x_{(3)} = x_6 = 22.30 ,$$
$$\beta_1(x_{(1)}) = 2.25, \qquad \beta_1(x_{(2)}) = 2.90, \qquad \beta_1(x_{(3)}) = 3.38 , \tag{9.46}$$

and

$$x_{(1)} = x_2 = 7.73, \qquad x_{(2)} = x_5 = 18.67, \qquad x_{(3)} = x_9 = 33.24 ,$$
$$\beta_1(x_{(1)}) = 2.25, \qquad \beta_1(x_{(2)}) = 3.19, \qquad \beta_1(x_{(3)}) = 3.84 . \tag{9.47}$$

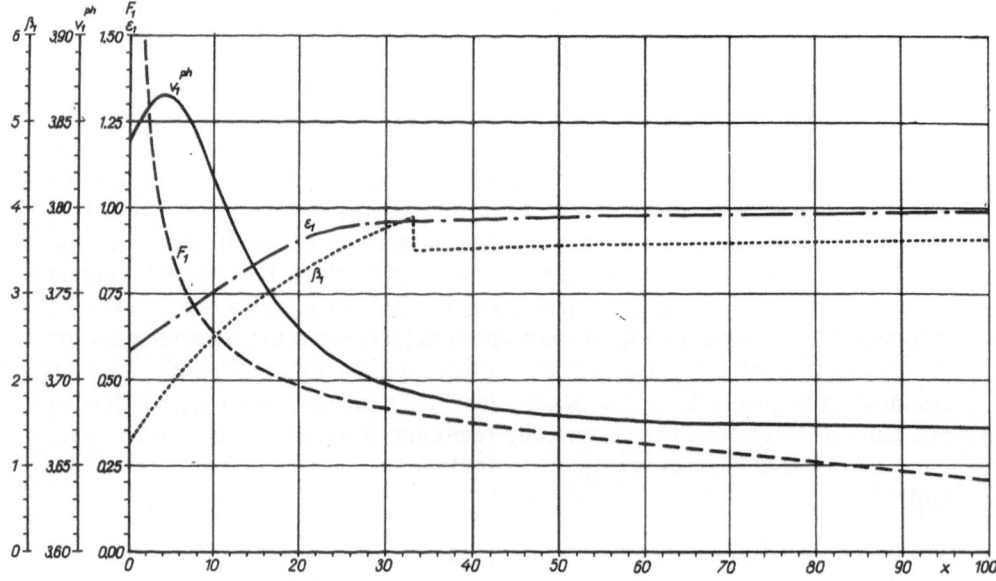

Fig. 9.8. The synthesized approximate dependences for the description of the S_0 wave contribution [9.65, Fig. 8]

By solving two systems of algebraic equations, we obtain the following values of the constants

$$K_1 = 6.65, \qquad K_2 = 6.53, \qquad K_3 = 34.4 ,$$
$$K_4 = 3.72, \qquad K_5 = 2.28, \qquad K_6 = 2.39 .$$

(9.48)

It is appropriate to fix the parameter x_{00} so that the approximate values of $\beta_1(x)$ calculated according to (9.44) and (9.45) would differ little from each other. The position of the 9th resonance of the S_0 wave satisfies this condition

$$x_{00} = 33.24 .$$

(9.49)

In Fig. 9.8, the plots of $v_1^{ph}(x)$, $F_1(x)$, $\varepsilon_1(x)$ and $\beta_1(x)$, computed according to the approximate formulae, are presented. The positions $\tilde{x}_n$ and the values $|P_1(\tilde{x}_n)|$ of the function $|P_1(x)|$ are presented in Fig. 9.9 and in Table 9.1.

Now, all the functional dependences of the quantities characterizing the contribution of the S_0 wave are described by approximate formulae, and it is

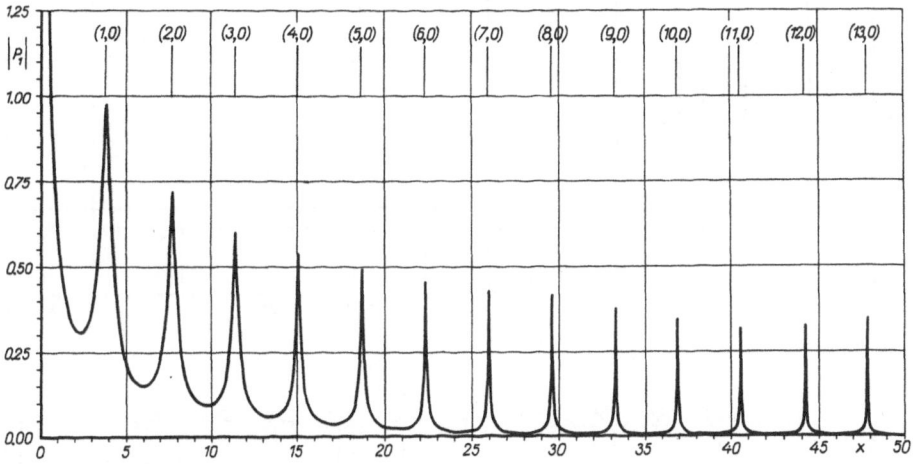

Fig. 9.9.1

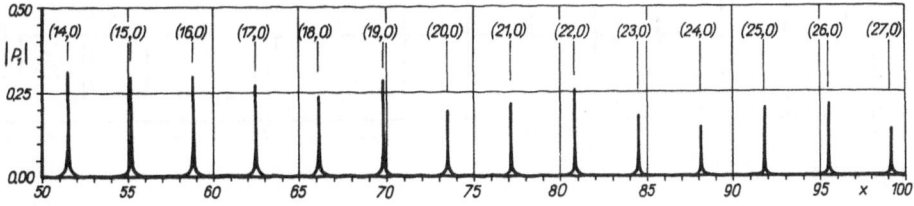

Fig. 9.9.2

Figs. 9.9.1 and 9.9.2. The modulus of $|P_1(x)|$, see (9.21), representing the contribution of the S_0 wave to the backscattered form function [9.65, Fig. 9]

possible to realize the synthesis of the backscattered far field form function in which only the contribution of the specular reflection (the refraction in the layer included) and the re-radiation induced by the S_0 wave would be taken into account. To this degree of approximation, the form function is described by

$$f_3(x) = f_{10}(x) + P_1(x) . \tag{9.50}$$

In Fig. 9.10, the plot of $|f_3(x)|$ is presented. The values of $|f_3(\hat{x}_n)|$ at the resonance points $\hat{x}_n$ are given in Table 9.1. The comparison of the plots in

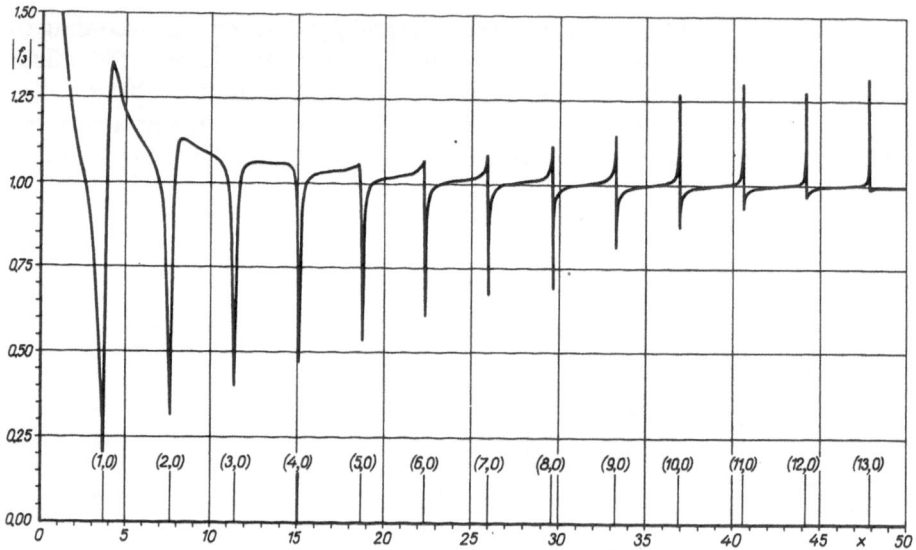

Fig. 9.10.1

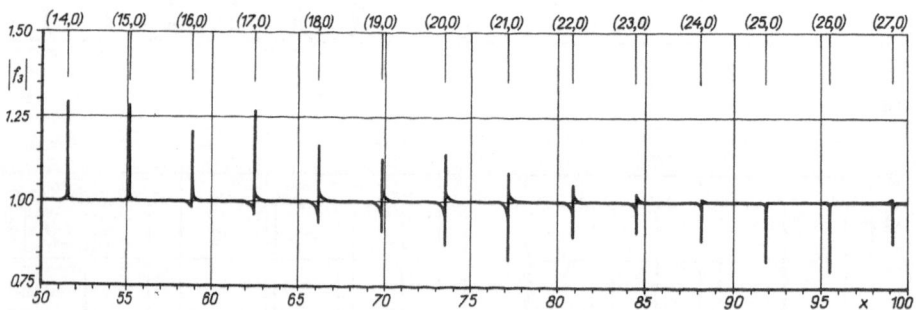

Fig. 9.10.2

Figs. 9.10.1 and 9.10.2 The modulus of $|f_3(x)|$, see (9.50). This is the modulus of the synthesized backscattered form function in which only the contribution of the specular reflection (the refraction in the layer included) and the re-radiation induced by the S_0 wave has been taken into account [9.65, Fig. 10]

Figs. 9.1 and 9.10 and the appropriate columns in Table 9.1 verify that, outside the strong-bending domain, $|f_3(x)|$ describes the modulus of the form function well.

9.2.3. The A Wave

To describe $v_2^{ph}(x)$, we shall use the extrema positions of the curve $|f_{00}(x)|$ (see Fig. 9.7), corresponding to the A wave resonances. According to the results of the resonance scattering theory (see Fig. 9.3 and Table 9.2), in the strong-bending domain the relative phase velocity of the A wave $v_2^{ph}(x)$ changes in rather narrow limits: from 0.83 to 0.99. Let us assume that the formula for $v_2^{ph}(x)$ may be represented as

$$v_2^{ph}(x) = L_0 + L_1 r_0 + L_2 r_0^2 , \tag{9.51}$$

$$r_0 = \tfrac{1}{2} \ln x , \tag{9.52}$$

and L_0, L_1 and L_2 are the free parameters. The resonances of the A wave in the strong-bending domain should be used as the reference points. They should be selected so that the influence of the S_0 wave on y_n becomes as small as possible; that is, the resonances y_n would be situated farthest away from the resonances at x_n. We shall choose the resonances 63, 73 and 83 of the A wave as the reference points

$$x_{(1)} = x_{63} = 54.16, \qquad x_{(2)} = x_{73} = 67.97, \qquad x_{(3)} = x_{83} = 81.52 ,$$
$$v_2^{ph}(x_{(1)}) = 0.8597, \qquad v_2^{ph}(x_{(2)}) = 0.9311, \qquad v_2^{ph}(x_{(3)}) = 0.9822 . \tag{9.53}$$

By solving the system of algebraic equations, we obtain

$$L_0 = -1.7066, \qquad L_1 = 4.40, \qquad L_2 = -1.66 . \tag{9.54}$$

Using (9.51) to calculate the successive resonance position of the A wave,

$$y_n = n v_2^{ph}(y_n) , \tag{9.55}$$

and comparing them with the inputs, the accuracy of (9.51) may be estimated.

It should be noted that, since we have a few reliable reference points that are necessary in order to obtain the parameters in $F_2(x)$ using only the $|f_{00}(x)|$ function, we have introduced

$$f_*(x) = f(x) - \tilde{f}(x) . \tag{9.56}$$

We shall define the positions of $|f_*(x)|$, corresponding to the A wave resonances, by $\tilde{y}_n$ and insert in Table 9.2 the values $\tilde{y}_n$ and $|f_*(\tilde{y}_n)|$. Now we shall use the averaged values of the positions and values, corresponding to $|f_{00}(y)|$ and $|f_*(y)|$, as the reference points. Using the averaged magnitudes at the 63, 68 and 73 resonances of the A wave as the reference points

$$x_{(1)} = y_{63} = 54.16, \qquad x_{(2)} = y_{68} = 61.35, \qquad x_{(3)} = y_{73} = 67.97 ,$$
$$|f_{0*}(x_{(1)})| = 0.161, \qquad |f_{0*}(x_{(2)})| = 0.299, \qquad |f_{0*}(x_{(3)})| = 0.257 \tag{9.57}$$

in the approximate formula for $F_2(x)$

$$F_2(x) = \exp(-M_2 x + M_1 x^{1/2} - M_0) \tag{9.58}$$

and solving the system of algebraic equations, we obtain

$$M_0 = 100.4, \qquad M_1 = 24.95, \qquad M_2 = 1.568 . \tag{9.59}$$

We shall use the same approach for describing $\varepsilon_2(x)$ as in the $\varepsilon_1(x)$ case. In the strong-bending domain, far away from the S_0 wave resonances, we shall assume that

$$|f_{00}(x)| = |P_2(x)| . \tag{9.60}$$

According to (9.18–20), we obtain

$$E_2(y_{n+1/2}) = |f_{00}(y_{n+1/2})|/F_2(y_{n+1/2}) , \tag{9.61}$$

and by the definition (9.19), we calculate

$$\varepsilon_2(y_{n+1/2}) = [1 - E_2(y_{n+1/2})]/[1 + E_2(y_{n+1/2})] . \tag{9.62}$$

Similarly to (9.35), the formula for $\varepsilon_2(x)$ may be written in the form

$$\varepsilon_2(x) = (N_2 x^2 - N_1 x + N_0)^{-1} . \tag{9.63}$$

We shall use the antiresonances 64, 67 and 72 of the A wave as the reference points

$$x_{(1)} = x_{64+1/2} = 56.36, \quad x_{(2)} = x_{67+1/2} = 60.65, \quad x_{(3)} = x_{72+1/2} = 67.29 ,$$

$$\varepsilon_2(x_{(1)}) = 0.8297, \qquad \varepsilon_2(x_{(2)}) = 0.6650, \qquad \varepsilon_2(x_{(3)}) = 0.3389 . \tag{9.64}$$

Then the solution of the algebraic equation system is

$$N_0 = 44.57, \qquad N_1 = 1.548, \qquad N_2 = 1.381 \times 10^{-2} . \tag{9.65}$$

For an appropriate description of $\beta_2(x)$, we propose the formula

$$\beta_2(x) = Q_0 - Q_1/x + Q_2/x^2 . \tag{9.66}$$

We shall use the positions of the resonances 62, 73 and 81 as the reference points

$$x_{(1)} = y_{62} = 52.68, \qquad x_{(2)} = y_{73} = 67.97, \qquad x_{(3)} = y_{81} = 78.83 ,$$

$$\beta_2(x_{(1)}) = 2.55, \qquad \beta_2(x_{(2)}) = 2.04, \qquad \beta_3(x_{(3)}) = 4.82 . \tag{9.67}$$

Then we obtain

$$Q_0 = 65.81, \qquad Q_1 = 7787, \qquad Q_2 = 234661 . \tag{9.68}$$

Figure 9.11 gives plots of $v_2^{\text{ph}}(x)$, $F_2(x)$, $\varepsilon_2(x)$ and $\beta_2(x)$ calculated according to the approximate formulae. The positions $\tilde{y}_n$ and values $|P_2(\tilde{y}_n)|$ of the function $|P_2(x)|$ at the resonance frequencies are given in Table 9.2. The $|P_2(x)|$ plot in the strong-bending domain is presented in Fig. 9.12.

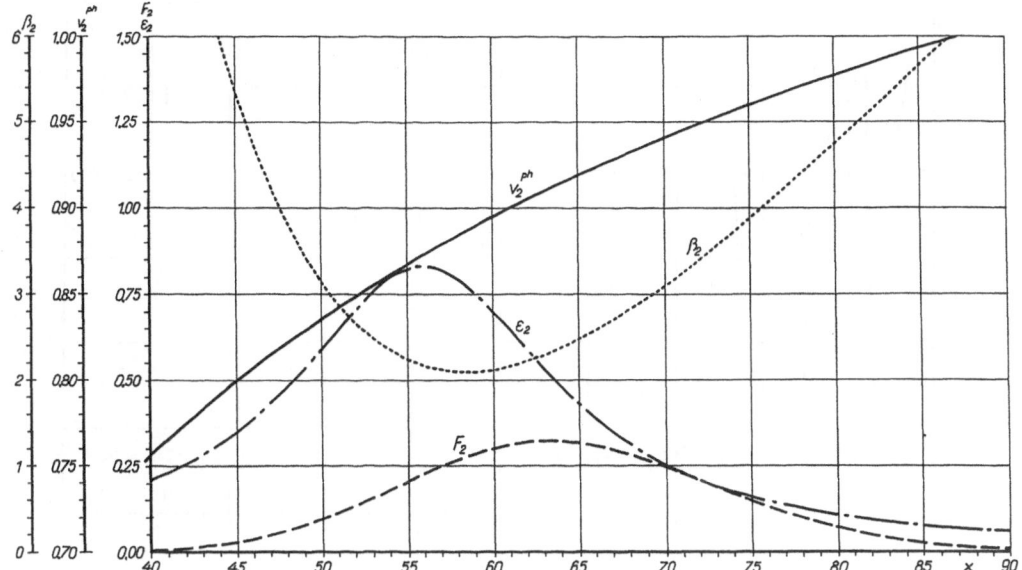

Fig. 9.11. The synthesized approximate dependences for the description of the A wave contribution [9.65, Fig. 11]

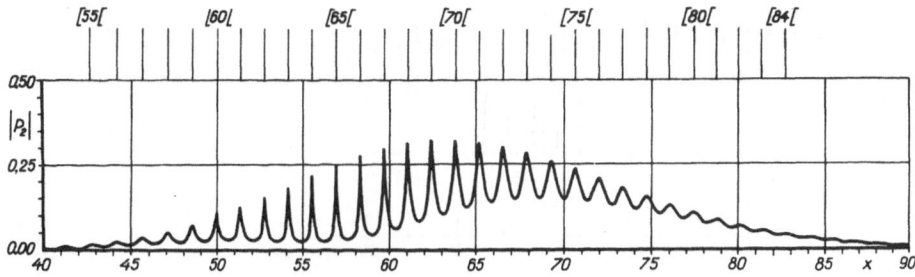

Fig. 9.12. The modulus of $|P_2(x)|$, see (9.21), representing the contribution of the A wave to the backscattered form function [9.65, Fig. 12]

Using the functional dependences of the quantities characterizing the A wave, the quantity $f_3(x)$, (9.50), may be corrected by including the contribution of the A wave. To this degree of approximation, the backscattered far field form function is given by

$$f_4(x) = f_{10}(x) + P_1(x) + P_2(x) . \tag{9.69}$$

Figure 9.13 gives the plot of $|f_4(x)|$. The $|f_4(\hat{x})|$ values at the resonance points $\hat{y}_n$ in the strong-bending domain are presented in Table 9.2.

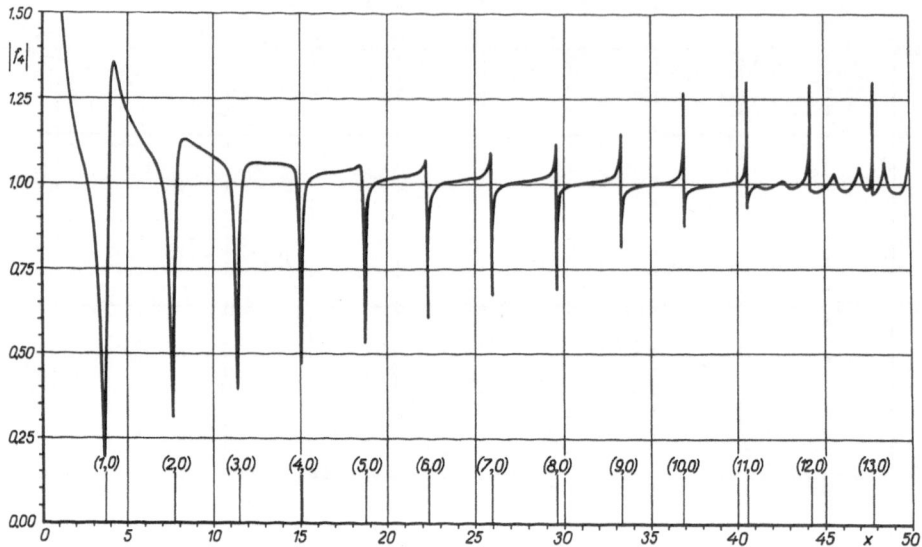

Fig. 9.13.1

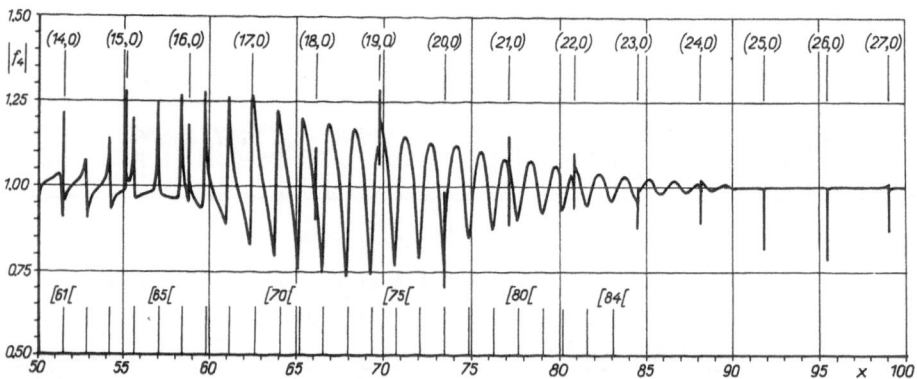

Fig. 9.13.2

Figs. 9.13.1 and 9.13.2. The modulus of $|f_4(x)|$, see (9.69). This figure is similar to Fig. 9.10 except that here the contribution of the specular reflection and re-radiation induced by the S_0 and A waves has been taken into account [9.65, Fig. 13]

The backscattered far field form function also may be represented by the formula

$$f_5(x) = \tilde{f}(x) + P_2(x) .\tag{9.70}$$

In Fig. 9.14, the plot of $|f_5(x)|$ is presented. The comparison of Figs. 9.1, 9.13 and 9.14 and the corresponding columns of Table 9.2 shows that the form function is well described by both $f_4(x)$ and $f_5(x)$.

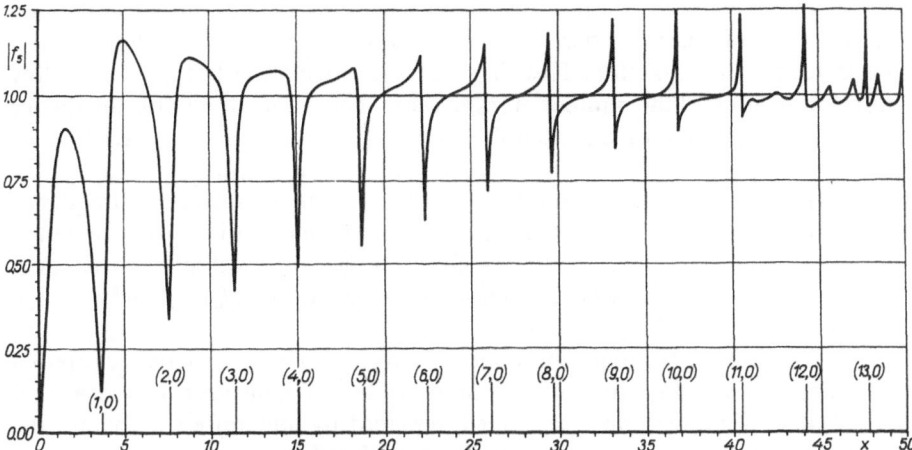

Fig. 9.14.1

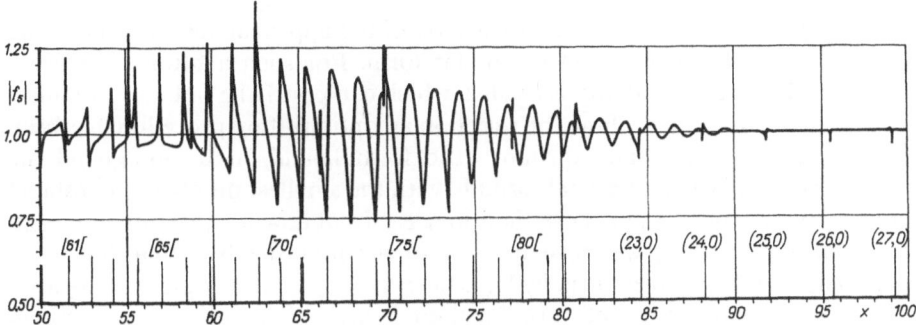

Fig. 9.14.2

Figs. 9.14.1 and 9.14.2. The modulus of $|f_s(x)|$, see (9.70). This figure is similar to Fig. 9.13 except that here the contribution of the A wave has been added to the solution from the partial-wave series when the motion of the shell is governed by the equations of the bending-free (membrane) thin shell theory [9.65, Fig. 14]

9.3 Discussion

In contrast to the acoustic rigid and soft bodies, for which the scattered acoustic pressure depends only on the parameters of the liquid surrounding the body, the scattered acoustic pressure for the case of an elastic scatterer is related to the material parameters of the body. Therefore, for each case, the calculation bears the imprint of the chosen specific material parameters. The above-considered situation of a metal shell immersed in water is a typical case in hydroelasticity.

Unlike the resonance scattering theory where the background is computed for every term of the series, here a single background, (9.10), is used for the total backscattered form function. This simplifies the calculation but does not allow a direct indication of the number of resonances. The resonance scattering theory or the method of isolation and identification of vibrational modes (MIIR) can be used for this purpose. When the MIIR is chosen, the angular diagram of the re-radiation (after the end of the forced vibration) at the resonance frequency should be computed. The comparison of our results with those computed according to the resonance scattering theory verifies the validity of the background used. This can be seen from the columns $\tilde{y}_n$ and y_n, $|f_*(\tilde{y}_n)|$ and $\xi_n(\tilde{y}_n)$ in Table 9.2 and Figs. 9.7, 9.2 and 9.3.

While describing the main components of S_0 and A waves, we have proposed twelve approximate formulae, each containing several constants. It is clear that each of them, if needed, may be refined, replaced by one more suitable for the description of the function considered and, therefore, containing a smaller number of constants. In fact, all of the formulas presented are themselves in some way the result of iteration.

Actually, the above-used approach of the total backscattered form function decomposition lets us reduce the problem of describing the functions $v_j^{ph}(x)$, $F_j(x)$, $\varepsilon_j(x)$ and $\beta_j(x)$ ($j = 1, 2$) to a selection of the approximate functions using the computed data, presented in tabular form. For such a fitting, of course, a certain element of creativity is needed, which can poorly be put into formulae. Essentially, during such a function selection an optimization procedure has been carried out for, on one hand, the most exact formula and, on the other hand, the most simple and clear empirical formula with the smallest number of constants chosen. We try not to put more than four easily obtained constants into each formula. To do that, we have taken into account the fact that the input parameters of the problem usually have the accuracy of three significant figures. The calculation of the "exact" form function is carried out with some step size in x and, therefore, the values that must be approximated themselves contain some error. Certainly, the calculation step size could be decreased, thus refining the input data.

By using the model of a thin liquid layer to describe the process of specular reflection and refraction, the difference $f_{00}(x)$, (9.12), was established; its modulus extrema are slightly shifted from the $p(x)$ extrema. This means that a certain error is permitted not only in the extrema positions but also in their values. Then, all of our subsequent operations consist of extracting information from these extrema. From their positions, the velocities $v_j^{ph}(x)$ and the total phase jumps $\beta_j(x)$ are determined; from the values at the resonance points, the amplitudes $F_j(x)$ are defined; and from the values at the antiresonance points, the damping coefficients $\varepsilon_j(x)$ are evaluated. In the approach used, the type of the approximation function $F_j(x)$ directly influences the choice of the functions $\varepsilon_j(x)$ and $\beta_j(x)$. Formalizing the procedure and transferring it entirely to the computer has not been successful as yet, because it contains the creative element of the selection of the empirical functions.

As we have ascertained, the requirements for the accuracy of the description of the functions $v_j^{\text{ph}}(x)$, $F_j(x)$, $\varepsilon_j(x)$, and $\beta_j(x)$ ($j = 1, 2$) are different. The velocities $v_j^{\text{ph}}(x)$ and the total phase jumps $\beta_j(x)$ must be described as accurately as possible while the amplitudes $F_j(x)$ and the damping coefficients $\varepsilon_j(x)$ may be described with moderate accuracy. This experience is reflected in the structure of all the above presented empirical dependences.

It should be noted that, for the selection of the $v_j^{\text{ph}}(x)$ ($j = 1, 2$) form, we used the rich experience of our predecessors. For the function $F_j(x)$, we had little experience of our own concerning the description of $F_1(x)$ in the case of a rather thin shell. The functional dependences of the $\varepsilon_j(x)$ and $\beta_j(x)$ were entirely unknown to us at the beginning of our investigation. Here, we keep in mind the form of the dependences in a broad x range.

It is clear that the combined formula for $\beta_j(x)$ [(9.44) and (9.45)] cannot be justified physically. The exact function $\beta_j(x)$ is smooth but in our approximation it has a jump at $x = x_{00}$, which causes the jumps in $|f_3(x)|$, (9.55), and $f_4(x)$, (9.69). Therefore, our approximation is not applicable either at $x = x_{00}$ or in its vicinity.

A number of such imperfections of our approach may be indicated. We accept these imperfections because we intend to give the qualitative description of the backscattered form function, but not its strict mathematical form. The approach described for the analysis and synthesis of the backscattered form function can be easily transferred to the problem of acoustic plane wave scattering by elastic spherical shells.

In a similar way to that described above, the contribution of the specularly reflected wave (the refraction in the layer included) can be separated from the contribution of the re-radiated waves, caused by the peripheral waves in the more complicated two-dimensional problems concerning plane acoustic wave scattering on a cylindrical shell with a smooth boundary curve, and on shells of revolution with a smooth meridian curve. The synthesis of the backscattered form function can be carried out only after the functions $v_j^{\text{ph}}(x)$, $F_j(x)$, $\varepsilon_j(x)$ and $\beta_j(x)$ are defined for every peripheral wave. These functions can be found from the solution of a model problem of eigenvibrations of a fluid-loaded elastic body. In the low-frequency range, corresponding to the first resonances of the A wave, the synthesized form function can be checked by comparison with the exact form function evaluated by numerical methods.

The material stated above is based on results first available as a preprint [9.64] and afterwards as a paper [9.65]. I wish to acknowledge stimulating remarks made by Dr. Werner G. Neubauer.

10 Peripheral Waves in the Scattering by Spherical Shells

We present here the results of applying resonance scattering theory to the problem of a plane acoustic wave scattering from a spherical shell with moderate thickness (with $h = 1/32$) and from a thick-walled shell (with $h = 1/10$). The computations have been carried out and analyzed in a broad frequency band $(0 \leqslant x \leqslant 400)$ and for large orders of resonances ($n \lesssim 100$). Attention is focused on the Lamb-type peripheral waves S_j and A_j ($j = 0, 1, 2, 3, 4$). The role of the so-called hidden resonances is elucidated.

10.1 Form Function and Isolated Modal Resonances

Let us consider a thin empty spherical shell submerged in an unbounded acoustic medium. A plane harmonic pressure wave strikes the shell and is scattered by it. The motion of the shell is described by the equations of linear elasticity theory and that of the liquid is described by the Helmholtz equation. It is a two-dimensional steady-state problem. The secondary (scattered by the shell) field of the acoustic pressure is analyzed. Using the method of separation of variables, one can obtain the exact solution in series form [10.1]. It is valid for arbitrary values of spherical coordinates r, θ. For simplicity, we shall examine the pressure at a fixed observation point situated in the far field at backscattering ($r/a = 10^4$, $\theta = \pi$). We shall consider a typical case in hydroelasticity, the case of an aluminum shell immersed in water. The physical parameters of the shell material and the liquid are given by (8.1). The relative thickness of the shell $h = 1 - b/a = 1/32$. The calculation is carried out in the range $0 \leqslant x \equiv ka \leqslant 400$ with the computation step size $l_x = 10/256$. Both limiting cases of a thick ($h = 1/10$) and a thin ($h = 1/100$) shell have been investigated in detail (see, for example [10.2–4]). The presented material is similar to that considered in Chap. 8. Therefore, only the essential features are discussed below.

We shall use the results presented in Chap. 3: the exact solution of the problem given by (3.37), the form function defined by (3.8) and (3.9), the difference between the pressure scattered by the elastic shell p_s and an acoustic rigid sphere $p_s^{(r)}$, see (3.24–26). For the sake of simplicity we shall use the notations introduced by (7.1).

In Fig. 10.1, the computed form function is presented. The resonance frequencies of partial modes are indicated on it. The extrema of the form

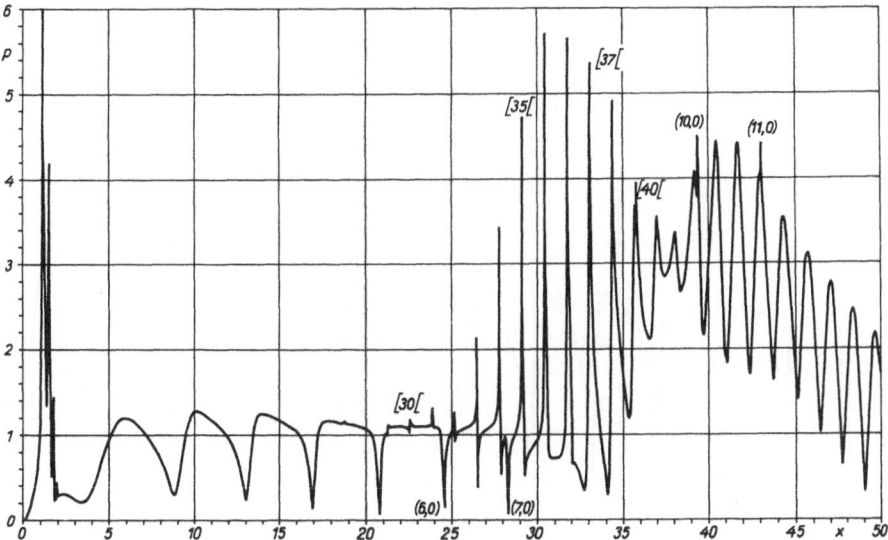

Fig. 10.1.1

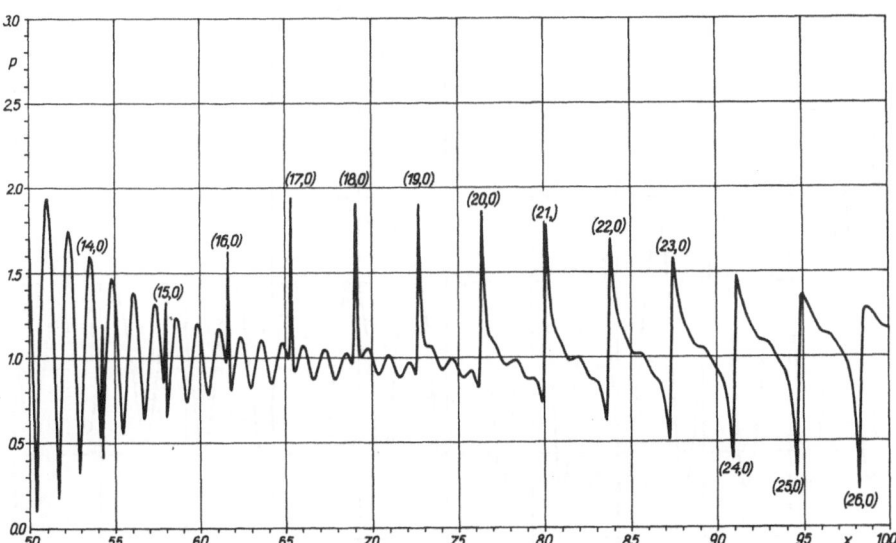

Fig. 10.1.2

Figs. 10.1.1–10.1.8 (p. 187–190). The form function of the acoustic pressure scattered by aluminum shell (with $h = 1/32$) immersed in water as a function of $x = ka$ for the parameters defined in (8.1). The isolated modal resonances of the Lamb-type symmetric waves are labeled by (n, l) and those of anti-symmetric waves by $(n, l($. The resonances of the fluid-borne bending A wave are designated by $[n[$

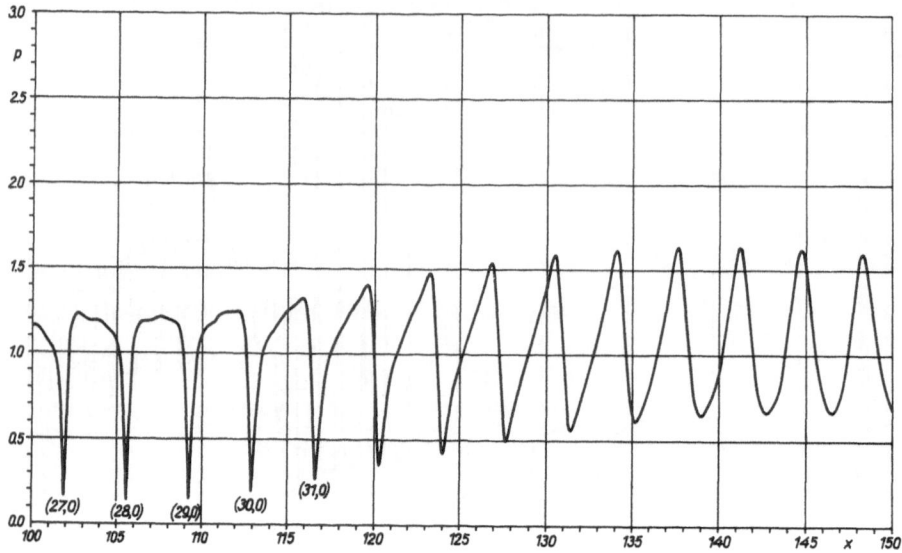

Fig. 10.1.3

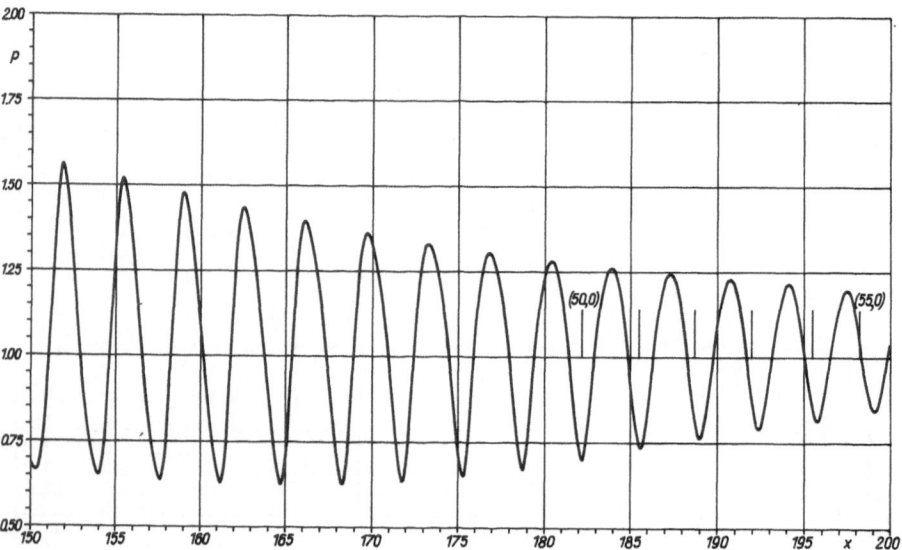

Fig. 10.1.4

function curve correspond to the resonance frequencies of the partial modes.
Usually with the frequency increasing the form function curve becomes more
discontinuous. This becomes particularly clearly apparent in the vicinity of the
cutoff frequencies of the Lamb-type waves, after which the resonances of the
newly generated wave can be observed.

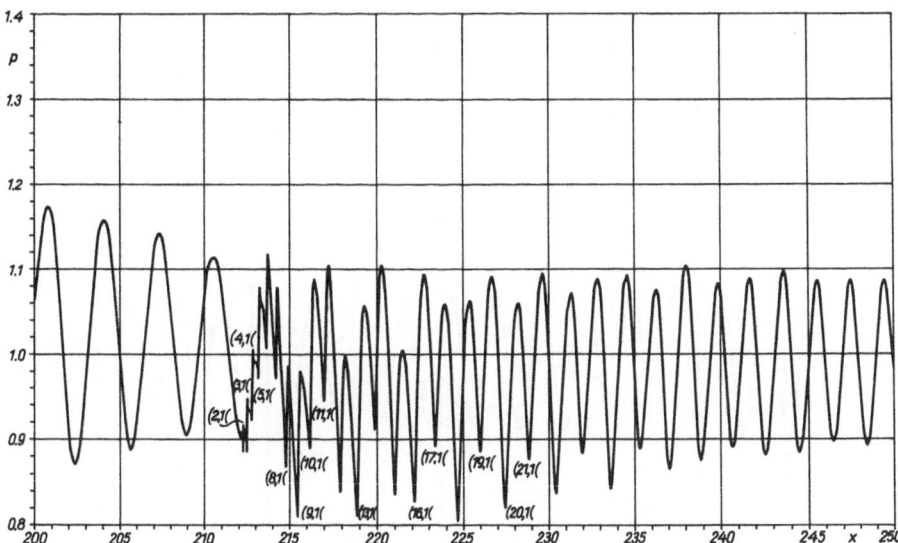

Fig. 10.1.5

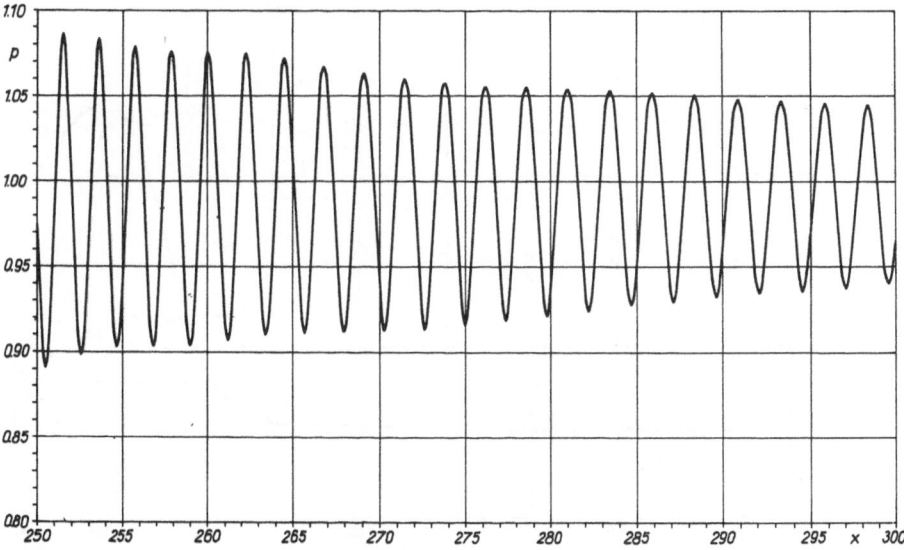

Fig. 10.1.6

The resonances of the fluid-borne bending wave A can be clearly observed in the strong-bending domain. In the example considered this is the range $20 \lesssim x \lesssim 60$. The resonances with $33 \lesssim n \lesssim 61$ are the most prominent ones. One can see the influence of the S_0 wave in the range $0 \lesssim x \lesssim 210$ (at $1 \lesssim n \lesssim 59$), i.e. up to the cutoff frequency of the A_1 wave ($x = 212.11$). The

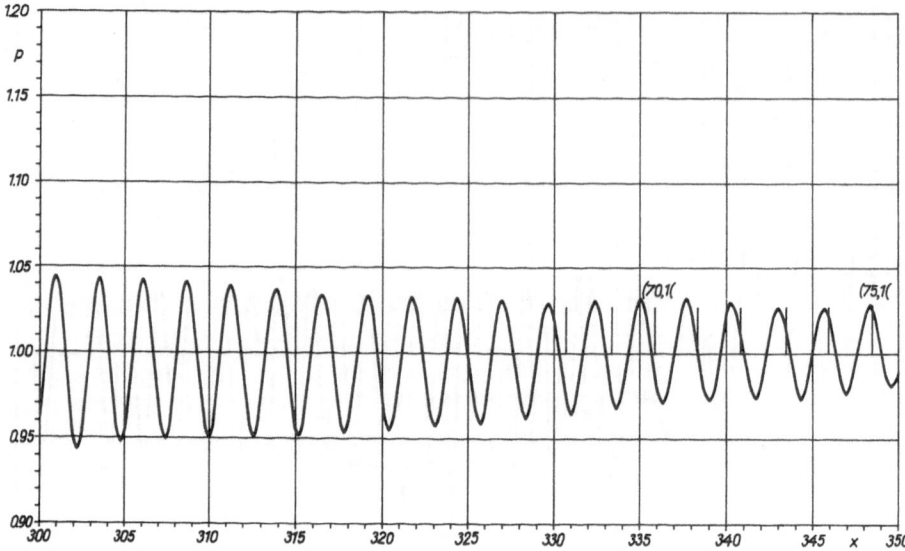

Fig. 10.1.7

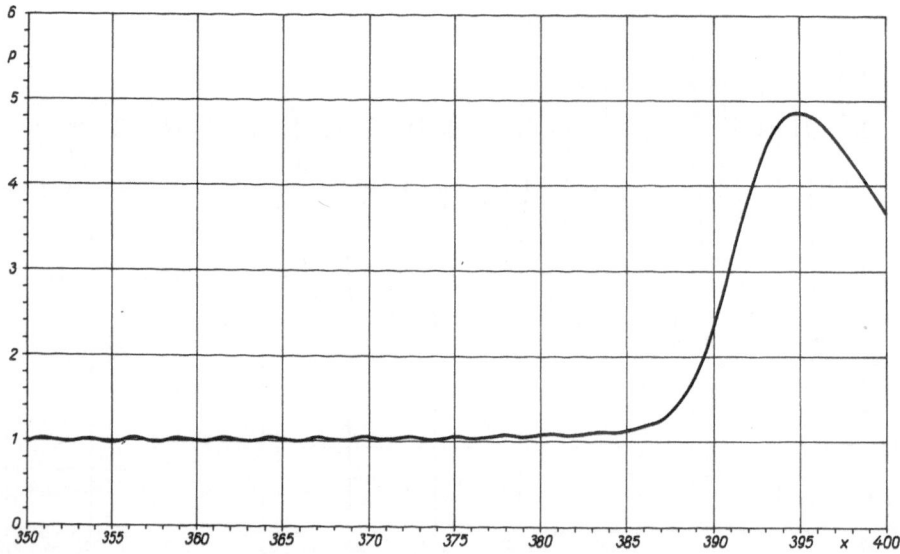

Fig. 10.1.8

successive resonances of this wave can be observed only as a superposition with
the A_1 wave resonances. At small n ($n \lesssim 30$), the latter have a small magnitude,
but possess a rather high Q factor. One can see the influence of the A_1 wave
resonances up to $x \lesssim 350$ (at $n \lesssim 75$). This influence decreases when x and
n increase. At $x \gtrsim 385$, one can see on the form function curve a high and broad

peak. It is caused by the resonances of the S_1 wave. The cutoff frequency of this wave is $x = 423.75$. With n increasing, the first fifty resonance frequencies of this wave move on the x axis from the right to the left, and the successive ones, as usual, from the left to the right. On the form function curve, the influence of the resonances of this wave is noticeable up to the cutoff frequency of the S_2 wave ($x = 436.37$).

It is clear that an explanation for the form function curve can be given only after the computation of the isolated modal resonances. The width (on the x axis) q_n of the partial mode resonance curve at its half-magnitude level will be used for a qualitative description of the Q factor.

10.1.1   The A Wave

At $26 \lesssim n \lesssim 36$, with n increasing, the resonance magnitude quickly grows from $\zeta_{26} = 0.0065$ to $\zeta_{36} = 4.7035$ (see Fig. 10.2). At $n > 36$ the magnitude decreases slowly to $\zeta_{51} = 4.1104$. The Q factor changes in the same way. The procedure of the resonance scattering theory allows one to separate the resonances only for $n \lesssim 50$.

10.1.2   The S_0 Wave

Generally speaking, with n increasing, the resonance magnitude grows slowly changing from $\zeta_1 = 0.9139$ to $\zeta_{100} = 1.3312$ (see Fig. 10.3). At $1 \lesssim n \lesssim 15$, with n increasing, the Q factor grows quickly; after this, at $15 \lesssim n \lesssim 75$, it decreases slowly. For big n values the Q factor, for example, may be characterized by $q_{100} = 21.6$.

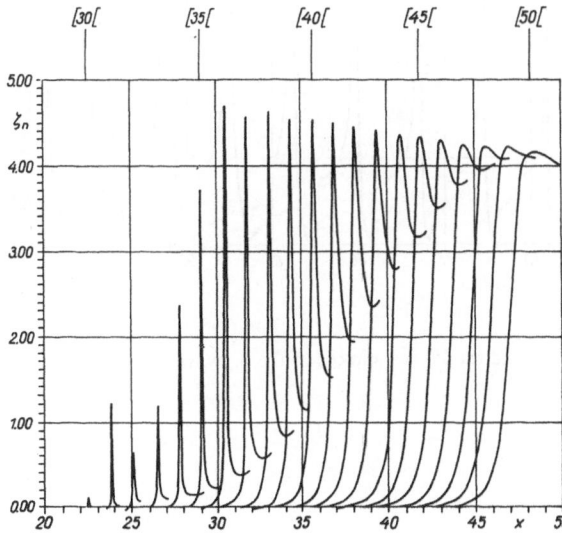

Fig. 10.2. The isolated modal resonances of the A wave

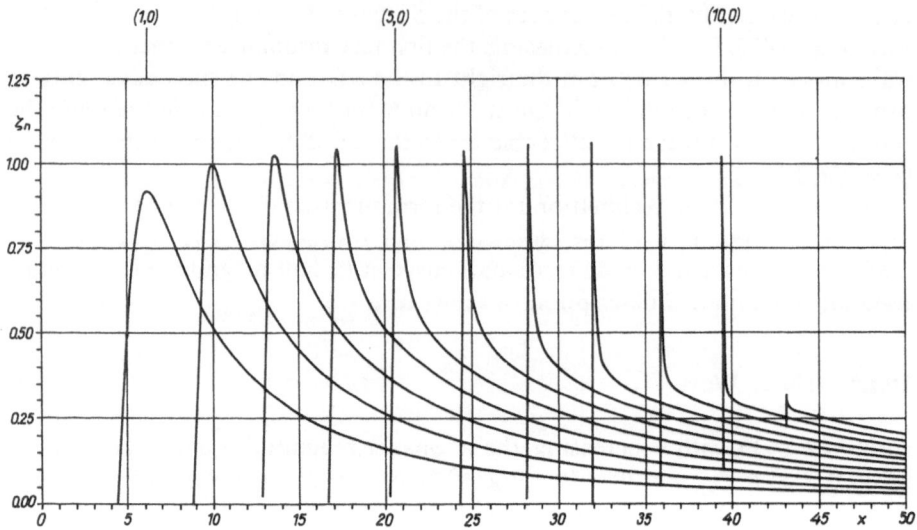

Fig. 10.3.1

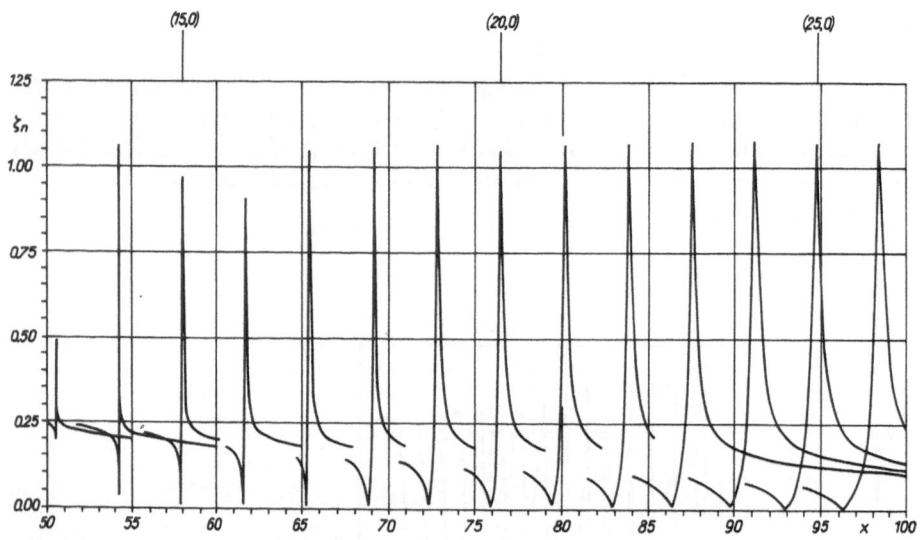

Fig. 10.3.2

Figs. 10.3.1–10.3.6 (p. 192–194). The isolated modal resonances of the S_0 wave

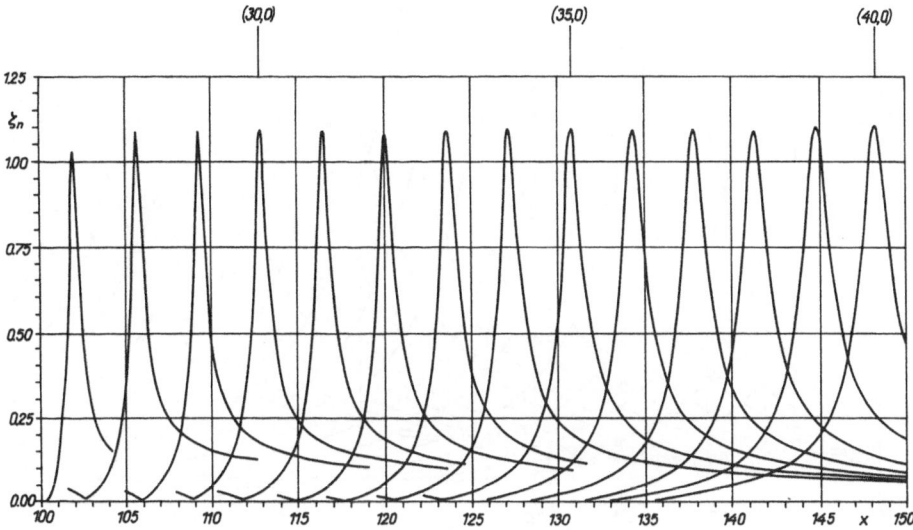

Fig. 10.3.3

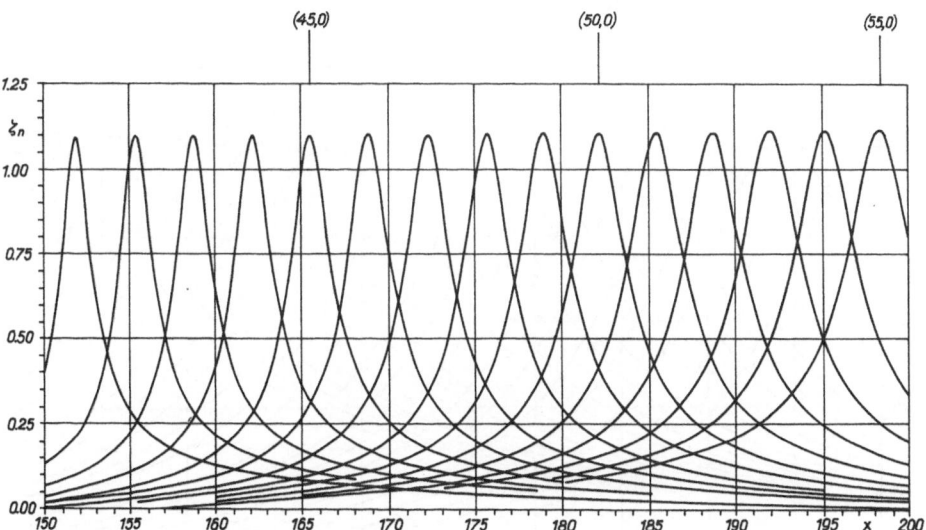

Fig. 10.3.4

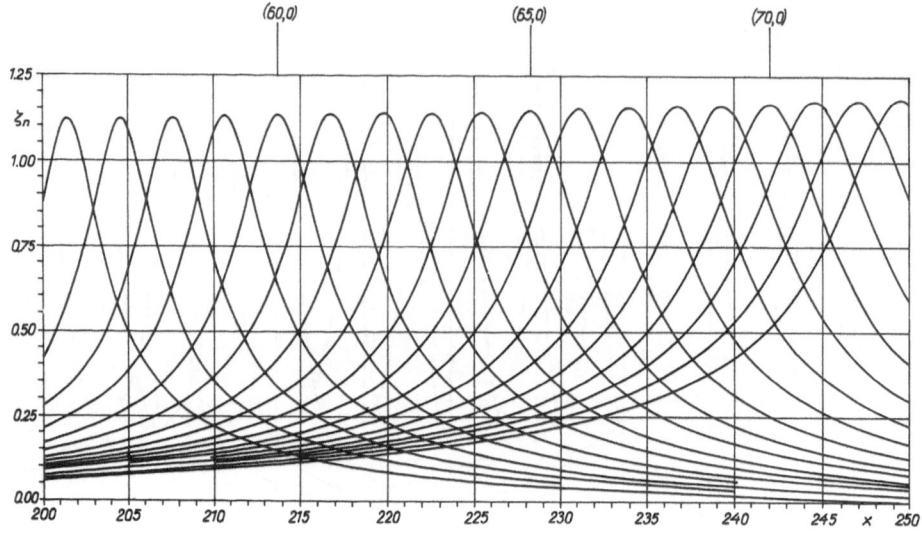

Fig. 10.3.5

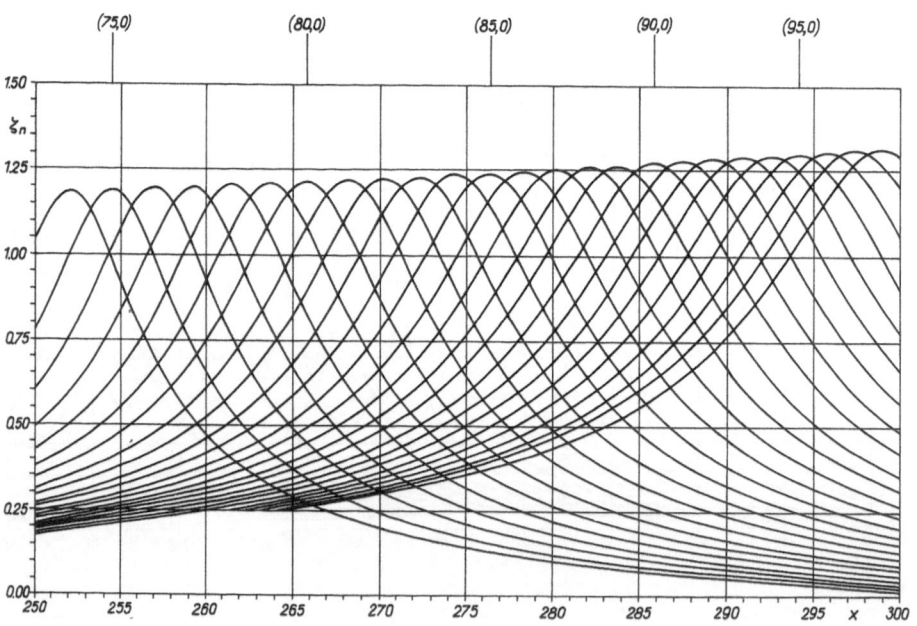

Fig. 10.3.6

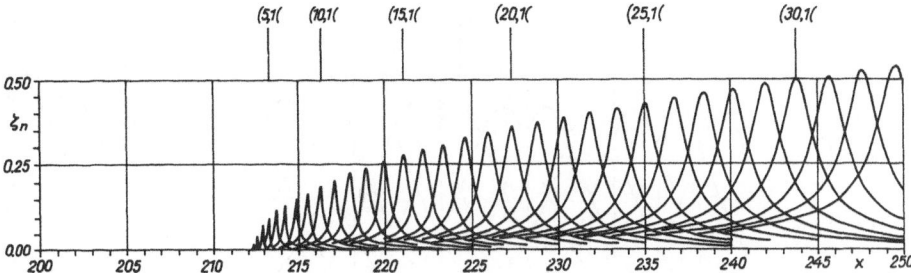

Fig. 10.4.1

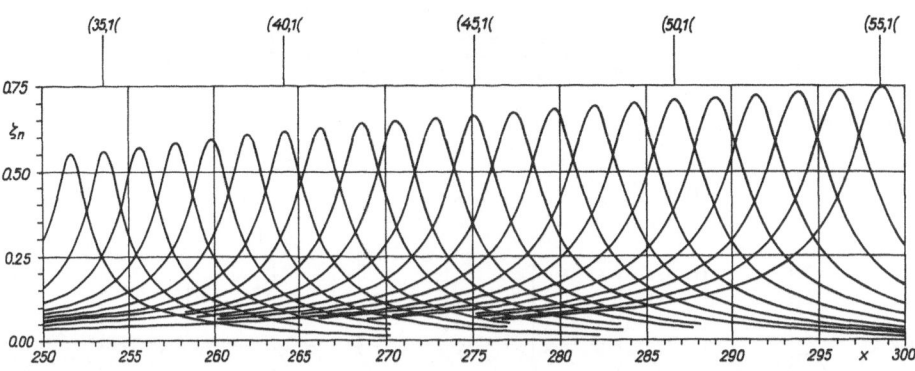

Fig. 10.4.2

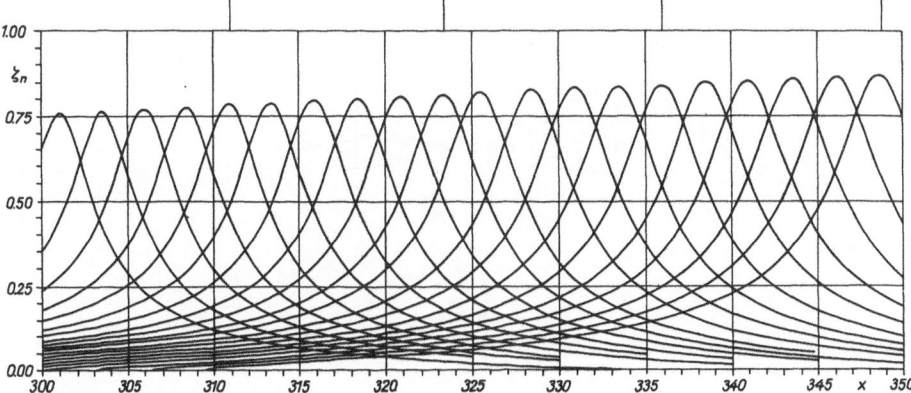

Fig. 10.4.3

Figs. 10.4.1–10.4.5 (p. 195–196). The isolated modal resonances of the A_1 wave

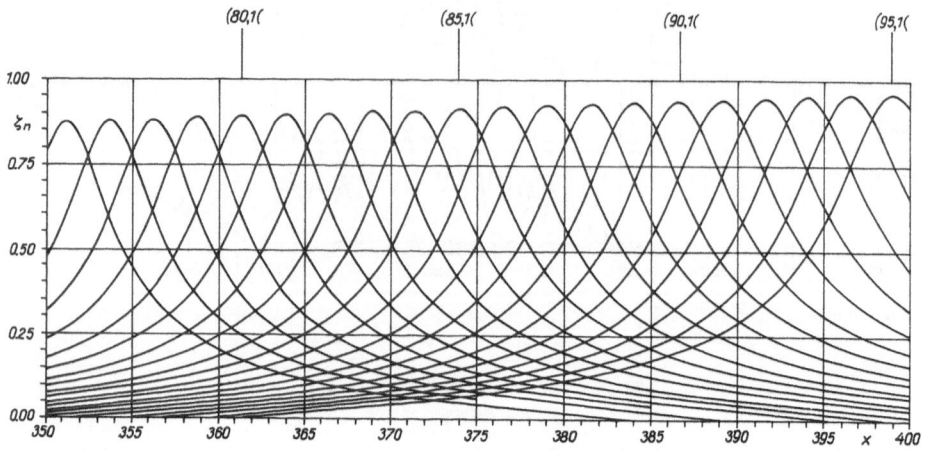

Fig. 10.4.4

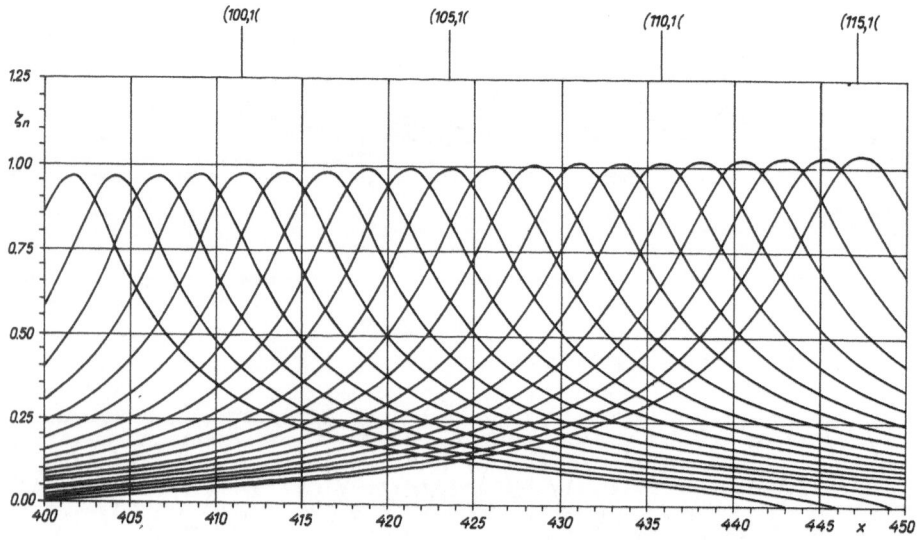

Fig. 10.4.5

10.1.3 The A_1 Wave

As one can see from Fig. 10.4, with n increasing, the resonance magnitude grows slowly from $\zeta_1 = 0.0042$ to $\zeta_{100} = 0.9770$. At $1 \lesssim n \lesssim 20$, with n increasing, the Q factor almost does not change. With n further increasing, it slowly decreases: $q_{60} = 6.4$, $q_{80} = 9.2$, $q_{100} = 11.6$.

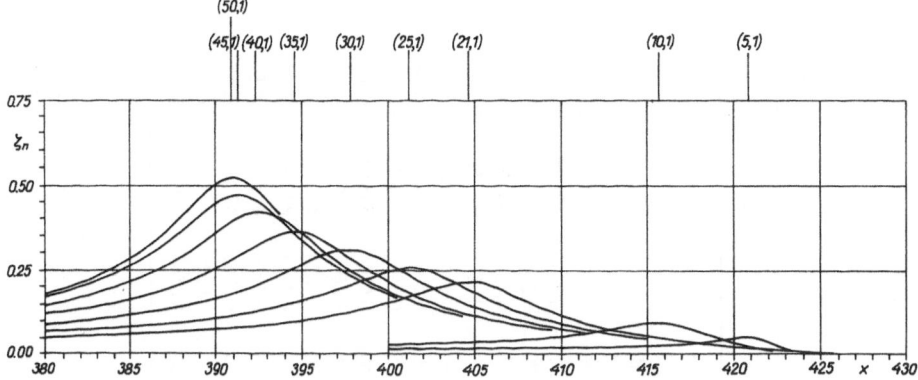

Fig 10.5.1

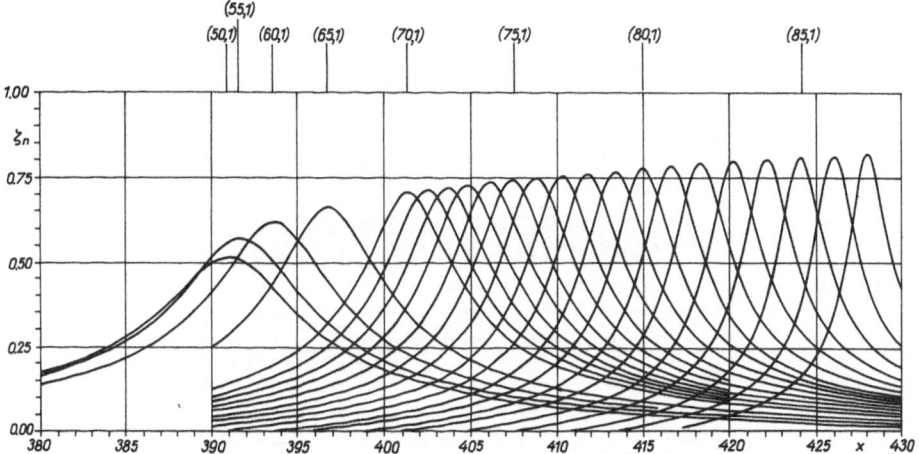

Fig. 10.5.2

Figs. 10.5.1 and 10.5.2. The isolated modal resonances of the S_1 wave

10.1.4 The S_1 Wave

As one can see from Fig. 10.5, with n increasing, the resonance magnitude grows slowly from $\zeta_1 = 0.0142$ to $\zeta_{98} = 0.8682$. The Q factor is small at $n \lesssim 12$. With n increasing, it grows: $q_{40} = 14$, $q_{60} = 11$, $q_{80} = 5.4$.

Table 10.1. The exact (x_n) and approximate (x_n^*) positions of modal resonances, and their magnitudes ζ_n computed for the case of an aluminum shell (with $h = 1/32$) immersed in water

A				S_0				A_1				S_1			
n	ζ_n	x_n	x_n^*	n	ζ_n	x_n	x_n^*	n	ζ_n	x_n	x_n^*	n	ζ_n	x_n	x_n^*
31	1.2321	23.87	23.36	1	0.9139	6.25	5.61	1	0.0042	212.11	212.09	1	0.0142	423.75	423.69
32	0.6392	25.20	24.69	2	0.9947	10.00	9.36	2	0.0204	212.27	212.25	5	0.0523	420.70	420.68
33	1.1969	26.48	26.03	3	1.0285	13.59	13.11	3	0.0486	212.50	212.48	10	0.1011	415.51	415.49
34	2.3386	27.81	27.38	4	1.0441	17.23	16.84	4	0.0780	212.81	212.79	15	0.1511	410.31	410.31
35	3.7359	29.14	28.74	5	1.0524	20.50	20.59	5	0.1022	213.20	213.17	20	0.2022	405.51	405.53
36	4.7035	30.47	30.11	6	1.0474	24.61	24.33	6	0.1215	213.67	213.63	25	0.2542	401.25	401.29
37	4.5803	31.80	31.48	7	1.0555	28.28	28.07	7	0.1369	214.22	214.17	30	0.3068	397.66	397.68
38	4.6195	33.09	32.85	8	1.0606	31.99	31.81	8	0.1574	214.80	214.78	35	0.3597	394.73	394.78
39	4.5477	34.38	34.21	9	1.0610	35.70	35.54	9	0.1750	215.51	215.46	40	0.4126	392.62	392.65
40	4.5369	35.66	35.57	10	1.0177	39.41	39.27	10	0.1942	216.25	216.21	45	0.4650	391.33	391.37
41	4.4816	36.95	36.91	20	1.0454	76.41	76.38	20	0.3606	227.34	227.30	50	0.5166	391.02	391.04
42	4.4499	38.20	38.25	30	1.0800	112.89	112.89	30	0.5004	243.79	243.72	55	0.5667	391.76	391.76
43	4.4102	39.45	39.57	40	1.0921	148.32	148.38	40	0.6137	263.98	263.87	60	0.6148	393.63	393.63
44	4.3731	40.70	40.88	50	1.1086	182.19	182.29	50	0.7047	286.64	286.49	65	0.6602	396.80	396.78
45	4.3378	41.95	42.17	60	1.1323	213.71	213.85	60	0.7785	310.82	310.68	70	0.7026	401.37	401.32
46	4.3054	43.16	43.44	70	1.1655	241.91	242.11	70	0.8395	335.90	335.76	75	0.7412	407.42	407.33
47	4.2797	44.41	44.69	80	1.2103	266.02	266.21	80	0.8913	361.29	361.23	80	0.7759	415.00	414.86
48	4.2437	47.50		90	1.2666	285.74	285.96	90	0.9364	386.56	386.64	85	0.8065	424.06	423.87
49	4.2099	46.99	47.14	100	1.3312	301.91	302.12	100	0.9770	411.41	411.64	90	0.8331	434.53	434.29
50	4.1687	48.32	48.34									95	0.8560	446.25	445.95

The resonance frequencies x_n and magnitudes ζ_n of some isolated resonances are given in Table 10.1. We did not round off the resonance positions here and give them with two significant digits after the decimal point for the purpose of actual comparison, although the computation step size is only $l_x = 10/256$.

10.2 Dispersion Curves

It is easy to find the phase and group velocities of a peripheral wave when the position of the resonance is known. The resonance frequency of a peripheral (propagating) wave coincides with the resonance frequency of a partial mode (standing wave). The resonance occurs when exactly $n + \frac{1}{2}$ wavelengths fit the meridian circle length

$$2\pi a = (n + \tfrac{1}{2})\lambda_{nl} \ . \tag{10.1}$$

Here n defines the ordinal number of the resonance and l indicates its family. From condition (10.1) the phase velocity (c^{ph}) and the group velocity (c^{gr}) may

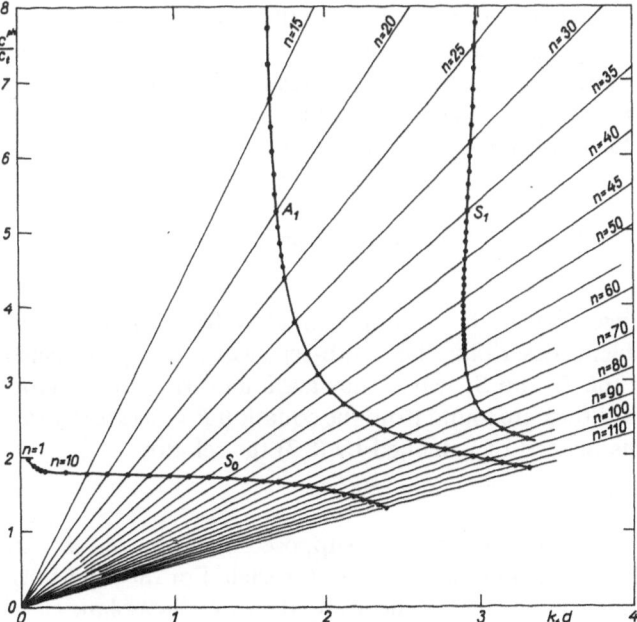

Fig. 10.6. The dispersion curves of the relative phase velocities of the Lamb-type peripheral waves S_0, A_1 and S_1 revolving around the aluminum shell (with $h = 1/32$) immersed in water

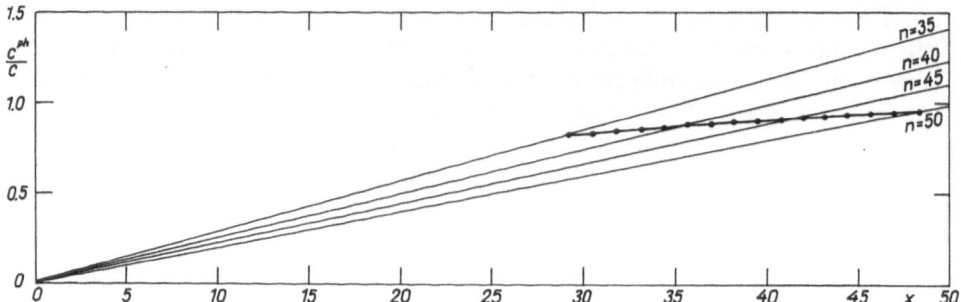

Fig. 10.7. The dispersion curve of the relative phase velocity of the fluid-borne bending A wave generated in an aluminum shell (with $h = 1/32$) immersed in water

be found

$$c^{ph}(x_{nl}) = c\frac{x_{nl}}{(n+\frac{1}{2})}, \qquad c^{gr}(x_{nl}) = c[x_{(n+1)l} - x_{nl}] . \tag{10.2}$$

The dispersion curves of the peripheral waves generated in shells with different relative thickness h can be compared by calculating the dependence of y on z. Here y and z are nondimensional quantities defined by (8.3). In Fig. 10.6, the dispersion curves of the relative phase velocities of the Lamb-type peripheral waves S_0, S_1 and A_1 are presented. In the z–y plane the resonances of the same order n, but for different dispersion curves l, are situated on the rays which pass through the origin. The angle between the ray corresponding to the nth resonance and the z axis is $\arctan 2 \{[(n + \frac{1}{2})h]^{-1}\}$. In Fig. 10.7, the dispersion curve of the relative phase velocity of the A wave is shown.

10.3 Approximation of Resonance Positions

The position of resonances can be found approximately from the dispersion equations of model problems describing the Lamb-type waves in a plane elastic layer (see Chap. 8). For the S_0 and S_1 waves they are found from the equation $E = 0$, and for the A_1 wave they are found from the equation $F = 0$ [see (8.2) and (8.3)]. In these equations instead of y defined by (8.3) the relation

$$y^* = 2(1 - h/2)z[(n + \frac{1}{2})h]^{-1} \tag{10.3}$$

should be used. In the model problem it is supposed that the S_0, S_1 and A_1 waves are propagating on the middle surface of the shell. For the fluid-borne A wave, we should consider the problem concerning the waves in a plane elastic layer, one side of which is in contact with the liquid and the other is free, as the model problem. The resonance frequencies are found from the equation [10.5, 6]

$$E(F + \psi) + F(E + \psi) = 0 , \tag{10.4}$$

where

$$\psi = (\rho/\rho_1)[(1 - \gamma_0^2 y^2)(1 - \gamma_1^2 y^2)^{-1}]^{1/2}, \qquad \gamma_0 = c_t/c_1, \qquad \gamma_1 = c_t/c \quad (10.5)$$

in which instead of $y = c^{ph}/c_t$ the relation

$$y^* = 2z[(n + \tfrac{1}{2})h]^{-1} \tag{10.6}$$

should be used. In the model problem it is supposed that the A wave is propagating on the outer surface of the shell. In contrast with the model problems concerning waves in a plane layer, here the dispersion equations must be solved for every value of n ($n = 1, 2, 3, \dots$). The resonance frequencies found from the dispersion equations for model problems are determined as x_{ni}^* and are listed in Table 10.1. There they are designated by x_n^* because the value of l is evident from the table caption.

The comparison of the results shows that, at not very small values of n, the coincidence is rather good. Qualitatively it can be described as: a long wave (with $\lambda \sim 4a$) strongly 'feels" the curvature of the scatterer; a wave with wavelength commensurable with the typical size of the scatterer (with $\lambda \sim a$) is not very "sensitive" to the curvature; a short wave (with $\lambda \sim \tfrac{1}{2}a$) is almost insensitive to the curvature. At small values of n the accuracy of determinating the resonance frequencies can be improved by using the procedure outlined in [10.7, 8]. A corresponding asymptotic formula is given by (8.10) and was verified by comparison with the exact solution for the case of scattering by a circular–cylindrical shell.

It should be mentioned that the approximation outlined above for the resonance frequencies is also valid for the case of scattering by a circular–cylindrical shell (at normal incidence), but n should be inserted in (10.3) and (10.6) instead of $n + \tfrac{1}{2}$. For the spherical case the summand $\tfrac{1}{2}$ in the $n + \tfrac{1}{2}$ term is caused by focusing.

It is astonishing that the approximation of the resonance frequency position given above can be used even in the case of a very thick shell. Earlier, we have computed the position of resonances for the case of an empty aluminum shell with $h = 1/10$ [10.2]. The corresponding data are listed in Tables 10.2 and 10.3 and the dispersion curves of the relative phase velocities are shown in Fig. 10.8. Here, as for the case of scattering by a tube (see Fig. 8.2), Lamb-type waves of higher order are generated in the shell by the incident plane wave. For small n values the principal error in the appoximation of the resonance position is caused by the curvature of the shell; for large n values the error is mainly caused by the difference in path. In Table 10.4 we give the number n_* which denotes the largest value of n for the case in which the approximate position of the resonance on the x axis does not differ from the exact one by more than 0.20.

One more observation should be mentioned here. It is clear from the comparison of Figs. 10.8 and 8.2 that the dispersion curves of the Lamb-type waves practically coincide for waves generated in spherical and cylindrical

Table 10.2. The positions of the isolated modal resonances of the Lamb-type waves generated in a spherical aluminum shell (with $h = 1/10$) immersed in water

n	S_0	S_1	S_2	S_3	S_4	A_0	A_1	A_2	A_3	A_4
1	7.50	131.76	137.34		397.50		66.68	198.83	272.77	331.33
2	10.55	130.70	138.67	265.08	397.46		67.19	198.83	272.93	331.37
3	14.10	129.57	140.20	265.16	397.38		68.01	198.87	273.16	331.45
4	17.77	128.52	141.80	265.23	397.31		69.02	198.87	273.44	331.52
5	21.48	127.46	143.48	265.31	397.19		70.27	198.91	273.79	331.64
6	25.23	126.52	145.23	265.43	397.07		71.72	198.98	274.22	331.80
7	28.98	125.63	147.07	265.59	396.95		73.36	199.02	274.73	331.95
8	32.70	124.84	148.95	265.74	396.84		75.12	199.10	275.27	332.15
9	36.37	124.16	150.86	265.94	396.72		77.07	199.18	275.90	332.38
10	40.04	123.52	152.81	266.13	396.64		79.10	199.30	276.60	332.66
11	43.67	123.05	154.80	266.37	396.52		81.25	199.41	277.34	332.93
12	47.23	122.66	156.88	266.60	396.45		83.52	199.57	278.16	333.20
13	50.74	122.38	158.95	266.88	396.37		85.86	199.73	279.02	333.55
14	54.22	122.27	161.09	267.15	396.29		88.24	199.92	279.96	333.91
15	57.62	122.27	163.24	267.90	396.25	20.00	90.70	200.16	280.94	334.30
16	60.90	122.46	165.43	267.81	396.21	22.23	93.20	200.39	281.95	334.73
17	64.14	122.77	167.70	268.20	396.21	24.41	95.74	200.66	283.05	335.16
18	67.23	123.28	169.96	268.59	396.21	26.60	98.28	200.98	284.14	335.66
19	70.23	123.95	172.23	269.02	396.25	28.79	100.86	201.33	285.31	336.17
20	73.13	124.84	174.57	269.49	396.29	31.02	103.44	201.72	286.52	336.72
21	75.86	125.94	176.91	269.96	396.33	33.24	106.02	202.15	287.77	337.30
22	78.48	127.27	179.30	270.47	396.45	35.47	108.59	202.62	289.02	337.93
23	80.94	128.79	181.72	271.02	396.52	37.70	111.13	203.13	290.35	338.59
24	83.28	130.55	184.14	271.60	396.64	39.92	113.67	203.71	291.68	339.30
25	85.47	132.54	186.60	272.19	396.80	42.15	116.17	204.34	293.05	340.00
26	87.50	134.69	189.10	272.85	396.95	44.41	118.67	205.00	294.41	340.78
27	89.41	137.07	191.56	273.52	397.15	46.64	121.13	205.74	295.82	341.60
28	91.17	139.57	194.10	274.22	397.38	48.91	123.63	206.56	297.27	342.50
29	92.85	142.23	196.64	275.00	397.58	51.13	126.09	207.42	298.67	343.40
30	94.45	145.00	199.18	275.78	397.85	53.40	128.52	208.36	300.16	344.38
31	95.98	147.85	201.76	276.60	398.13	55.63	130.97	209.34	301.60	345.35
32	97.42	150.82	204.34	277.50	398.40	57.89	133.40	210.43	303.09	346.45
33	98.33	153.83	206.91	278.44	398.71	60.16	135.82	211.64	304.57	347.54
34	100.20	156.88	209.45	279.38	399.06	62.38	138.20	212.93	306.05	348.71
35	101.52	159.96	211.91	280.39	399.41	64.61	140.59	214.38	307.54	349.92
36	102.85	163.05		281.45	399.77	66.88	142.93		309.06	351.21
37	104.14	166.17		282.58	400.20	69.10	145.23		310.55	352.54
38	105.43		220.86	283.75	400.59	71.33	147.50	217.81	312.03	353.91
39	106.76	172.42	223.52	284.96	401.05	73.55	149.77	219.49	313.52	355.39
40	108.05	175.51	226.21	286.21	401.48	75.78	151.95	221.17	315.00	356.88
41	109.38	178.59	228.94	287.54	401.99	78.01	154.14	222.93	316.48	358.44
42	110.70	181.64	231.76	288.95	402.50	80.23	156.25	224.77	317.97	360.08
43	112.07	184.69	234.57	290.39	403.01	82.46	158.32	226.64	319.41	361.76
44	113.44	187.66	237.42	291.88	403.56	84.65	160.35	228.59	320.90	363.48
45	114.80	190.63	240.27	293.48	404.14	86.88	162.34	230.59	322.34	365.27
46	116.21	193.52	243.16	295.08	404.73	89.06	164.30	232.62	323.83	367.11
47	117.66	196.41	246.09	296.80	405.35	91.25	166.21	234.73	325.27	369.02
48	119.10	199.22	249.06	298.55	405.98	93.48	168.09	236.84	326.76	370.98
49	120.55	201.99	252.03	300.35	406.64	95.66	169.02	239.02	328.20	372.97
50	122.03	204.73	255.04	302.57	407.34	97.85	171.72	241.21	329.69	375.04
51	123.55	207.38	258.09	304.22	408.05	100.04	173.52	243.48	331.13	377.12

Table 10.2. Continued

n	S_0	S_1	S_2	S_3	S_4	A_0	A_1	A_2	A_3	A_4
52	125.08	210.04	261.17	306.25	408.79	102.19	175.23	245.70	332.62	379.26
53	126.64	212.58	264.26	308.32	409.57	104.38	176.95	248.01	334.10	381.48
54	128.24	215.12	267.34	310.51	410.35	106.52	178.63	250.27	335.59	383.71
55	129.80	217.58	270.47	312.73	411.13	108.71	180.31	252.58	337.11	385.98
56	131.45	220.00	273.59	315.00	411.99	110.86	181.95	254.92	338.63	388.28
57	133.09	222.38	276.76	317.38	412.85	113.01	183.59	257.23	340.16	390.63
58	134.73	224.69	279.92	319.81	413.75	115.16	185.20	259.53	341.72	393.05
59	136.41	226.95	283.09	322.31	414.65	117.31	186.80	261.88	343.32	395.43
60	138.13	229.18	286.21	324.88	415.63	119.42	188.40	264.18	344.92	397.89
61	139.84	231.37	289.38	327.50	416.60	121.56	189.96	266.52	346.52	400.39
62	141.56	233.52	292.50	330.20	417.58	123.67	191.52	268.83	348.20	402.89
63	143.32	235.59	295.63	332.97	418.63	125.82	193.09	271.13	349.84	405.39
64	145.08	237.66	298.71	335.78	419.69	127.93	194.65	273.44	351.56	407.97
65	146.88	239.65	301.76	338.67	420.78	130.04	196.21	275.74	353.28	410.55
66	148.67	241.64	304.81	341.64	421.91	132.15	197.77	278.01	355.08	413.13
67	150.51	243.55	307.81	344.65	423.09	134.22	199.34	280.27	356.84	415.74
68	152.34	245.47	310.78	347.73	424.34	136.33	200.90	282.54	358.67	418.36
69	154.18	247.34	313.71	350.86	425.59	138.40	202.46	284.81	360.55	420.94
70	156.05	249.18	316.64	354.06	426.95	140.51	204.02	287.03	362.42	423.55
71	157.93	251.02	319.49	357.30		142.58	205.59	289.26	364.34	
72	159.84	252.77	322.31	360.59		144.65	207.15	291.45	366.33	
73	161.76	254.57	325.04	363.91	430.18	146.72	208.71	293.63	368.62	432.46
74	163.67	256.29	327.77	367.23	431.68	148.79	210.31	295.78		435.00
75	165.59	258.01	340.47	370.47	433.20	150.82	211.88	297.93		437.62
76	167.54	259.73	333.13		434.73	152.89	213.48	300.04		440.31
77	169.49	261.41	335.70		436.29	154.92	215.08	302.11		443.01
78	171.48	263.09	338.28	381.76	437.85	156.99	216.68	304.18	378.13	
79	173.44	264.73	340.82	385.27	439.45	159.02	218.28	306.25	380.31	
80	175.43	266.37	343.28	388.87	441.13	161.05	219.92	308.24	382.46	
81	177.42	268.01	345.74	392.50		163.09	221.52	310.27	384.65	
82	179.45	269.65	348.16	396.13		165.12	223.16	312.23	386.80	
83	181.48	271.25	350.55	399.80		167.15	224.80	314.18	388.98	
84	183.48	272.85	352.93	403.48		169.14	226.45	316.13	391.17	
85	185.55	274.45	355.23	407.15		171.17	228.13	318.05	393.36	
86	187.58	276.06	357.54	410.82		173.20	229.77	319.02	395.59	
87	189.61	277.66	359.81	414.49		175.20	231.45	321.76	397.77	
88	191.68	279.26	362.07	418.13		177.19	233.25	323.63	399.96	
89	193.75	280.82	364.26	421.80		179.22	234.81	325.43	402.19	
90	195.82	282.92	366.45	425.39		181.21	236.52	327.27	404.38	
91	197.89	283.98	368.63	428.98		183.20	238.24	329.02	406.60	
92	199.96	285.59	370.78	432.50		185.20	239.92	330.82	408.79	
93	202.07	287.15	372.89	436.02		187.19	241.67	332.54	410.98	
94	204.14	288.75	374.96	439.45		189.18	243.40	334.30	413.16	
95	206.25	290.31	377.03	442.85		191.18	245.12	336.02	415.35	
96	208.36	291.91	379.10			193.16	246.88	337.70	417.50	
97	210.47	293.52	381.13			195.16	248.63	339.41	419.69	
98	212.58	295.08	383.13			197.15	250.39	341.09	421.84	
99	214.69	296.68	385.09			199.14	252.15	342.77	423.98	
100	216.84	298.28	387.07			201.13	253.95	344.41	426.13	
101	218.95	299.84	388.98			203.13	255.70	346.05	428.24	
102	221.06	301.45	390.90			205.08	257.50	347.70	430.39	

Table 10.3. The positions x_n and magnitudes ζ_n of the isolated modal resonances of the fluid-borne A wave generated in a spherical aluminum shell (with $h = 1/10$) immersed in water

n	x_n	ζ_n	n	x_n	ζ_n
2	1.76	5.60	7	5.16	5.51
3	2.27	6.16	8	6.25	5.32
4	2.77	6.33	9	7.46	5.09
5	3.40	6.44	10	8.71	4.80
6	4.22	5.60	11	10.08	4.55

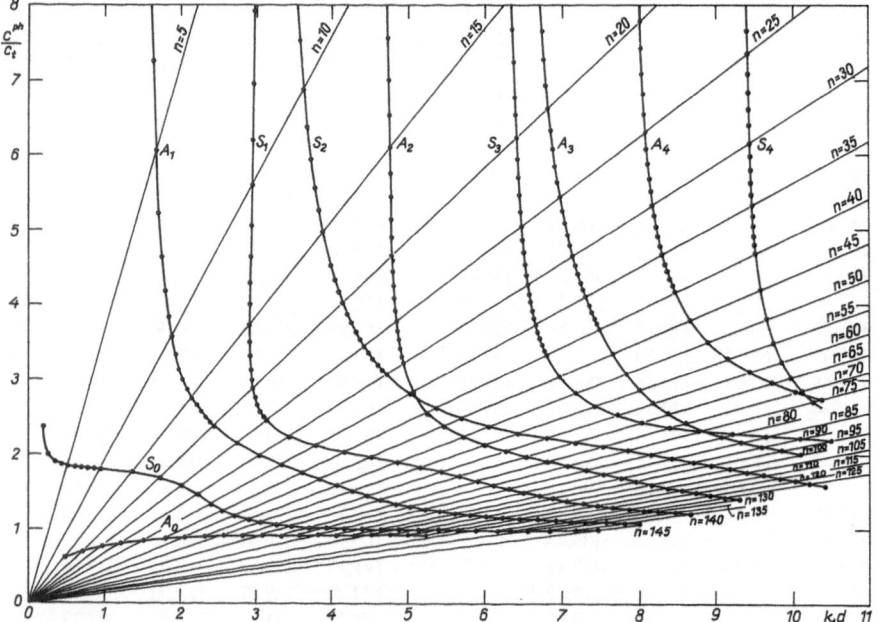

Fig. 10.8. The dispersion curves of the relative phase velocities of the Lamb-type peripheral waves revolving around an aluminum shell (with $h = 1/10$) immersed in water

Table 10.4. The number n_* of the properly approximated resonances of the Lamb-type waves for the case of the shell with $h = 1/10$

Wave	S_0	S_1	S_2	S_3	S_4	A_0	A_1	A_2	A_3	A_4
n_*	40	20	50	40	55	40	10	25	50	50

shells. This holds even for the most sensitive parts of the dispersion curves (in the vicinity of the cutoff frequencies and the points of intersection of dispersion curves, for example, of the waves S_l and A_l, $l = 2, 3, 4$).

10.4 Hidden Resonances

As the results of the computations have shown (see, for example, Fig. 10.1) not all of the resonances become apparent in the total form function. Particularly, this is related to the resonances with low Q factor which are overlapping in frequency. As a rule, the resonances neighboring in n are in anti-phase. The magnitude of the superposed resonances with low Q factor is smaller, and sometimes considerably smaller (by an order of magnitude) than every summand. The extrema of this superposition do not coincide with the resonance frequencies. In such a case the extrema of the form function curve are also shifted with respect to the resonance frequencies. Sometimes, this shift attains $\frac{1}{2}(x_{n+1} - x_n)$. In the example considered for a spherical aluminum shell with $h = 1/32$ immersed in water, the S_0 wave has such properties at $35 \lesssim n \lesssim 70$, the A_1 wave at $25 \lesssim n \lesssim 95$, and the S_1 wave at $1 \lesssim n \lesssim 80$.

Two examples of such kind of superposition of isolated modal resonances are shown in Fig. 10.9. In every case the contribution of all the resonances in the

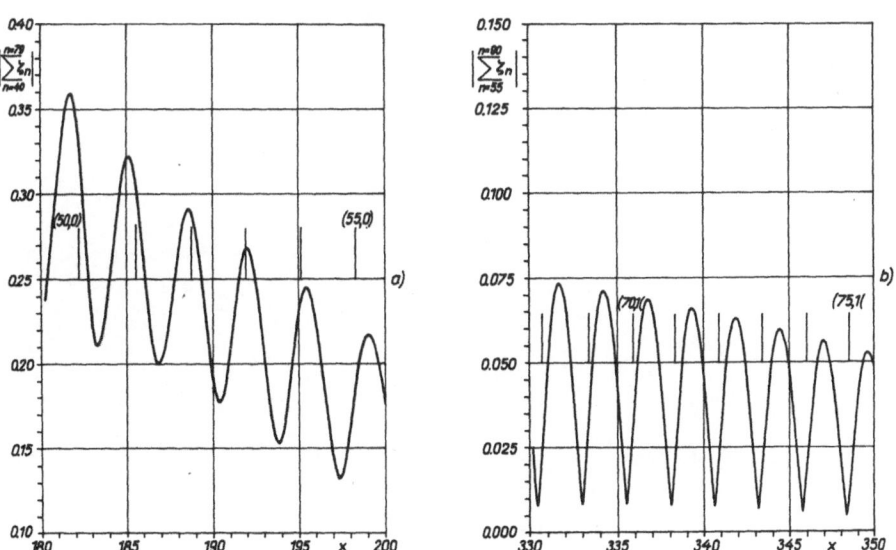

Fig. 10.9. The contribution of the isolated modal resonances in the restricted x domain **a** of the S_0 wave, **b** of the A_1 wave. The computation was carried out in case of an aluminum shell (with $h = 1/32$) immersed in water. The exact positions of the resonances are labeled by bars

restricted x range are taken into account. At $40 \lesssim n \lesssim 80$ ζ_n is of the order 1.10–1.20 for the S_0 wave, but the magnitude of the modulus of the superposition is of the order 0.35; at $55 \lesssim n \lesssim 90$ ζ_n is of the order 0.75–0.95 for the A_1 wave, but the magnitude of the modulus of the superposition is of the order 0.075 and decreases with x increasing. The exact positions of modal resonances are labeled by bars. As a rule, they coincide neither with the extrema positions of the superposition (see Fig. 10.9) nor with the extrema of the form function (see Fig. 10.1).

The anti-phase superposition of the successive (in n) isolated resonances, which have a low Q factor and which overlap on the x axis, is typical for both the scattering by cylindrical and spherical shells. For the latter case the result of such a superposition is easier to show, because the value of ζ_n is sufficiently large.

The results described above were partly reported at the Spring Conference of the Institute of Acoustics [10.9].

11   Peripheral Waves in the Scattering
of a Plane Acoustic Wave Obliquely Incident
on a Solid Elastic Cylinder

In contrast to the limiting problem of the scattering of a plane acoustic wave at normal incidence, the problem of an obliquely incident wave has not been investigated in detail. The formulation of the problem and its solution in the Rayleigh series form using the eigenfunctions is given in [11.1] (see Chap. 1). There the form function was computed at various angles of incidence (see Fig. 1.2). In [11.2], for the analysis of the peripheral waves that propagate along helical paths over the cylinder surface, the procedure of the resonance scattering theory has been used, and formulae are presented for the calculation of the critical entrance-reradiation angles, resonance frequencies, magnitudes and Q factors. There, a qualitative view of the dispersion curves of the phase velocities is presented. In a series of papers [11.3–5] the observation of supplementary (guided) waves that propagate along helical paths has been described and their behavior has been interpreted.

Here we again analyze the form function. As in [11.1], the form function computation is carried out by the direct summation of the Rayleigh series. For its interpretation, the resonance scattering theory is used, but, in contrast to [11.2], here we use a computational algorithm rather than the formal descriptive approach. Each term of the Rayleigh series for the scattered pressure is separated into two components: the resonances and the background. The modal resonance components are computed and the resonances of single kind are joined into families. The computation has been carried out for the case of an aluminum cylinder immersed in water, for the range $0 \leqslant x \equiv ka \leqslant 50$. Besides the creeping waves, nineteen types of peripheral waves propagate along helical paths over the cylinder surface and affect the form function. Only nine of them are generated by the plane wave at normal incidence. The influence of the angle of incidence on the form function and the isolated modal resonances is investigated. A correspondence is established between the extrema of the form function and the modal resonances of the peripheral waves.

11.1   Form Function

A solid circular cylinder of infinite extent is immersed in an unbounded acoustic medium. A plane acoustic pressure wave whose direction of propagation forms an angle α with the normal to the longitudinal axis of the cylinder is incident on it, and is scattered by it. The pressure in the incident wave p_i may be written in the form of (1.7). The (secondary) pressure p_s scattered by the cylinder may be presented in a form similar to that given in (1.23)

$$p_s = p_0 \sum_{n=0}^{\infty} \varepsilon_n(-i)^n R_n H_n^{(1)}(\delta r) \cos n\theta , \qquad (11.1)$$

where R_n is a coefficient which should be found from the boundary conditions on the surface of cylinder at $r = a$ (see (1.44)).

For each value of the summation index n, the coefficient R_n should be found from a system of four algebraic equations and may be written as

$$R_n = P_n/Q_n , \qquad (11.2)$$

where P_n, Q_n are the determinants of fourth rank, see (1.66) and (1.58–60). In the far field (at $\delta r \gg 1$), instead of $H_n^{(1)}(\delta r)$, its asymptotic representation (1.61) can be used. Then p_s becomes

$$p_s \cong p_* \left(\frac{a}{2r}\right)^{1/2} \exp[i(\delta r + \gamma z' - \omega t)] \sum_{n=0}^{\infty} 2(i\pi\delta a)^{-1/2} \varepsilon_n R_n \cos n\theta . \qquad (11.3)$$

If the partial scattering function

$$S_n = 2R_n + 1 \qquad (11.4)$$

and the partial form function

$$f_n(\theta) = (i\pi\delta a)^{-1/2} \varepsilon_n (S_n - 1) \cos n\theta \qquad (11.5)$$

are introduced, then the scattered pressure takes the form

$$p_s \cong p_* \left(\frac{a}{2r}\right)^{1/2} \exp[i(\delta r + \gamma z' - \omega t)] f(\theta) , \qquad (11.6)$$

where

$$f(\theta) = \sum_{n=0}^{\infty} f_n(\theta) . \qquad (11.7)$$

In the limiting case when $\rho/\rho_1 \to 0$, one can obtain the partial form function corresponding to an acoustically rigid cylinder

$$p_s^{(r)} = - [J_n'(\delta a)]/[H_n^{(1)'}(\delta a)] , \qquad (11.8)$$

where the prime denotes the derivative with respect to the argument.

In the far field the difference between the acoustic pressure scattered by an elastic cylinder p_s and an acoustically rigid cylinder $p_s^{(r)}$ may be written in the form

$$p_s - p_s^{(r)} = p_* \left(\frac{a}{2r}\right)^{1/2} \exp[i(\delta r + \gamma z' - \omega t)] \psi^{(r)}(\theta) , \qquad (11.9)$$

where

$$\psi^{(\mathrm{r})}(\theta) = \sum_{n=0}^{\infty} \psi_n^{(\mathrm{r})}(\theta) , \tag{11.10}$$

$$\psi_n^{(\mathrm{r})}(\theta) = 2(i\pi\delta a)^{-1/2} \varepsilon_n \left(R_n + \frac{J_n'(\delta a)}{H_n^{(1)'}(\delta a)} \right) \cos n\theta . \tag{11.11}$$

At a fixed observation point, for the sake of brevity, we introduce the following notations

$$p = |f|, \qquad \zeta_n = |\psi_n^{(\mathrm{r})}|, \qquad x = ka . \tag{11.12}$$

We shall call $p(x)$ the form function and $\zeta_n(x)$ the modal resonance component.

The computation of $p(x)$ and $\zeta_n(x)$ was carried out in the far field on the ray $\alpha = 0$ in the range $0 \leqslant x \leqslant 50$. The computation step size was $l_x = 10/256$. The calculation was performed for the case of an aluminum cylinder immersed in water (see (7.2)).

The angle of incidence is one of the principal parameters of the problem considered. In order to investigate the influence of this angle on the form function and the modal resonances, a series of computations has been carried out for $\alpha = 0°$, $1°$, $2.5°$, $5°$, $7.5°$ and $10°$. As typical examples, form functions at $\alpha = 0°$ and $\alpha = 10°$ are shown in Figs. 7.1 and 11.1, respectively. Curve $p(x)$ is

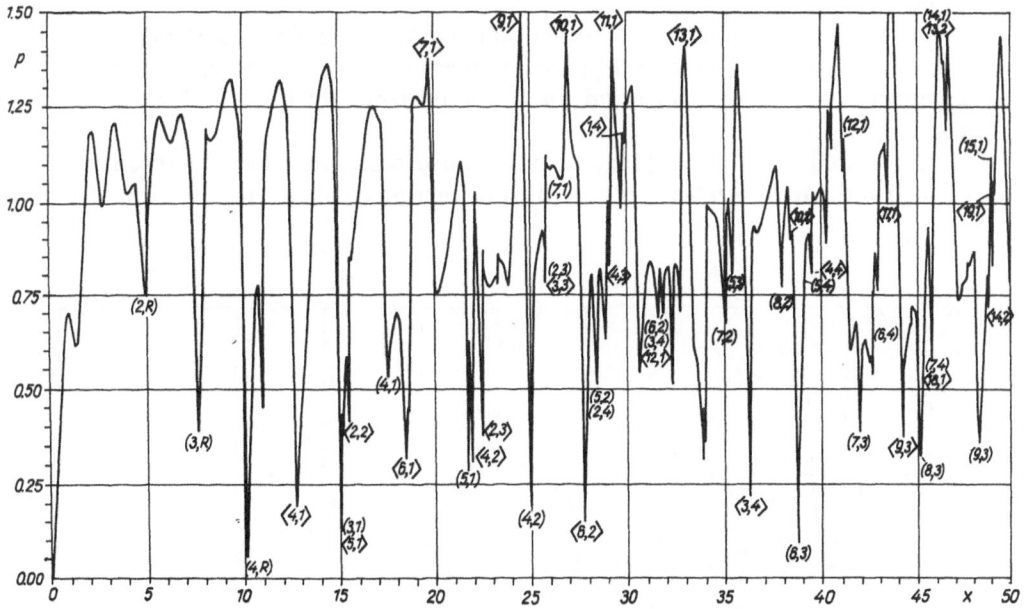

Fig. 11.1. The form function of the acoustic pressure scattered by a circular cylinder at oblique incidence, at $\alpha = 10°$. The main modal resonances are labeled by (n, l) and the supplementary ones by $\langle n, l \rangle$ [11.10, Fig. 2]

strongly discontinuous and irregular. Except in the narrow range $0 \leqslant x \lesssim 5$, it is difficult to connect the extrema of the form function with a resonance component of some partial mode and, even more, to trace the shift of some resonance position on the x axis as the α angle varies.

11.2 Modal Resonances

At a fixed angle of incidence modal resonances are calculated in order to interpret the behavior of the form function. Firstly, at chosen n, all the resonances present in the x domain considered are isolated. Then, the resonances of single type are joined into families. The resonance is denoted by two symbols n, l. The first of them indicates the ordinal number of the resonance, and the second defines its family. Two kinds of families are observed. The resonances of the first kind, which also occur at $\alpha = 0$, are labeled by (n, l), and the resonances of the second kind, which are generated only at $\alpha \neq 0$, are denoted by $\langle n, l \rangle$. Here, according to the notations used in [11.6], the family of the Rayleigh wave resonances is distinguished by a letter symbol $l = R$, and the families of the Whispering Gallery wave resonances by indices $l = 1, 2, 3, \ldots$. In [11.3] only one family of $\langle n, l \rangle$ resonances was observed, namely the first one, $\langle n, 1 \rangle$. There this family has been named "supplementary", and its resonances are denoted by $(n, m = 2)$.

As typical examples, the curves of isolated modal resonances of the Rayleigh wave (n, R), two transverse Whispering Gallery waves $(n, 1)$ and $(n, 4)$, and two complementary waves $\langle n, 1 \rangle$ and $\langle n, 4 \rangle$ are presented in Figs. 11.2–6. All the curves were computed at $\alpha = 10°$.

As is well known, for plane acoustic wave scattering the first resonance $(n = 1)$ of the Rayleigh wave is generated neither on a sphere, nor on a cylinder.

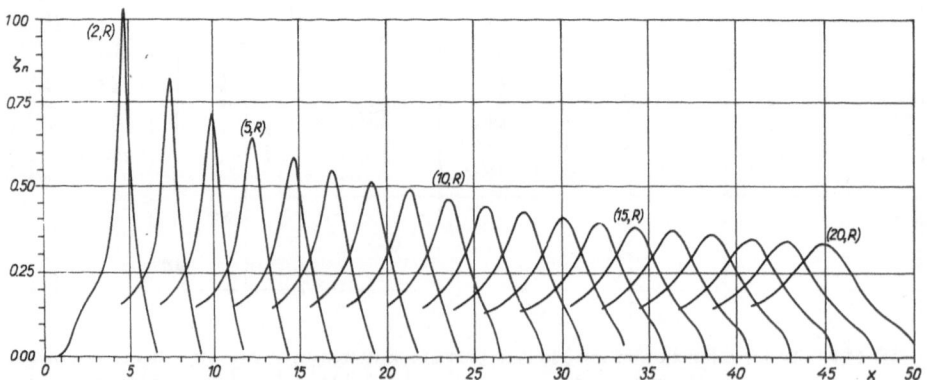

Fig. 11.2. Modal resonances of the Rayleigh wave at $\alpha = 10°$ [11.10, Fig. 3]

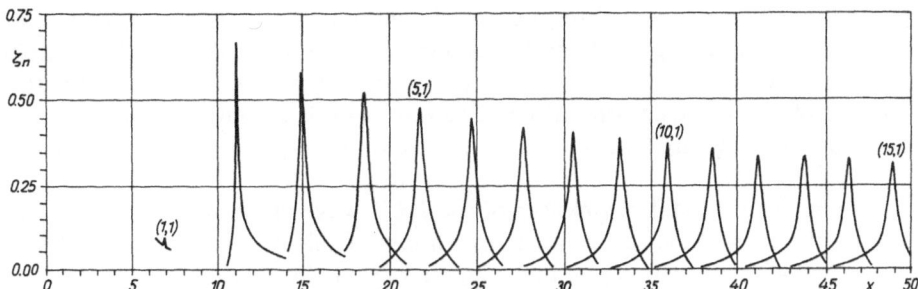

Fig. 11.3. Modal resonances of the $l = 1$ wave at $\alpha = 10°$ [11.10, Fig. 4]

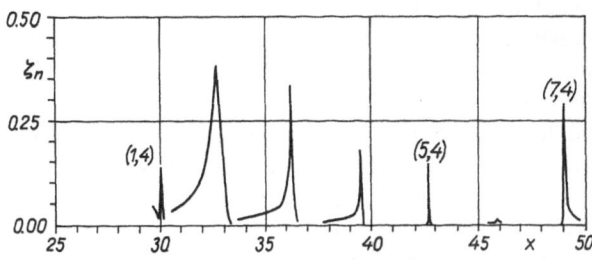

Fig. 11.4. Modal resonances of the $l = 4$ wave at $\alpha = 10°$ [11.10, Fig. 5]

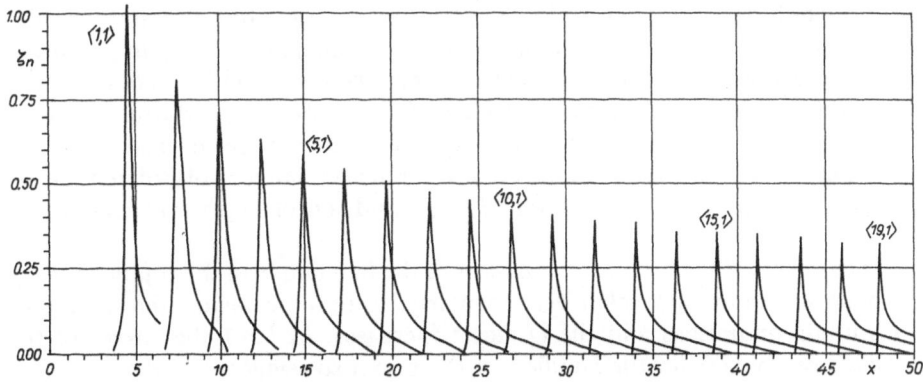

Fig. 11.5. Modal resonances of the $\langle l \rangle = 1$ wave at $\alpha = 10°$ [11.10, Fig. 6]

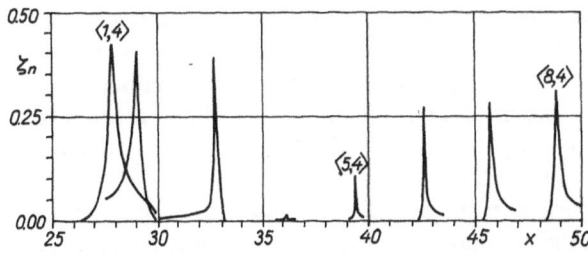

Fig. 11.6. Modal resonances of the $\langle l \rangle = 4$ wave at $\alpha = 10°$ [11.10, Fig. 7]

The family of isolated modal resonances of the Rayleigh wave differs from all the other families in its regularity and large magnitude at the resonance frequency. With n increasing, the value of the resonance magnitude decreases and the Q factor of the resonance diminishes. From the graph, the latter can be gleaned from the opening angle of the resonance curve at the maximum, or from the width of the resonance at $2^{-1/2}$ of its peak value.

The isolated modal resonances of the $l = 1$ wave are also very regular. At small n the magnitudes of isolated resonances of this wave are somewhat less compared with the corresponding magnitudes of the Rayleigh wave, and at $n \geqslant 15$ they are commensurable with them. The Q factor of isolated modal resonances of this wave is higher than that of the Rayleigh wave. With n increasing, the Q factor slowly diminishes.

The family of the isolated resonances for the $l = 4$ wave is typical in the sense of its irregularity. With n increasing, the value of the resonance magnitude first increases, then it attains a maximal value, decreases and again increases. The Q factor first increases and then decreases.

Using the analytical formulae given in [11.2], one can calculate the resonance frequency, magnitude, Q factor and phase for every isolated resonance from any family. The results obtained in such a way coincide with the directly computed ones, but are restricted in the frequency domain: they can be used only in the vicinity of the resonance. Both the computed and analytical results describe each resonance separately. They cannot be used to judge the behavior of the whole family of the isolated modal resonances. For this purpose the geometrical theory of diffraction in the form presented in [11.7–9] may be used.

However, we cannot now give explicit approximate formulae to describe the form function over all the range considered $0 \leqslant x \leqslant 50$ for an arbitrary angle of incidence α. In such a formula, besides the contributions furnished by the creeping waves, the contributions of all the nineteen families of isolated modal resonances, nine of which are of (n, l) type and ten of $\langle n, l \rangle$ type, should be present.

The curves of the isolated resonances of the first family of the supplementary resonances follow each other very regularly. With n increasing, the resonance magnitude diminishes and the Q factor increases. At $l > 1$ the curves of the isolated resonances for the families $\langle n, l \rangle$ are not so regular. For these families the behavior of the isolated resonances presented in Fig. 11.6 is typical. With n increasing, the resonance magnitude first diminishes, then (after $n = 4$) it increases; the Q factor first increases and then diminishes.

After the isolated resonances are computed and joined into families, one can identify form function curve extrema, that is, one can correlate them with resonances of peripheral waves. The positions of the resonance frequencies are labeled in Figs. 7.1 and 11.1. Sometimes the position of the isolated resonance is shifted by several computation steps relative to the form function extremum. Only one third of the extrema of the form function curve is identified. This can be explained in the following ways. Firstly, some extrema of the form function curve are not connected with the resonances; moreover, they may be connected with antiresonances. Such a fact may be observed for the case when the peripheral wave possesses a large magnitude and low Q factor. The Rayleigh wave is a characteristic example of such a wave. Secondly, some extrema of the form function curve may be caused by the interference of waves and do not depend at all on the resonance frequencies. Thirdly, some resonances which are successive in n, may be superimposed almost in anti-phase. These are the so-called hidden resonances (see Chap. 10).

A part of the form function extrema from the so-called quasi-rigid region is connected not with the peripheral but with the creeping (diffracted) waves. Here these waves are not considered and their resonances are not indicated on the form function curve. The properties of these waves are investigated in [11.2]. The formulae presented in that paper permit the resonance frequencies of the creeping waves to be determined and the corresponding extrema in the form function curve to be indicated. The quasi-rigid region is bounded from above by the frequency corresponding to the $(2, R)$ resonance.

Taking into account the fact that the form function is formed by approximately twenty families of superimposed resonances, each of which having a specific frequency, magnitude, Q factor, and phase, it is no wonder that the form function curve has such a discontinuous appearance.

At small angles of incidence ($\alpha = 1°$, $2°$ and $2.5°$) the positions of the resonances $\langle n, 1 \rangle$ for $n < 9$ are in good agreement with the calculated and measured data presented in [11.3, Fig. 9]. It should be noted that at small α the resonances $\langle n, l \rangle$ are not as well observed as is shown in Fig. 11.5 and even in Fig. 11.6. They are observed in the form function curve as small "notches". With α increasing, they become more clearly identified.

11.3 Dispersion Curves of the Phase Velocities

As is well known, at normal incidence the dependence of the phase velocity of the peripheral wave on its resonance frequency $x(n, l)$ is determined by the formula

$$c^{\mathrm{ph}} = c \frac{x(n, l)}{n} . \tag{11.13}$$

At oblique incidence it is determined by

$$c^{\mathrm{ph}} = c \frac{\dfrac{x}{n}}{\sqrt{1 + \left(\dfrac{x}{n} \sin \alpha\right)^2}} \tag{11.14}$$

which follows from the formula for the addition of wave numbers

$$k^2 = k_{z'}^2 + k_\theta^2 , \tag{11.15}$$

where

$$k = \omega/c^{\mathrm{ph}}, \qquad k_{z'} = (\omega \sin\alpha)/c, \qquad k_\theta = n/a \,. \tag{11.16}$$

Here k is the total wave number; $k_{z'}$ and k_θ are the axial and tangential wave numbers, respectively.

In (11.14), $x(n, l)$ should be used for the main resonances and $x\langle n, l\rangle$ for the supplementary resonances.

Often it is more suitable to use the relative phase velocity c^{ph}/c_t and to investigate the dependence of this quantity on x. Here we just use the relative phase velocity. For completeness, in Fig. 11.7 the dispersion curves at normal incidence are presented for eight types of waves. In the $x - (c^{\mathrm{ph}}/c_t)$ plane, the resonances of the different waves of the same order lie on rays passing through the origin. This property follows from the definition of the resonance. The Rayleigh wave does not possess a critical frequency, but the Whispering Gallery waves do. The greater l, the larger is the critical frequency and the higher is the

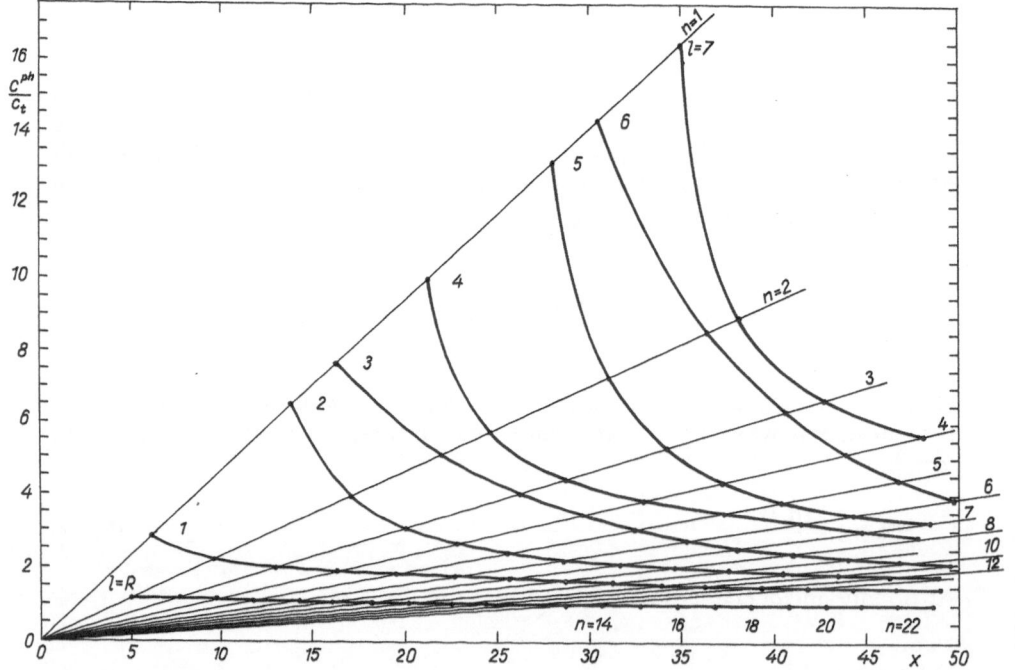

Fig. 11.7. Dispersion curves of the phase velocities for eight types of peripheral waves at normal incidence [11.10, Fig. 8]

Fig. 11.9. Dispersion curves for the phase velocities of the supplementary waves $\langle l\rangle = 1, 2, 3, 4$ at $\alpha = 1°$ [11.10, Fig. 10]

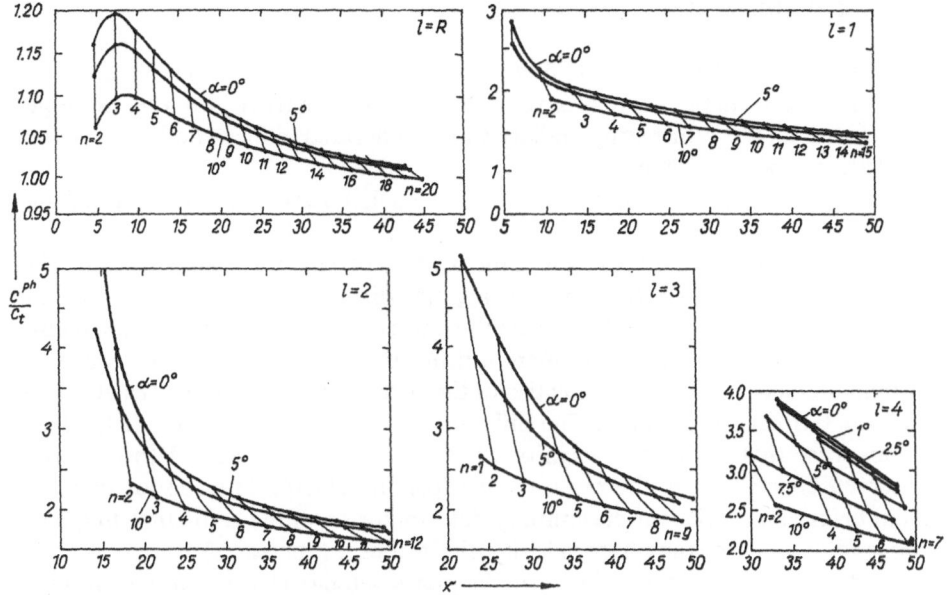

Fig. 11.8. Dispersion curves of the phase velocities for the first five peripheral waves ($l = R, 1, 2, 3, 4$) at different angles of incidence [11.10, Fig. 9]

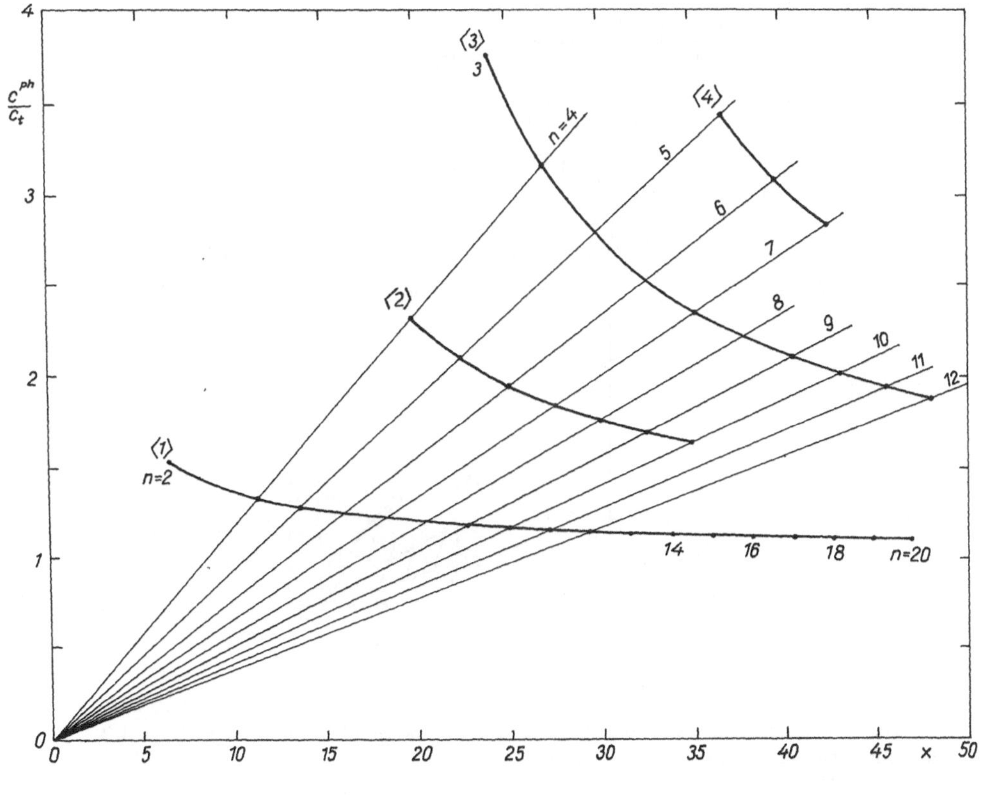

phase velocity of the wave at its first resonance. With n increasing, the beam of the dispersion curves of phase velocities becomes narrow. In the example considered, all the Whispering Gallery waves generated in the cylinder are of the transverse type (see Chap. 7).

In Fig. 11.8, the dispersion curves of the Rayleigh wave and the Whispering Gallery waves $l = 1, 2, 3, 4$ for different angles of incidence are shown. With α increasing, the resonances of the same order n of each wave are shifted to the right on the x axis. With n increasing, the dispersion curves calculated at different α move closer together. At small angles of incidence ($\alpha < 10°$), the phase velocity of the Rayleigh wave differs a little from the velocity of this wave at normal incidence, while the velocities of the Whispering Gallery waves at $\alpha = 0$ and $\alpha \neq 0$ differ significantly. The difference is particularly marked at the first resonances, that is, in the vicinity of the critical frequency. Although the observed effects follow directly from the formula (11.14), one cannot indicate them beforehand, without preliminary determination of the resonance frequencies, and to say how the phase velocity diminishes (or increases) as α is changed at fixed n or x. The larger α, the more the phase velocity changes at fixed n. This

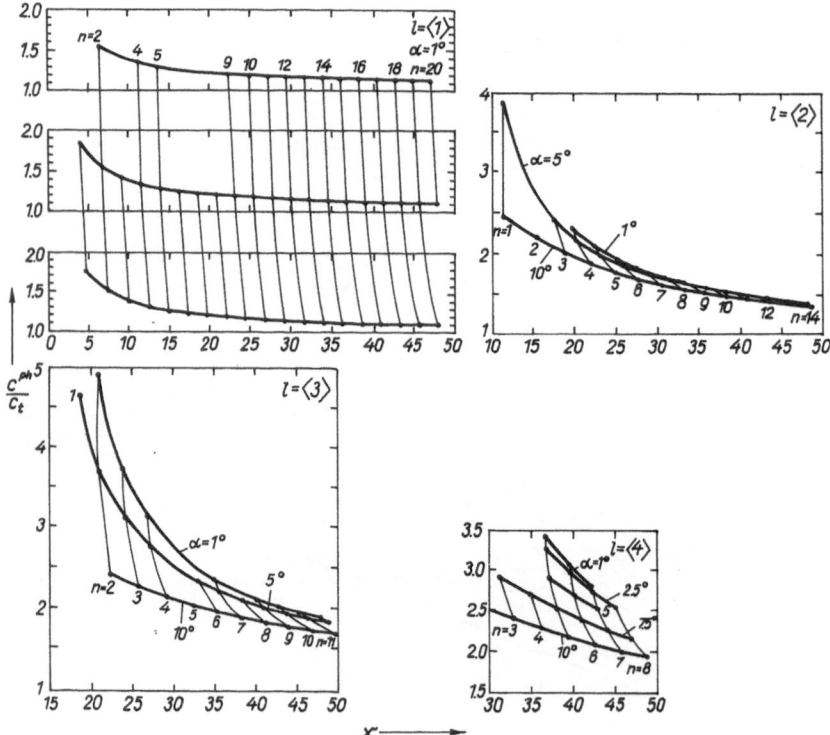

Fig. 11.10. Dispersion curves for the phase velocities of the supplementary waves $\langle l \rangle = 1, 2, 3, 4$ at different angles of incidence [11.10, Fig. 11]

is well observed in Fig. 11.8 for the $l = 4$ wave. For example, at $n = 4$ we obtain

α	$0°$	$5°$	$10°$
c^{ph}/c_t	3.89	3.33	2.30

that is

$$\left[\frac{c^{ph}}{c_t}(\alpha = 5°)\right]\Big/\left[\frac{c^{ph}}{c_t}(\alpha = 0°)\right] = 0.856 \ ,$$

while

$$\left[\frac{c^{ph}}{c_t}(\alpha = 10°)\right]\Big/\left[\frac{c^{ph}}{c_t}(\alpha = 5°)\right] = 0.691 \ .$$

In Fig. 11.9, at an angle of incidence $\alpha = 1°$, the dispersion curves of the phase velocities of the waves $\langle l \rangle = 1, 2, 3, 4$ are shown, while in Fig. 11.10 they are presented separately – for different angles of incidence for each wave. The

Table 11.1. The position on the x axis of modal resonances of the $\langle l \rangle = 1$ wave for shells with different relative thickness h at the angle of incidence $\alpha = 10°$ [11.10, Table I]

n	Relative thickness h						
	1	1/2	1/4	1/8	1/16	1/32	1/512
1	4.80	3.48	2.89	2.70	2.58	2.30	2.27
2	7.58	6.41	5.43	5.04	4.88	4.61	4.53
3	10.12	9.22	7.97	7.42	7.19	6.91	6.80
4	12.54	11.99	10.55	9.80	9.49	9.22	9.06
5	14.96	14.61	13.09	12.22	11.80	11.52	11.33
6	17.38	17.19	15.66	14.61	14.14	13.83	13.59
7	19.77	19.65	18.20	17.03	16.44	16.13	15.86
8	22.15	22.11	20.74	19.41	18.79	18.48	18.13
9	24.53	24.49	23.28	21.84	21.13	20.78	20.43
10	26.88	26.88	25.82	24.26	23.48	23.09	22.70
11	29.26	29.26	28.32	26.64	25.78	25.39	24.96
12	31.64	31.64	30.82	29.06	28.13	27.70	27.23
13	33.98	33.98	33.28	31.48	30.47	29.96	29.49
14	36.33	36.33	35.74	33.91	32.81	32.27	31.76
15	38.71	38.71	38.20	36.29	35.16	34.57	34.02
16	41.05	41.05	40.63	38.71	37.46	36.88	36.29
17	43.40	43.40	43.05	41.09	39.80	39.18	38.55
18	45.74	45.74	45.47	43.52	42.15	41.38	40.82
19	48.09	48.09	47.85	45.94	44.49	43.79	43.09
20				48.32	46.84	46.09	45.35

dispersion curves of the $\langle l \rangle$ waves are similar to the ones described above in the case of the (l) waves.

At different α values the dispersion curves of the $\langle l \rangle = 1$ wave practically coincide into the same curve, but the resonance position for a fixed order n is different along this curve. The larger α, the more the resonance is shifted to the right along the dispersion curve. At $\alpha = 1°$ the procedure used does not permit separation of the resonances with $n = 1, 3, 6, 7, 8$. At other α values, these resonances clearly manifest themselves.

The $\langle l \rangle = 1$ wave is a shear wave. This can be easily shown by calculating the dispersion curves for shells with reduced relative thickness $h = 1 - b/a$.

$$h = 2^{-m} \quad (m = 0, 1, 2, \ldots, 10) . \tag{11.17}$$

Here $h = 0$ corresponds to the solid cylinder.

As the results computed at $\alpha = 10°$ have shown (see Table 11.1), with h diminishing the phase velocity of the $\langle l \rangle = 1$ wave becomes less dependent on x, the wave becomes less dispersive and its phase velocity exceeds only slightly the c_t value.

Shear waves generated in cylindrical shells at oblique incidence are considered in the next chapter.

This material was published in [11.10].

12 Analysis of Peripheral Waves in the Scattering of a Plane Acoustic Wave Obliquely Incident on a Circular–Cylindrical Shell

The scattering problem of a plane acoustic wave from a shell, with the propagation direction making an angle α with the normal to the longitudinal axis of the shell, is a natural generalization of the corresponding problem at normal incidence. The exact solution of the problem considered is given in series form in Chap. 1. As to the formulation and the method of solution, the problem is similar to the corresponding problem of a solid elastic cylinder (see [12.1] and Chap. 11). Mathematically, these problems differ in the number of boundary conditions. The problem considered possesses in addition three boundary conditions on the inner surface of the shell. From the physical point of view, the case of scattering by the shell is characterized by the fact that one group of the peripheral waves generated in the shell is of Lamp-type and the other is of shear-type.

12.1 Form Function Analysis

We shall here use the notations defined in Chap. 11. The form function is computed according to formula (11.7) in which the coefficient R_n should be defined as

$$R_n = A_n[\varepsilon_n(-i)^n]^{-1} \tag{12.1}$$

with A_n given by (1.50–53). The computation of the form function is carried out for the case of an aluminum shell immersed in water, see (7.2). The computation range and step size are $0 \leqslant x \leqslant 400$ and $l_x = 10/256$, respectively. The step size chosen allows the determination of all the essential properties of the form function. The computation has been carried out at $\alpha = 0°$, $1°$, $5°$ and $10°$. The observation point is situated in the far field at $\alpha = 0°$.

In Fig. 12.1, the form function $p(x)$ computed at $\alpha = 10°$ is shown. In order that the extrema of the form function could be well observed in the range $200 \leqslant x \leqslant 250$, the scale on the ordinate axis is enlarged in comparison with that used in other ranges. The positions of the resonances are indicated in Fig. 12.1. As usual, the first index, n, indicates the ordinal number of the resonance and the second, l, defines its family. The resonances of the S_0, A, T_0 and T_1 waves become very clear. At $\alpha = 10°$, the A_1 wave resonances are of a low Q factor and cannot be distinguished on the form function. At normal

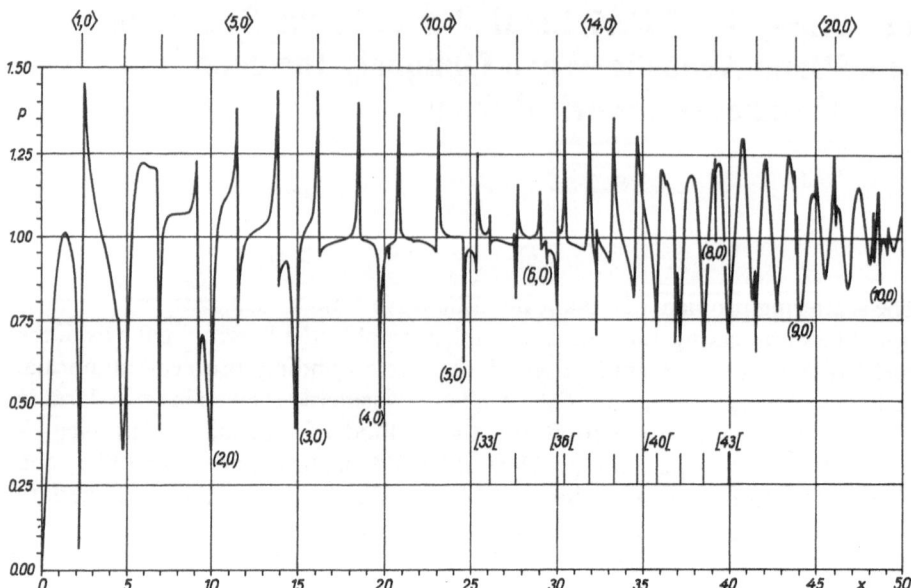

Fig. 12.1.1

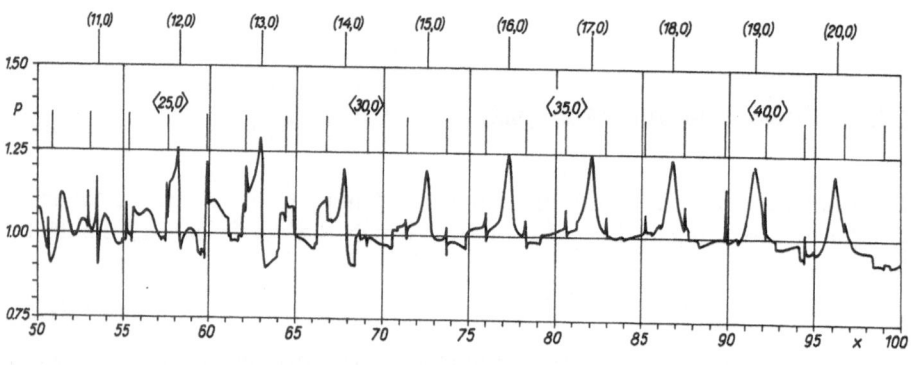

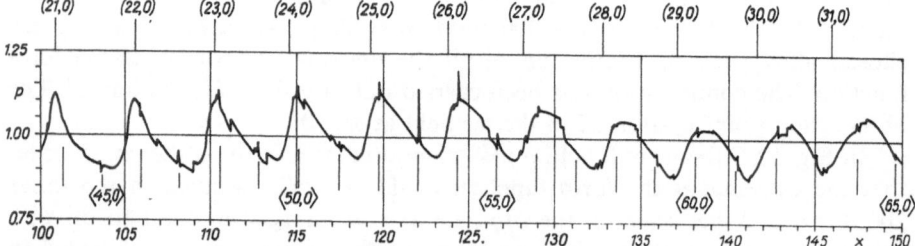

Fig. 12.1.2

Figs. 12.1.1–12.1.6 (p. 220–222). The form function $p(x)$ of the acoustic pressure scattered by a circular–cylindrical shell (with $h = 1/32$) immersed in water for the parameters defined in (7.2). Incidence angle $\alpha = 10°$, observation angle $\alpha = 0°$. The S_0 wave resonances are denoted by $(n, 0)$; the A wave by $[n[$; the A_1 wave by $(n, 1($; the T_0 wave by $\langle n, 0 \rangle$; the T_1 wave by $\langle n, 1 \langle$ [12.8, Fig. 1]

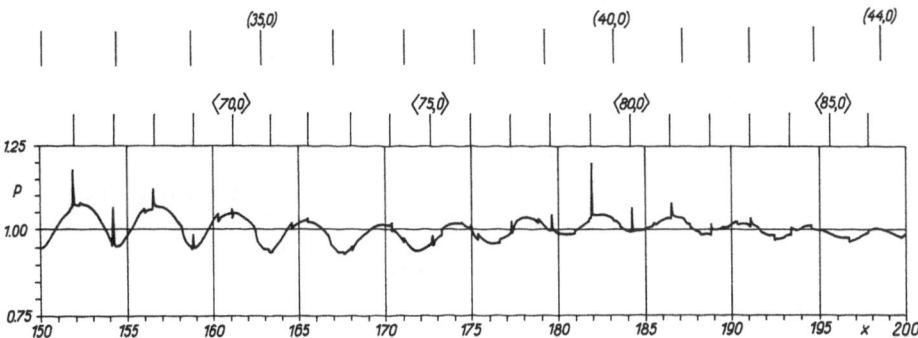

Fig. 12.1.3

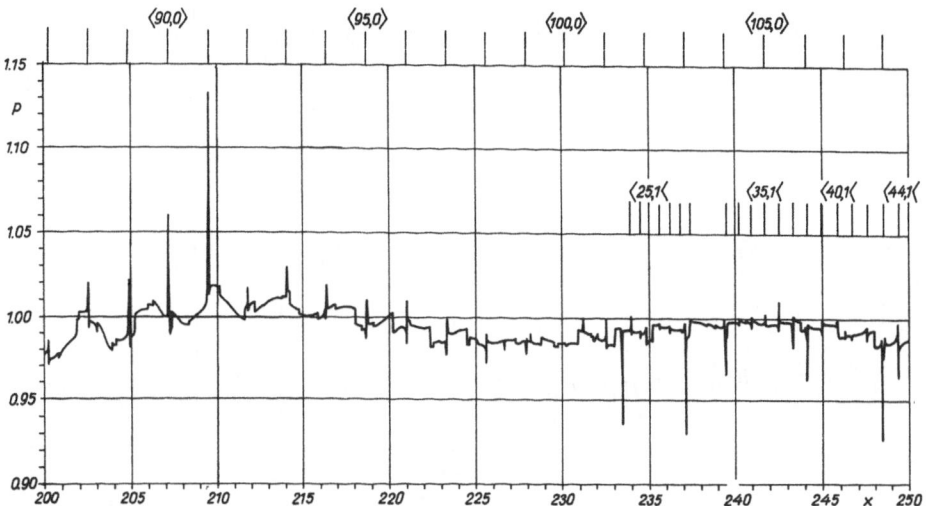

Fig. 12.1.4

incidence, when they are of a high Q factor, their contribution in the form function can be clearly observed, especially at small values of n. In Fig. 12.1, not all the resonances, but only the typical ones are labeled. In comparison with the case of normal incidence, when shear waves are not generated in the shell, the graph of the form function becomes more discontinuous (see Fig. 12.1 and Figs. $1c_1$, $1c_2$ in [12.2]). This is especially noticeable at $0 \leqslant x \leqslant 100$, where the contribution of the T_0 wave in the form function is essential, and also in the range $200 \leqslant x \leqslant 250$, where the shape of the form function is determined mainly by the contributions of the T_0 and T_1 waves. The resonances of the T_0 wave are clearly manifested in the range $250 \leqslant x \leqslant 400$. The appearance of new extrema on the form function curve, caused by the resonances of the shear waves, strongly complicates its shape. When these extrema are identified, i.e. put in

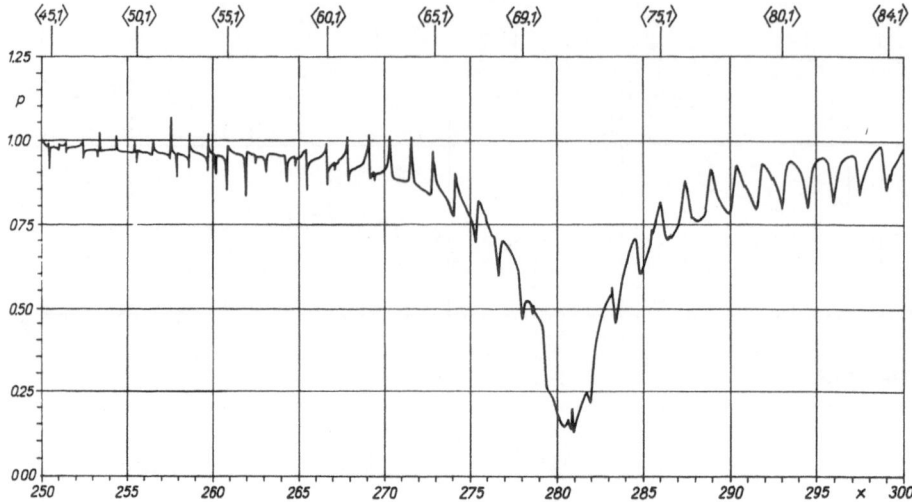

Fig. 12.1.5

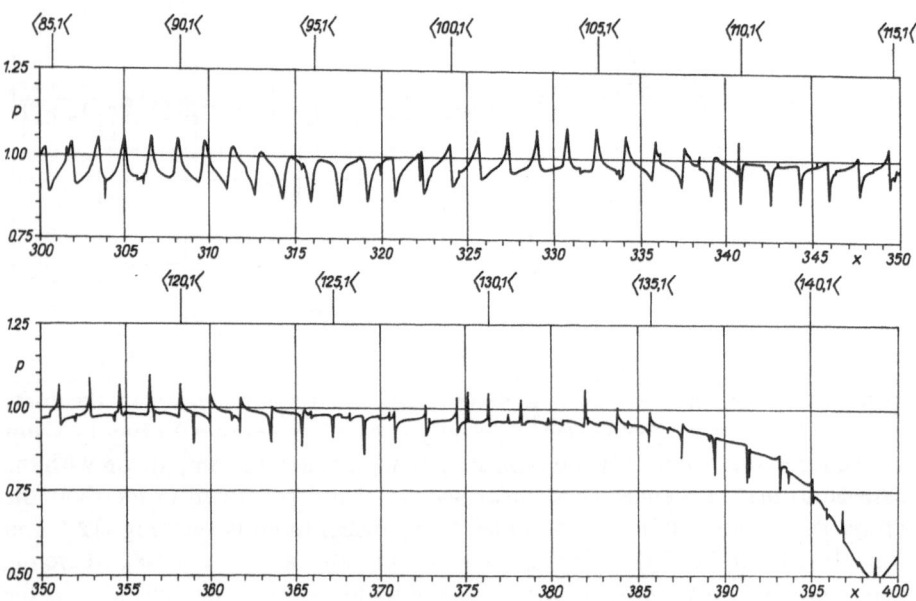

Fig. 12.1.6

correspondence with the relevant resonances, the curve of the form function may be used to obtain information about the elastic shell. Besides the spectrogram, the form function contains information on the magnitude and the Q factor of resonance.

The deep and broad "dip" in the form function curve in the range $270 \leqslant x \leqslant 290$ is apparently caused by the contribution of the A_1 wave. The resonance magnitudes of the partial modes of this wave are rather small here (~ 0.15) and Q factor is also small. Assuming that at small n the contributions of the A_1 wave resonances are almost in phase with each other and in anti-phase with the specularly reflected wave, the superposition of these waves will be of a shape exhibited by the curve in Fig. 12.1. With n increasing, the resonances of the A_1 wave change towards being in anti-phase, and the "dip" gradually disappears.

12.2 Modal Resonances

The computation of the resonance components of partial modes is carried out according to the procedure of the resonance scattering theory [12.3]. Soft and hard backgrounds are used according to the wave family, angle of incidence α, and the ordinal number of the resonance. At the angle of incidence $\alpha = 10°$ the curves of the resonance components of the partial modes are shown in Fig. 12.2.

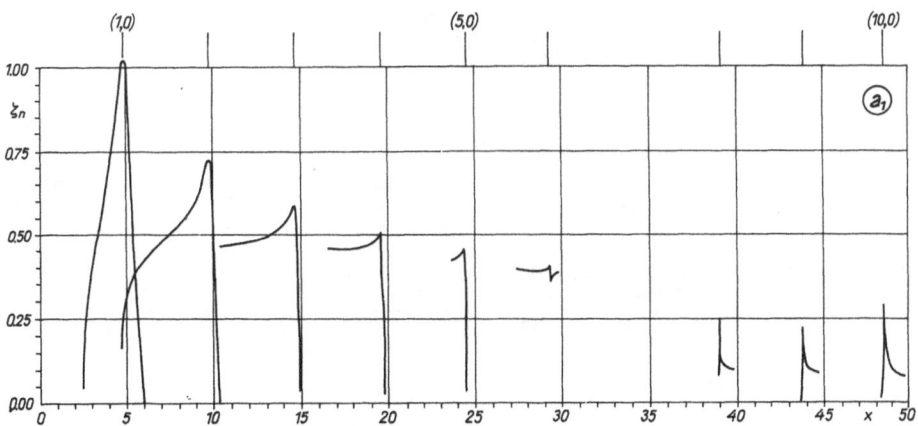

Fig. 12.2.1

Figs. 12.2.1–12.2.6 (p. 223–225). Partial mode resonance components for: **a** the S_0 wave; **b** the A wave; **c** the A_1 wave; **d** the T_0 wave; **e** the T_1 wave. Incidence angle $\alpha = 10°$, observation angle $\alpha = 0°$ [12.8, Fig. 2]

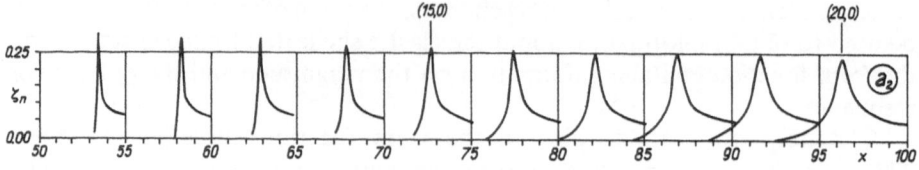

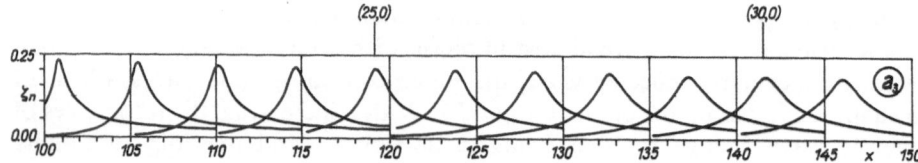

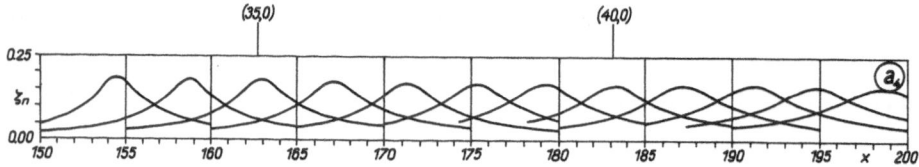

Fig. 12.2.2

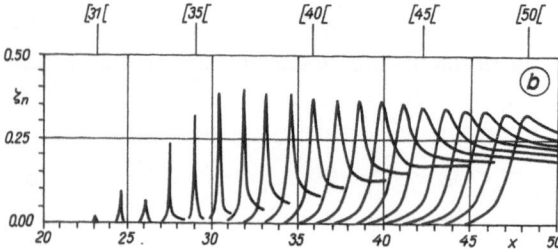

Fig. 12.2.3

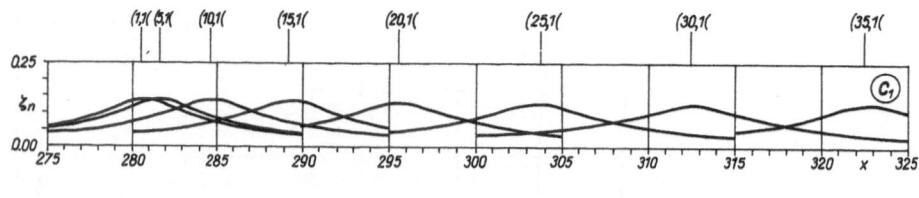

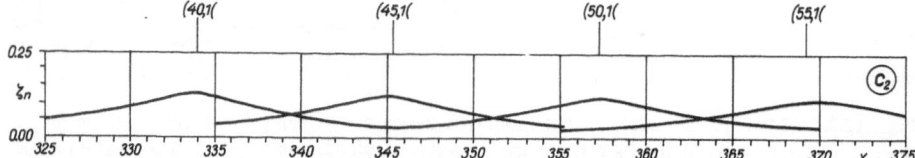

Fig. 12.2.4

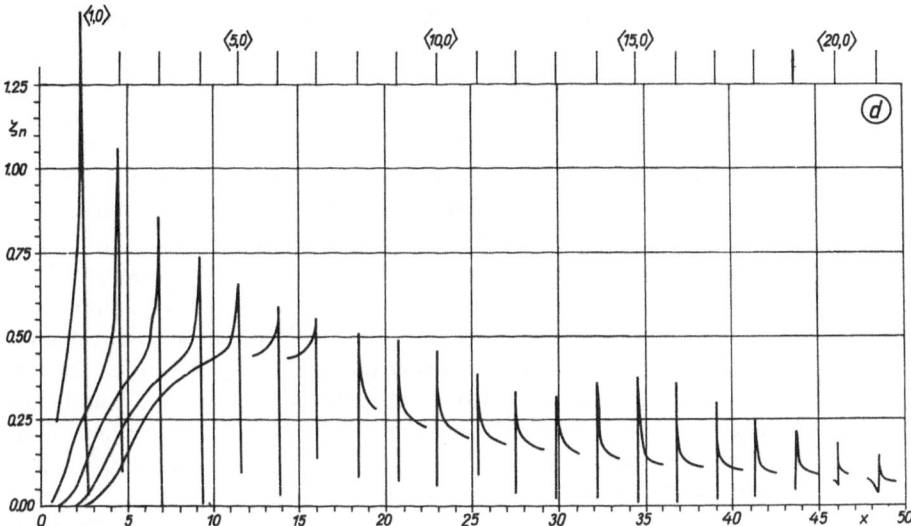

Fig. 12.2.5

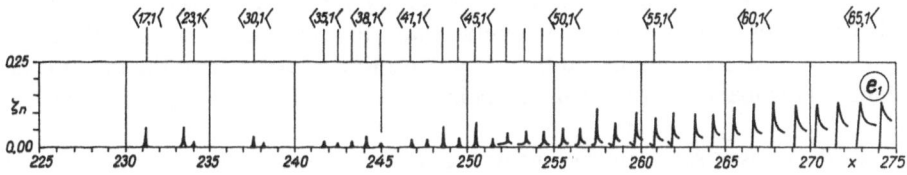

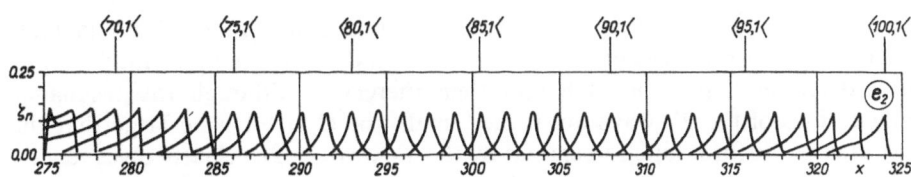

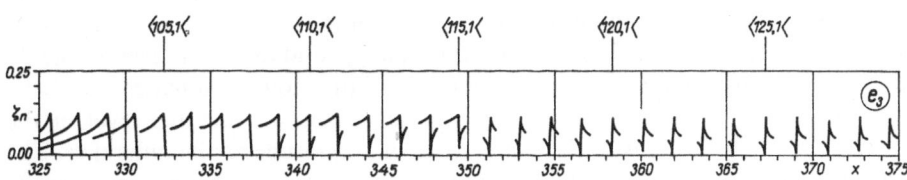

Fig. 12.2.6

The modal resonance components of the S_0, A and A_1 wave look like the corresponding ones at normal incidence (at $\alpha = 0$).

The properties of the S_0 and A waves at normal incidence have been investigated in detail (see Chap. 9). Thus, we shall not describe here the modal resonance components of these waves. We shall only note that with n increasing, the resonance magnitude and Q factor of the S_0 wave decrease monotonically. One can judge the Q factor by the width q of the isolated resonance curve at its half-magnitude level. The greater q, the smaller is the Q factor. Even the resonances of the S_0 wave which possess a moderate Q factor, for example those with $n \sim 40$, can be clearly distinguished on the form function curve. The incident plane wave generates the A wave in the shell only in a very restricted range of frequency ($20 \lesssim x \lesssim 50$) and of order ($30 \lesssim n \lesssim 50$). With n increasing, the resonance magnitudes of the A wave first increase (up to $n \sim 37$) and then decrease. With n increasing, the Q factor of the A wave resonance decreases. Due to the superposition of this wave onto the S_0 and T_0 waves, one can discern in the form function only the resonances with $33 \leqslant n \leqslant 43$.

The Q factor of the modal resonances of the A_1 wave for the same order decreases with α increasing. At $\alpha = 10°$ the curves of isolated resonances of this wave are so gently sloping that in order to distinguish them, we have presented only every fifth curve.

At small n the magnitudes of modal resonances of the T_0 wave are rather large. With n increasing, they decrease gradually. On the x axis the resonances of this wave follow each other very regularly and the relevant extrema of the form function are well distinguished from $n = 1$ up to $n \sim 100$, where the magnitude of modal resonances is rather small. This is caused by the high Q factor of the T_0 wave.

In the chosen scale on the ordinate axis the curves of modal resonances of the T_1 wave can be shown only beginning with $n = 17$. Here the general rule becomes apparent that the smaller the angle of incidence, the higher is the ordinal number of the first observable T_1 wave resonance. With n increasing, the resonance magnitudes first increase, then attain a maximal value and later decrease. With n increasing, the Q factor of modal resonances of the T_1 wave first decreases (up to $n \sim 80$) and then increases. Although the resonance magnitudes of the T_1 wave are rather small, the Q factor is so large that the extrema of the form function caused by the T_1 wave can be clearly observed. At $n \lesssim 100$ the modal resonances of the T_1 and A_1 waves are well separated and here the chosen background is adequate. Conversely, at $n \gtrsim 100$ the resonances of the T_1 wave can only be seen superimposed onto the resonances of the A_1 wave. Here the curve of the modal resonances of the A_1 wave is very smoothly sloping (almost a straight line), and on this background the resonances of the T_1 wave can be observed. So here the magnitude of the modal resonance should be defined as the difference between the actual magnitude and the level of the background. The latter is of the same order of magnitude as the difference. The background component caused by the resonance curve of the A_1 wave can be eliminated by a computer program, but we did not use this opportunity here.

12.3 Phase Velocities of Peripheral Waves

It is easy to find the phase velocity of a peripheral wave when the position of the resonance is known. For each wave the phase velocity c^{ph} on the outer surface of the shell (at $r = a$) is defined by the formula (11.14). This formula has been used in [12.4] for the calculation of the phase velocities of the peripheral waves revolving around an acoustically rigid cylinder.

The position of the resonance frequencies are listed in the first columns of Tables 12.1–5. We did not round off the resonance positions, and give them with two significant digits after the decimal point for the purpose of actual comparison, although the computation step size is $l_x = 10/256$. The dispersion curves of the relative phase velocity of the A, A_1 and T_1 waves are given in Figs. 12.3–5. For convenience of comparison, we used on the ordinate axis the variable $k_t d$ [where $k_t = \omega/c_t$, $d = (a - b)/2$] and on the abscissa axis the variable c^{ph}/c_t.

At various angles of incidence the dispersion curves of the phase velocities of peripheral waves are situated one below the other. When the scale on the ordinate axis is small, the dispersion curves for various values of α practically

Table 12.1. The positions on the x axis of the resonance frequencies of the S_0 wave at various angles of incidence. First column: exact computation according to the resonance scattering theory; second column: approximation [12.8, Table I]

n	$\alpha = 0°$		$\alpha = 5°$		$\alpha = 10°$	
	1	2	1	2	1	2
1	3.79	3.74	4.02	3.95	4.96	4.88
2	7.54	7.49	8.00	7.91	9.84	9.75
3	11.32	11.23	11.95	11.86	14.77	14.62
4	15.25	14.98	16.21	15.81	19.65	19.49
5	19.06	18.72	20.12	19.76	24.53	24.35
6	22.77	22.46	23.98	23.71	29.38	29.21
7	26.45	26.20	27.89	27.66		34.06
8	30.16	29.94	31.84	31.61	39.06	38.91
9	33.87	33.67	35.74	35.55	43.91	43.75
10	37.62	37.40	39.65	39.49	48.71	48.58
11		41.13	43.59	43.42	53.52	53.40
12	45.00	44.86	47.50	47.35	58.32	58.21
13	48.71	48.59	51.41	51.28	63.09	63.00
14	52.42	52.31	55.31	55.20	67.85	67.78
15	56.13	56.02	59.22	59.12	72.62	72.55
16	59.84	59.74	63.13	63.07	77.38	77.31
17	63.52	63.44	67.03	66.94	82.11	82.04
18	67.23	67.15	70.90	70.84	86.80	86.76
19	70.94	70.84	74.80	74.73	91.48	91.46
20	74.61	74.54	78.67	78.62	96.17	96.14

Table 12.2. The position on the x axis of the resonance frequencies of the A wave [12.8, Table II]

n	$\alpha = 0°$		$\alpha = 5°$		$\alpha = 10°$	
	1	2	1	2	1	2
31	22.73	22.70	22.81	22.78	23.16	23.04
32	24.18	24.02	24.30	24.11	24.65	24.39
33	25.66	25.36	25.74	25.96	26.09	25.76
34	27.11	26.71	27.23	26.81	27.58	27.15
35	28.52	28.06	28.63	28.16	29.02	28.54
36	29.96	29.43	30.04	29.55	30.43	29.94
37	31.33	30.79	31.45	30.93	31.84	31.34
38	32.70	32.16	32.81	32.30	33.24	32.74
39	34.02	33.53	34.14	33.68	34.57	34.14
40	35.35	34.89	35.47	35.05	35.90	35.53
41	36.64	36.24	36.76	36.41	37.22	36.91
42	37.93	37.58	38.05	37.76	38.52	38.28
43	39.18	38.91	39.30	39.09	39.80	39.64
44	40.47	40.23	40.55	40.41	41.05	40.97
45	41.72	41.53	41.75	41.72	42.30	42.29
46		42.81	43.05	43.00	43.51	43.59
47		44.07	44.18	44.27	44.73	44.87
48		45.31	45.39	45.51	45.94	46.13
49		46.54	46.56	46.74	47.15	47.36
50		47.74	47.73	47.94	48.32	48.58

Table 12.3. The position on the x axis of the resonance frequencies of the A_1 wave [12.8, Table III]

n	$\alpha = 0°$		$\alpha = 5°$		$\alpha = 10°$	
	1	2	1	2	1	2
1	212.07	212.04	225.78	225.78	280.63	280.59
10	215.86	215.82	229.61	229.60	284.49	284.49
20	226.64	226.60	240.55	240.51	295.68	295.67
30	242.85	242.80	256.99	256.95	312.67	312.62
40	262.89	262.79	277.38	277.30	333.67	333.63
50	285.43	285.32	300.35	300.25	357.23	357.18
60	309.57	309.45	324.88	324.79	379.53	379.53
70	334.61	334.50	350.27	350.20		

coincide. With α increasing, the resonance of the peripheral wave with fixed n slides down along the disperison curve. It is evident from formula (11.14) that the influence of the angle of incidence is stronger for larger x/n and α. Thus, the resonance frequencies with small values of n are shifted more to the right than those with larger values of n. This shift is especially noticeable for the A_1 and T_1 waves, which possess a cutoff frequency at a large x value.

Table 12.4. The position on the x axis of the resonance frequencies of the T_0 wave [12.8, Table IV]

n	$\alpha = 5°$		$\alpha = 10°$		n	$\alpha = 10°$	
	1	2	1	2		1	2
1	2.19	2.19	2.30	2.30	30	69.10	69.07
2	4.38	4.36	4.61	4.60	40	92.11	92.09
3	6.52	6.54	6.91	6.91	50	115.12	115.11
4	8.75	8.72	9.22	9.21	60	138.16	138.13
5	10.94	10.94	11.52	11.51	70	161.17	161.16
6	13.13	13.08	13.83	13.81	80	184.18	184.18
7	15.27	15.26	16.13	16.12			
8	17.46	17.44	18.48	18.42			
9	19.65	19.62	20.78	20.72			
10	21.84	21.79	23.09	23.02			
11	23.98	23.97	25.39	25.32			
12	26.17	26.15	27.70	27.63			
13	28.38	28.33	29.96	29.93			
14	30.54	30.51	32.27	32.23			
15	32.73	32.69	34.57	34.53			
16	34.88	34.87	36.88	36.84			
17	37.07	37.05	39.18	39.14			
18	39.26	39.23	41.38	41.44			
19	41.45	41.41	43.79	43.74			
20	43.63	43.59	46.09	46.06			

Table 12.5. The position on the x axis of the resonance frequencies of the T_1 wave [12.8, Table V]

n	$\alpha = 5°$		$\alpha = 10°$	
	1	2	1	2
1		215.69		227.84
10		216.77	229.02	228.98
20		224.98	232.46	232.43
30		230.43	238.13	238.06
40	233.52	233.45	245.78	245.73
50	241.72	241.64	255.35	255.25
60	252.30	252.22	266.56	266.43
70	264.25	264.18	279.10	279.06
80	277.46	277.33	293.01	292.96
90	291.60	291.53	308.01	307.95
100	306.68	306.62	323.98	323.89
110	322.54	322.47	340.74	340.64
120	339.17	338.99	358.20	358.09
130	356.17	356.08	376.25	376.10
140	373.79	373.65	394.84	394.72
150	391.80	391.65	413.87	413.72

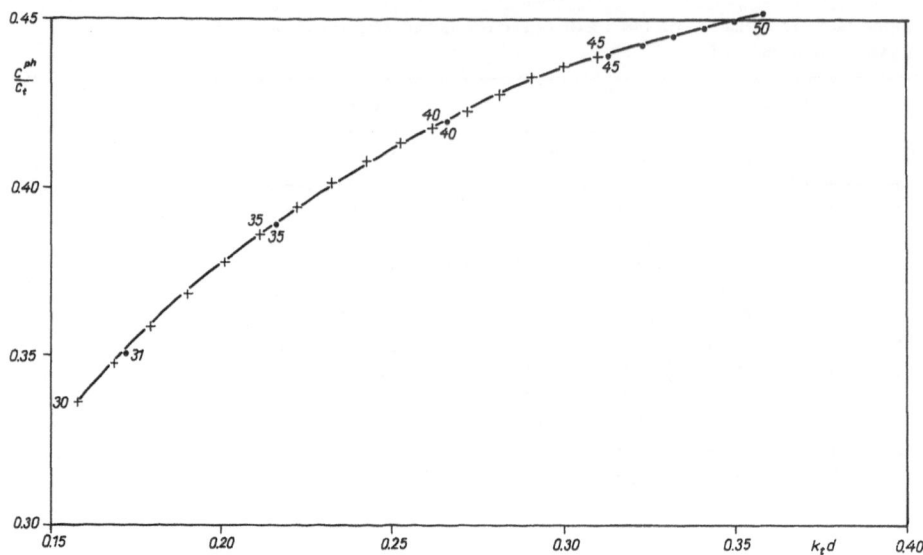

Fig. 12.3. The dispersion curve of the relative phase velocity of the A wave. Crosses denote $\alpha = 0°$, solid circles denote $\alpha = 10°$ [12.8, Fig. 3]

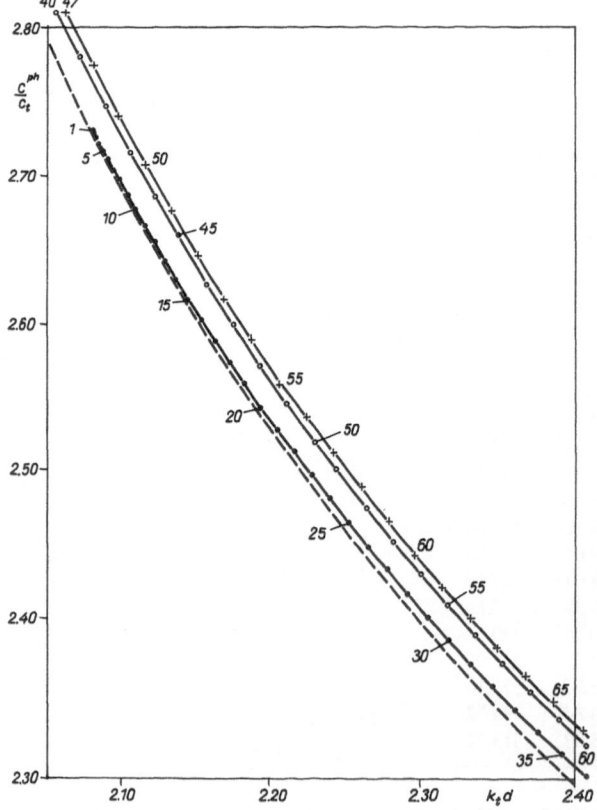

Fig. 12.4. The dispersion curves of the relative phase velocity of the A_1 wave at various incidence angles. Crosses denote $\alpha = 0°$, open circles denote $\alpha = 5°$, solid circles denote $\alpha = 10°$. The dispersion curve of the A_1 wave in the case of "dry" plane layer is shown by the dashed line [12.8, Fig. 4]

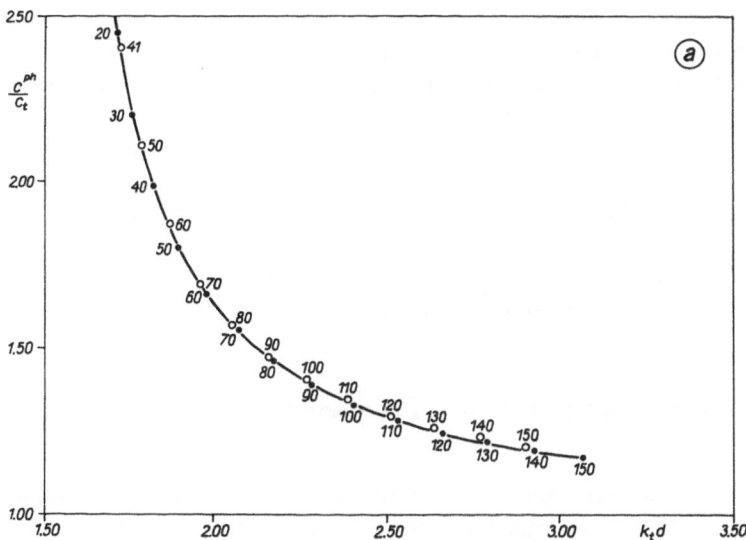

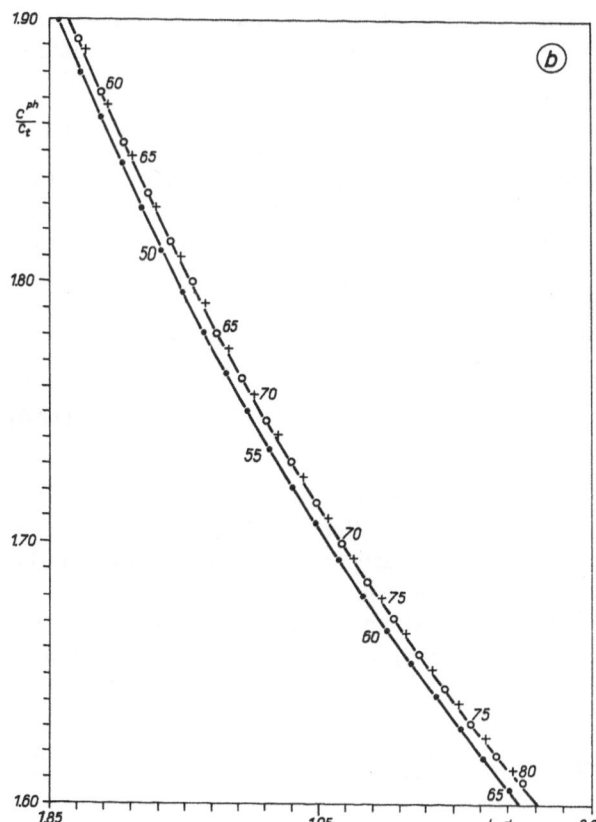

Figs. 12.5. The dispersion curve of the relative phase velocity of the shear wave T_1 at various incidence angles: **a** in a small scale and broad range ($1.5 \leqslant k_t d \leqslant 3.5$); **b** in a large scale and narrow range ($1.85 \leqslant k_t d \leqslant 2.05$). Crosses denote $\alpha = 1°$, open circles denote $\alpha = 5°$, solid circles denote $\alpha = 10°$ [12.8, Fig. 5]

It should be noted that at $\alpha \to 0$ both the A_1 and T_1 waves possess the same cutoff frequency ($k_t d = \pi/2$) and therefore, in the first computations, we do not know beforehand with which wave resonances we are dealing with. Only some experience, investigation of dispersion curves and systematic analysis of modal resonances allow us to unambiguously define the family of the observed resonance. In particular, at a fixed α, the dispersion curve of the A_1 wave is situated high above the corresponding curve of the T_1 wave, and the T_1 wave resonance possess higher Q factors than the A_1 wave resonances. The A_1 wave resonances can be observed beginning with $n = 1$, while the T_1 wave resonances only start at some value of n which corresponds to a two-digit number.

12.4   Estimation of the Resonance Frequency Positions

At oblique incidence, a plane acoustic pressure wave generates in a circular–cylindrical shell vibrations in which the parameters of its stress–strain state change according to the law

$$\exp\{\mathrm{i}[n\theta + (x \sin \alpha)/(z'/a)]\} \, , \tag{12.2}$$

where $0 \leqslant \theta \leqslant 2\pi$ is the polar angle and z'/a is a dimensionless coordinate along the longitudinal axis of the shell. If the stress–strain state of the shell even in one direction changes rapidly, i.e.

$$\max(n, x \sin \alpha) \gg 1 \, , \tag{12.3}$$

then in determining the resonance frequency positions, as a first approximation the curvature of the shell can be neglected and a plane elastic layer of the same thickness $2d$ can be used as a model. Here it is assumed that the stress–strain state along the Cartesian coordinates in the middle surface of the layer changes according to the exponential law corresponding to (12.2).

The influence of the liquid on the position of the resonance frequencies can be neglected if the condition

$$x \cos \alpha \gg n \tag{12.4}$$

is fulfilled.

At the small incidence angle α as considered here, the inequality (12.4) becomes $x \gg n$. It is easy to convince oneself that all the resonances of the peripheral waves generated in the shell satisfy condition (12.4), except the resonances of the A wave, for which $x \cos \alpha < n$.

Now we shall prove the foregoing statement. Using the asymptotic representation of the Hankel function $H_n^{(1)}(x \cos \alpha)$, when condition (12.4) is fulfilled, the following expression is obtained for the acoustic pressure scattered by the shell

$$p_s \sim p_s(a)(a/r)^{1/2} \exp\{\mathrm{i}x \cos \alpha[(r/a) - 1]\} \, . \tag{12.5}$$

Here $p_s(a) = p_s(r = a, \theta, z')$.

The boundary conditions in the direction of the normal to the shell on its outer surface have the form

$$v_r = -i\omega u_r, \qquad \sigma_{rr} = -[p_i(a) + p_s(a)] \quad \text{at } r = a , \qquad (12.6)$$

where v_r is the radial component of the velocity vector in the liquid, and u_r is the radial displacement of the shell.

According to the Euler equation, we have

$$v_r = \frac{-i}{\rho\omega} \frac{\partial}{\partial r} (p_i + p_s) . \qquad (12.7)$$

Now we express v_r in (12.7) according to (12.6), and $p_s(\alpha)$ according to (12.5). In the latter case, when differentiating, we shall take into account that x is large. Then we obtain

$$p_s(a) \sim \frac{i\rho c\omega}{\cos\alpha} u_r + \frac{ic}{\omega\cos\alpha} \frac{\partial p_i}{\partial r}\bigg|_{r=a} . \qquad (12.8)$$

Substituting (12.8) into the second condition (12.6), we get

$$\sigma_{rr} + \frac{i\rho c\omega}{\cos\alpha} u_r = -\left(p_i + \frac{ic}{\omega\cos\alpha} \frac{\partial p_i}{\partial r}\right)\bigg|_{r=a} . \qquad (12.9)$$

When $u_r \gtrsim \max(u_\theta, u_{z'})$ and condition (12.4) is fulfilled, the asymptotic analysis of the equations of motion of the shell, performed similarly to [12.5], allows us to obtain the following estimate

$$\sigma_{rr} \sim \rho_1 c_t \omega u_r . \qquad (12.10)$$

By introducing this estimate into (12.9), it is easily seen that the imaginary part of the expression on the right-hand side of the equality has the order $0(\varepsilon)$ with respect to its real part. Here $\varepsilon = \rho c/(\rho_1 c_t)$ is the relative impedance.

Thus we conclude that when condition (12.4) is fulfilled, then in determining the position of the resonance frequencies, the complex nature of the left-hand side of (12.9) may be neglected, i.e. the shell may be considered as "dry". However, the second term on the left-hand side of (12.9) plays an important role in the determination of the magnitudes of the resonances and in the estimation of their Q factors. From (12.9) and (12.10) it follows that resonances of the peripheral waves at which $u_r \gtrsim \max(u_\theta, u_{z'})$, are effectively damped by the liquid and possess a small Q factor. For example, the resonances of the S_0 wave with $n \gtrsim h^{-1}$, which are situated on the "shelf" of the initial part of the dispersion curve, behave in this way at $\alpha = 10°$.

In the case when $u_r \ll \max(u_\theta, u_{z'})$, i.e. when the vibrations of the shell are quasitangential, the observable resonances possess a high Q factor. This can be explained in the following way. For quasitangential vibrations the shell effectively "slips" in the liquid, and is practically unaware of its resistance.

Three types of quasitangential vibrations may be distinguished:

(i) vibrations analogous to the vibrations of a thin plate in the case of a generalized plane stress; to these correspond vibrations with resonances situated on the "shelf" of the initial part of the S_0 wave dispersion curve at $1 \ll n \ll h^{-1}$;

(ii) vibrations corresponding to the shear waves T_0 and T_1;

(iii) high-frequency, long-wave vibrations in the vicinity of the cutoff frequencies, coinciding with the eigenfrequencies of the shear vibrations of the transverse fibre of the shell [12.6]. For an obliquely incident plane wave such vibrations become realizable at rather small angles of incidence α which, near the cutoff frequency, fulfill the condition

$$x \sin \alpha \ll h^{-1} . \tag{12.11}$$

This condition is a restriction on the wavelength along the longitudinal axis of the shell. With α increasing, condition (12.11) is no longer fulfilled, and the vibrations, corresponding to the first resonances of the A_1 wave at

$$(\pi c_t/2nc) \sin \alpha \sim 1 \tag{12.12}$$

are no longer quasitangential. This explains the above-mentioned reduction of the Q factor of the first resonances of the A_1 wave with increasing α.

Estimation of the resonance frequencies of the peripheral waves S_0 and A_1 with a rather high accuracy may be obtained from the dispersion equations of the Lamb-type waves, which correspond to the vibrations of the "dry" elastic layer. These equations emerge from the dispersion equations in the case of the plane strain state $E = 0$ and $F = 0$ (see (8.2) and (8.3)), respectively, by means of replacing y by y^*,

$$y^* = \{a_1[1 + (a_2nh/a_1z)^2]^{1/2}\}^{-1} \tag{12.13}$$

where

$$a_1 = (c_t/c) \sin \alpha, \qquad a_2 = (1 + b/a)^{-1} . \tag{12.14}$$

It follows from the results of [12.5, 7] that in the problems of high-frequency vibrations of "dry" thin shells, using the model of a plane layer, the phase velocity of the peripheral waves in the layer must be set equal to the velocity of the corresponding peripheral waves on the middle surface of the shell. It should be noted that in Sect. 12.3 the analysis of the phase velocities of peripheral waves was carried out on the outer surface of the shell.

The positions of the resonances of the peripheral waves T_0 and T_1 may be found from the dispersion equations of the shear waves corresponding to the vibrations of a "dry" plane elastic layer. Here the shear vibrations are understood as vibrations of the layer for which both the normal displacement and volume extension are equal to zero. These dispersion equations take the form

$$\sinh p_t = 0, \qquad \cosh p_t = 0 \tag{12.15}$$

for symmetric and antisymmetric waves, respectively. Here in p_t [see (8.3)] y^* must be used instead of y.

For the position of the T_0 and T_1 wave resonances, explicit approximate formulae follow directly from (12.15). The resonance frequencies of the T_0 wave are defined by the formula

$$z = a_2 nh(1 - a_1^2)^{-1/2} .$$ (12.16)

and those of the T_1 wave by the formula

$$z = \frac{\pi}{2} \sqrt{1 + \frac{\left(\frac{2}{\pi} a_2 nh\right)^2}{1 - a_1^2}} .$$ (12.17)

In the analysis of the A wave resonances it is already impossible to neglect the influence of the liquid, because for this wave $x \cos \alpha < n$ and the boundary layer of the liquid makes the shell "heavier". In fact, if for $x \cos \alpha < n$, using arguments analogous to those set forth above, we express the pressure $p_s(\alpha)$ through u_r, then in the left-hand side of (12.9) instead of the second, purely imaginary term, we shall obtain a term which can be expressed by the Hankel functions and their derivatives. The real part of it is, generally speaking, commensurable with σ_{rr}. In the case of the A wave, estimates of the resonance frequency positions can be found by solving the problem of the vibration of an elastic layer on a liquid half space. The corresponding dispersion equation can be obtained from (10.4) by replacing $y = c^{\mathrm{ph}}/c_t$ by y^*, defined by (12.13). In the latter expression one should set $a_2 \equiv 1$, which means that the velocity of the outer surface of the shell is considered. This gives an explicit account of the influence of the liquid on the vibrations of the shell.

The approximate positions of the resonance frequencies are listed in the second columns of Tables 12.1–5. One can see that the correspondence is good enough even for small values of x and n (as in the case of the S_0 wave).

This material was first published in [12.8].

13   On the Causes of Possible Errors in the Solution of Scattering Problems

The considerations that follow below deviate from the main theme of this book. The major portion of the book has so far dealt mainly with the physical essence of the phenomenon, but here the main attention is now paid to some details of obtaining and interpreting the solution. This may seem strange and the corresponding presentations inappropriate. However, our experience (as a reader, listener and participant of discussions) has shown that many publications and presentations contain systematic errors which sometimes manifest themselves in incorrect results and erroneous conclusions. We tried to recognize the typical errors committed, and to show how to eliminate them. Here we followed the well-known principle that the contra-indication is as much needed as the indication. This chapter is addressed to an inexperienced reader. However, some of the given examples may also be interesting for the experienced reader. As such examples, we shall use the solutions of the problem of scattering by elastic spheres and cylinders. Here the exact solution exists and therefore it becomes easy to explain the cause of the mistake. In more complicated scattering problems, for example, by a system of individual elastic bodies, by a body with parameters different in some subdomains, or by a body in the vicinity of a boundary, it is difficult to obtain the solution, but easy to commit errors, which reduce to zero all the expended efforts. We hope that the information given below will help to avoid such an error.

13.1   Application of a Model Outside its Region of Validity

The problem of acoustic wave scattering by an elastic body is a contact problem. The contact conditions connect the equations of motion of the elastic body to those of the medium surrounding it. The contact conditions change in correspondence with the model used to describe the motion of the elastic body. They allow us to connect the equation of motion of the liquid with substantially different equations of motion of the elastic body. In particular, one can prescribe certain contact conditions which will entirely "kill" the elasticity of the scatterer. These are the Neumann and Dirichlet conditions and the bodies corresponding to them are termed acoustically rigid and soft bodies, respectively.

In engineering practice solid elastic bodies are rarely encountered, but shells are often utilized. In order to describe the motion of an elastic body (thin-walled or not very thin-walled), the equations of thin shell theory are frequently used. Generally, linear theory is employed in scattering problems. The main advantage of this is that the three-dimensional problem is reduced to the two-dimensional one. For the small parameter – the relative thickness of the shell h – the passage to the limit is carried out and the motion of the shell is reduced to the motion of its middle surface. The contact conditions are also imposed on this surface.

The thin shell theory, as such, is a brilliant achievement of the mechanics of our century. It should be pointed out, however, that in scattering problems this theory should be used in a reasonable way.

Even the fact that, in modeling a three-dimensional elastic curved layer by using the theory of a thin shell, one cannot describe the refraction of waves between outer and inner surfaces, should make one cautious. The second warning of possible trouble is the fact that boundary conditions are imposed on the middle surface of the shell, while in the real problem they are defined on its outer and inner surfaces. The ratio h/λ can be considered the third physical caution. When it is small ($h/\lambda \ll 1$) the thin shell theory can be effectively used and with $h/\lambda \sim 1$ the thin shell theory will certainly break down. Here λ is the wavelength in the liquid surrounding the shell; it is used as a typical measure of size.

Depending on the thin shell theory chosen, one obtains the physical model in which some or other number of wave types can propagate. For example, the membrane (bending-free) thin shell theory can be used when solving the problem of acoustic wave scattering by a thin-walled elastic cylinder. In this case only one wave, namely the membrane wave, can be generated in the shell by the incident wave (in elasticity theory this wave is named the Lamb-type zero-order symmetric wave S_0). When the Timoshenko-type thin shell theory is used in the same problem, three types of waves can be generated (see Chap. 9).

Often the dynamical equations of two-dimensional thin shell theories are constructed so that the initial parts of the dispersion curves of some first modes of three-dimensional elasticity theory will be well modeled. It is clear that when the thin shell theories designed in such a way are used for solving a scattering problem, the peripheral waves of higher order, not described by these equations, could not be generated. We believe that the procedure of devising a two-dimensional thin shell theory when only the dispersion curves are used as a reference, is not well justified. In fact, aside from the velocity (usually the phase velocities are compared), the wave is specified by magnitude, phase and Q factor, and these quantities cannot be properly utilized, because in a two-dimensional theory the free parameters needed are simply missing. On the other hand, it is clear that if a more complicated thin shell theory is devised (according to which not one or two waves, but several waves could be generated in the shell), even these two-dimensional equations of the theory would define a system which in complexity would approach the elasticity theory. For this reason all the advantages of the two-dimensional theory would be lost.

Ordinarily, when going over from three-dimensional elasticity theory to the theory of two-dimensional thin shells, one does not use complicated theories, but the classical ones in the form presented by Lurie [13.1], Novozhilov [13.2] and Gol'denveizer [13.3]. The refined Timoshenko-type [13.4] and Kennard-type [13.5] theories are more rarely utilized. Until now the equations of the theory of thin shells presented by Berdichevskii [13.6] have not been used in the scattering problems.

The scattering problem, as a contact problem, is sufficiently difficult. Nobody wants to complicate it, but would rather like to simplify it, therefore the theories of thin shells are used. Here the first risk is hidden: using the model out of the range of its applicability. Restrictions on the frequency should be set when passing from the elasticity theory to the thin shell theory. To transcend the frequency domain where the thin shell theory is entirely justified is risky for two reasons. Firstly, the waves described by the theory at low frequencies would be badly continued. Secondly, the waves which are not supposed to appear in this two-dimensional theory would be not described at all. The largest error will be committed when the main lobe of the loading (the incident wave) moves into a domain not described by the two-dimensional theory. If the loading spectrum is broadband, it covers also the high-frequency range. Therefore, even in the case when the carrier frequency is situated in the range where the two-dimensional theory is entirely justified, on account of the broadband spectrum of the loading an error slips into the solution. The error can originate in the high-frequency range of the spectrum, where the two-dimensional theory inadequately describes the motion of the elastic body. The broader the spectrum of the loading, the larger is the error. We shall illustrate this by several examples.

In Fig. 9.1, the form function is given for case of scattering by an Armco iron shell with $h = 1/64$ immersed in water [see (9.3)]. Here the motion of the shell is described by the equations of linear elasticity theory. As is shown in Chap. 9, in the domain $0 \leqslant x \leqslant 100$ only two waves, namely the S_0 and A wave, furnish contributions to the form function. At $50 \lesssim x \lesssim 80$ the strong-bending domain can be clearly observed. If here the membrane thin shell theory would be used, the strong-bending domain will disappear (see Fig. 9.5). It is clear that the membrane shell theory does not describe the bending and the error caused in the solution by this can be large. Thus, in the strong-bending domain, the contributions furnished by both the S_0 and A waves are commensurable in order. Therefore in this domain the solution based on the membrane shell theory is wrong in principle. When using the membrane shell theory in the problem considered, one should exclude the domain $50 \lesssim x \lesssim 80$ from consideration.

It must be admitted that even for such an easy problem as the scattering by a circular–cylindrical shell (with given relative thickness and material parameters), we have until now no answer to a simple question: at which x values, depending on h, will the strong-bending domain be situated? This is a part of a more common question: how is the energy distributed between different peripheral waves generated in the shell by the incident wave, and at which frequencies is the contribution of a given peripheral wave essential? The same question may be formulated in another way: when can one neglect the contribution in the form function furnished by one or another peripheral wave?

Figure 13.9 (Sect. 13.5) shows the form function of the pressure scattered by an Armco iron circular–cylindrical shell with $h = 1/24$ immersed in water. The material parameters are given in (9.3). The equations of linear elasticity theory are used to describe the motion of the cylindrical layer. If instead of them the equations of the Timoshenko-type thin shell theory are used, then the form function takes the form presented in [13.7, Fig. 4.4]. We have compared these curves earlier [13.7, Chap. 4]. The main result of this comparison can be formulated as follows: at $x \gtrsim 30$ the thin shell theory is not justified, because here it describes both the contributions of the S_0 and A wave in an erroneous fashion.

In this problem it is hard to predict that the Timoshenko-type thin shell theory, which is intended for describing the motion of rather thick shells and is commonly and fruitfully used for that case, will lead to erroneous results for a shell with $h = 1/24$ at such low x values.

For a thicker shell with $h = 1/10$, as can be seen from Figs. 8.1 and 8.8 and Table 8.3, the A_2 wave contributes to the form function beginning from $x = 198.83$. This wave is not described by the Timoshenko-type thin shell theory and therefore at x bigger than this critical value thin shell theory cannot be used.

These considerations confirm this evident condition: the thin shell theory can be used only up to that frequency below which the theory is in principle able to describe all of the peripheral waves generated in the shell. This is a necessary condition, but not a sufficient one. Besides the correct number of waves, the properties of each of them should be adequately described. The latter condition restricts even more the frequency domain in which the thin shell theory can be used in scattering problems.

13.2 Application of the Procedure Outside its Region of Validity

One of the commonly used methods for solving the scattering problem is the Sommerfeld–Watson transformation, which transforms the series of partial modes into an integral on the complex plane. The main aim of such a transformation is to avoid the slow convergence of the series of eigenfunctions. Actually, as a result of such a transformation, the initial series is replaced by the series of residues of the poles, which converges more rapidly.

Quite often the Sommerfeld–Watson transformation is used when it formally cannot be applied. Essentially, this transformation is valid for the high-frequency (short-wavelength) case. For each peripheral wave, there is a lowest x value, onward from which the asymptotic description of the wave obtained by means of the Sommerfeld–Watson transformation is valid. Usually, this value can be obtained from the Sommerfeld–Watson transformation itself.

An unpleasant situation occurs when the Sommerfeld–Watson transformation is applied to the series in the problem where the equations of thin shell theory are used. Here the equations can be used only for $x < x_*$ and the transformation only for $x > x_0$. Particularly it can happen that $x_* < x_0$ and then, formally, the asymptotics obtained with the Sommerfeld–Watson transformation can absolutely not be used. Somewhat better is the situation when $x_0 < x_*$. In this case the asymptotics obtained are valid at $x_0 \leqslant x \leqslant x_*$. Therefore, when the equations of thin shell theory are used, the Sommerfeld–Watson transformation should be applied gingerly.

In the second typical situation one can obtain the asymptotic descriptions of two peripheral waves, but according to the asymptotic estimate, for one of them the description can be used when $x > x_{**}$, while for the other only when $x > x_{00}$. The values x_{**} and x_{00} may differ in order. The biggest of them can be situated outside the region which is the most interesting in the problem considered.

The third typical situation occurs when, using the Sommerfeld–Watson transformation, one can obtain a rather good description of a peripheral (or creeping) wave and not a very correct description of a geometrical wave (propagating on a straight line). A typical example is the result of applying the Sommerfeld–Watson transformation in the two-dimensional problem of a plane acoustic wave scattered by an acoustically rigid cylinder. In this case the scattered field is composed of waves of two types, namely the specularly reflected and the diffracted waves. From the latter ones, only one wave is significant, namely that with the smallest order, because the magnitude of this wave is much bigger than the magnitudes of all the other waves. With x increasing, the magnitude of the diffracted wave rapidly diminishes and for $x \gtrsim 10$ it becomes much smaller than the magnitude of the specularly reflected wave. Therefore, at $x \gtrsim 10$ the contribution furnished by this wave to the form function may be neglected.

In this problem, the most interesting domain is that where the specularly reflected wave interferes with the diffracted wave. Application of the Sommerfeld–Watson transformation permits us to obtain a rather good description of the diffracted wave [13.8, 9], but not a very successful description of the specularly reflected wave. The latter may be represented in the form

$$\bar{p}_1(x) = A_2(x) + iB_2(x) \, ,$$

$$A_2(x) = 1 - \alpha_2 x^{-2}, \qquad B_2(x) = - \alpha_1 x^{-1}, \tag{13.1}$$

$$\alpha_1 = \frac{11}{16}, \qquad \alpha_2 = \frac{353}{512} \, .$$

While obtaining the expression of (13.1) it was supposed that $x \gg 1$, but the formula (13.1) is often used even at $x \sim 1$. From the structure of this formula one can see that at $x \sim 1$ it cannot describe the reflected wave very well. Therefore, at $x \lesssim 10$ one should abandon the description given by (13.1) and replace it by a more realistic one. The formula given in [13.10] can be used for this purpose

$$\bar{p}_1(x) = A_1(x) + iB_1(x) \, ,$$

$$A_1(x) = \frac{1}{2}\left(\frac{x^2}{x^2 + a_1} + \frac{x^2}{x^2 + a_2}\right), \qquad B_1(x) = \frac{-a_3 x}{x^2 + a_4},$$

$$a_1 = 1.57, \qquad a_2 = -0.30, \qquad a_3 = 0.673, \qquad a_4 = 0.979 . \tag{13.2}$$

The geometrical divergence of the rays and the increase of the phase in the formulae of (13.1) and (13.2) should be taken into account by an additional factor.

In problems of acoustic wave scattering by elastic bodies, the application of the Sommerfeld–Watson transformation and the subsequent utilization of the saddle-point method (see [13.11, 12]) cause, at small x values, an error in the description of the specularly reflected wave. This error arises from the procedure itself. Qualitatively, the asymptotic approximation obtained in such a way is not better than the one given by (13.1).

One further remark should be made. As the analysis of the solutions of scattering problems has shown, the asymptotic estimates bounding the domain of validity of the asymptotic formula from below are frequently rather rigid (see [13.13]). The comparison with the independently obtained exact solutions suggests that the asymptotic formulae can be applied in domains broader than those given by the asymptotic estimates. However, in new problems where the solution is not known beforehand, the asymptotic solutions cannot be used with any justification outside the domains prescribed by the asymptotic estimates.

13.3 Boundedness of the Frequency Domain

The computation of the solutions for the scattering of acoustic waves by elastic bodies, even those of simple shape, is very time consuming. To obtain sufficient accuracy when x increases, the series must be summed up to rather large values of n. The computation time grows somewhat quicker than linearly in x. If the internal structure of the elastic scatterer is complicated, for example, when the cylinder or sphere is multilayered and the material of the layers is absorptive, the computation time may be twice or triple the time needed for the case of a solid cylinder or sphere.

When the internal structure of the scatterer is to be optimized in some way, and a series of computations have to be carried out in which, for example, the number of layers and the order of their priority, their material and their thickness are to be changed, the total computation time can become rather large.

Therefore, the computations are sometimes carried out in a rather narrow and non-representative frequency domain. The boundedness of the frequency domain at once leads to a filtration: the resonance frequencies situated outside the calculation domain disappear. For some types of waves one part of the resonances can be in the domain of calculation, and the other part outside; for waves of other types none of the resonances may be present in this domain.

Let us turn to the problem of scattering by an aluminum cylindrical shell with $h = 1/10$ (see Chap. 8). We shall suppose that the form function is computed in the domain $0 \leqslant x \leqslant 150$. In this domain we will find all the resonances of the Lamb-type waves up to n_* included

Wave type	S_0	A_0	S_1	A_1	S_2
n_*	67	74	32	39	9

In the more restricted domain $0 \leqslant x \leqslant 75$, only the resonances up to n_* will remain

Wave type	S_0	A_0	A_1
n_*	21	40	8

Our computational experience has shown that the cutoff frequencies for the case of a plane "dry" layer can be used as preliminary estimates of the cutoff frequencies of peripheral waves generated in cylindrical and spherical shells (see Chaps. 10–12). This is correct even for a rather thick shell with $h = 1/10$.

It is difficult to say beforehand which type of peripheral wave will be generated in an elastic scatterer of complicated inner structure, for example a multilayered one. Therefore, it is not clear which resonances will be filtered when a restriction on the frequency band is set.

13.4 Extremely Large Computation Steps

The form function is computed using a certain step size. The value of the step size must be chosen such that after the computation, all the principal information on the physical process can be obtained. Usually, the computation step size is obtained from the condition

$$ml_x = \lambda_i , \tag{13.3}$$

where l_x is the step size along the x axis, m is the number of steps over the wavelength, and λ_i is the wavelength of some wave, usually the smallest of all those occurring in the process. The value $m = 2$ provides the description of zeroes of the wave. Values $m \geqslant 4$ provide the description of its magnitude. The bigger m, the better all the properties of the wave are described. When $m \sim 16$ (or ~ 32), the wave is rather well described.

In the standard situation, when l_x is too large, the following error occurs in the form function: the resonance frequencies are shifted, which introduces errors in the phase and group velocities of the peripheral waves, and the magnitude of the peripheral waves is reduced. The peripheral waves with a large Q factor and magnitude, which quickly change along the x axis, are those most sensitive to the value of the step size. Usually, these properties are held by the waves in the vicinity of the cutoff frequencies, i.e. at small resonance orders n. Extremely large step sizes can bring about the situation where some wave(s) will be totally omitted in the form function. This can occur at

$$l_x \gtrsim \lambda_i \, . \tag{13.4}$$

When transient problems are considered, an additional condition on the computation step size must be posed. If the solution of the steady-state problem is considered as the Fourier transform with respect to time, then using its convolution with the Fourier transform of the incident wave, the solution of the transient problem can be obtained. When the Fourier transform of the incident wave is restricted to a narrow frequency band, as it is in the case of a sufficiently long harmonic pulse, one must take care that a sufficient number q of computational steps are placed on the width of the main lobe L

$$q \, l_x = L \, . \tag{13.5}$$

With l_x diminishing, the computation time increases, therefore the frequency band must be restricted. Such restriction can be clearly observed when the scattering problems are solved by numerical methods. When three-dimensional problems are considered, it is the restrictions on the computation step size which make the computations in the resonance domain extremely time-consuming. In such complicated problems the solution is often given in the form of angular diagrams, calculated at some not very high frequency and rarely as a form function. The latter quantity is often non-representative, because the frequency band is very restricted.

13.5 Series Truncation at a Small Number of Terms

Only a small number of problems in acoustic wave scattering by elastic bodies have a solution in series form. But even in these cases, before the analysis of the solution can be carried out, the series must be summed for each frequency value, which is considered as a parameter, in a broad frequency band. If the problem is examined as a transient one, this series is used as the Laplace or Fourier transform with respect to time, and the inverse transform should be carried out.

As an example, let us consider the problem of a plane acoustic wave scattered by a cylindrical shell. The solution of this problem is given by (2.54). The physical parameters are defined by (9.3) and the relative thickness of the shell is $h = 1/24$. The motion of the shell is described by the equations of linear elasticity theory. We have here the exact solution of the steady-state problem, which should be used for obtaining numerical values.

In the far field, the acoustic pressure scattered by the shell can be represented in the form

$$p_s = p_* \left(\frac{a}{2r} \right)^{1/2} \exp\left[i(kr - \omega t) \right] \sum_{n=0}^{\infty} b_n(x, \varphi) ,$$

$$(13.6)$$

$$b_n(x, \varphi) = \frac{\varepsilon_n}{(i\pi x)^{1/2}} R_n \cos n\varphi .$$

Let us denote a partial sum by S_N

$$S_N = \sum_{n=0}^{N} b_n(x, \phi) .$$

$$(13.7)$$

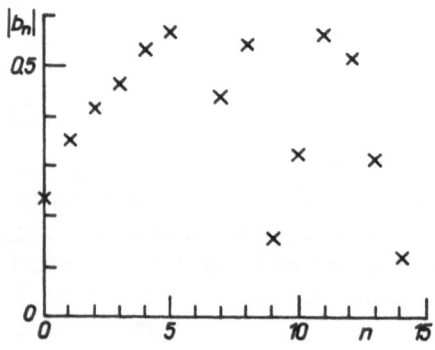

Fig. 13.1. The dependence of $|b_n|$ on n for backscattering by an Armco iron shell (with $h = 1/24$) immersed in water, at $x = 15.31$

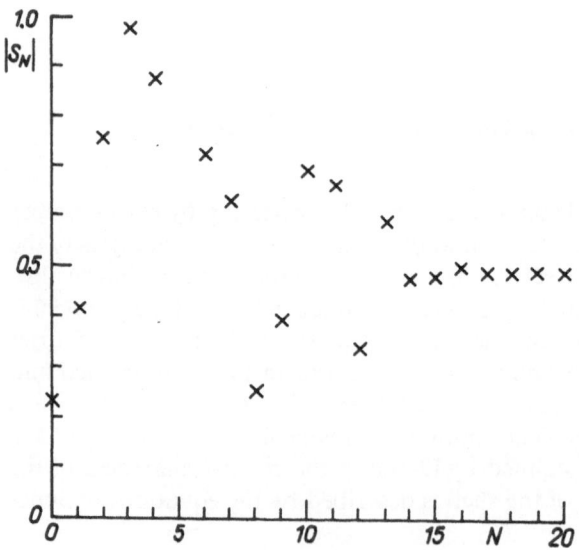

Fig. 13.2. The dependence of $|S_N|$ on N for backscattering by an Armco iron shell (with $h = 1/24$) immersed in water, at $x = 15.31$

Table 13.1. The dependence of $|b_n|$ on n [see (13.6)] and the dependence of $|S_N|$ on N [see (13.7)] for backscattering by an Armco iron shell (with $h = 1/24$) immersed in water

$n; N$	$|b_n|$	$|S_N|$	$n; N$	$|b_n|$	$|S_N|$
0	0.234	0.234	14	0.114	0.48017
1	0.352	0.416	15	0.531×10^{-2}	0.48545
2	0.419	0.757	16	0.171×10^{-1}	0.50241
3	0.464	0.984	17	0.788×10^{-2}	0.49458
4	0.534	0.876	18	0.656×10^{-3}	0.49523
5	0.571	0.871	19	0.297×10^{-2}	0.49819
6	0.142	0.729	20	0.385×10^{-2}	0.49436
7	0.441	0.636	21	0.148×10^{-2}	0.49584
8	0.543	0.256	22	0.298×10^{-3}	0.49613
9	0.153	0.399	23	0.234×10^{-4}	0.49610
10	0.328	0.695	24	0.253×10^{-5}	0.49611
11	0.566	0.668	25	0.276×10^{-6}	0.49611
12	0.514	0.341	26	0.284×10^{-7}	0.49611
13	0.311	0.594			

In Fig. 13.1, a plot of $|b_n|$ as a function of n is given and in Fig. 13.2, the dependence of $|S_N|$ on N is presented. The computation is carried out for backscattering ($\varphi = 180°$) and at fixed $x = x_0$. The value $x = 15.31$ corresponds to the fourth resonance of the S_0 wave. The numerical x value, and choosing it at a resonance position, are unessential for the following considerations.

One can see from Fig. 13.1 that at $n < x$ the function $|b_n|$ has large values and is strongly varying with n; above $n \sim x_0$ the $|b_n|$ value abruptly decreases (see Table 13.1). The partial sums change correspondingly. At $n \lesssim x_0$ the series does not yet converge; on the contrary, at $n \gtrsim x_0$ the series converges rapidly. It should be noted that for backscattering the series is an alternating one and therefore converges rapidly compared with the case of forward scattering.

In Figs. 13.3–8, the form functions of the considered problem are presented. In the series (13.7), aside from the zero term ($n = 0$), we preserve all the terms up to 8, 16, 32 and 64, respectively. The proper form function of the same problem, where the upper number n in the sum (13.7) is found from the condition

$$|b_n| < \varepsilon \tag{13.8}$$

is shown in Fig. 13.9. We have used $\varepsilon = 10^{-7}$.

At different levels of truncation the form function is correct only up to the following x values

N	8	16	32	64
x	0	16	27	60

or approximately, up to $x \sim N$.

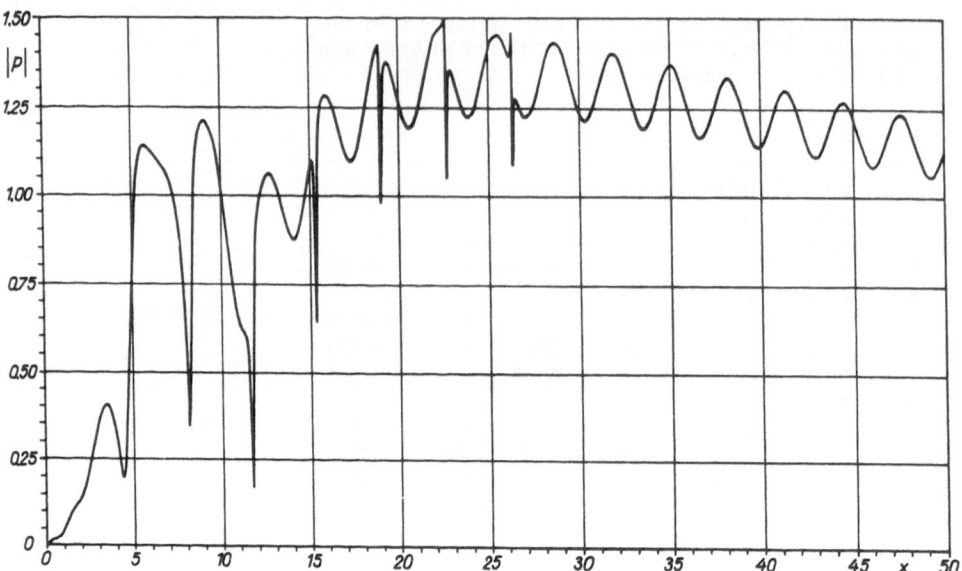

Fig. 13.3. The form function for backscattering by an Armco iron shell (with $h = 1/24$) when the series is truncated at $N = 8$

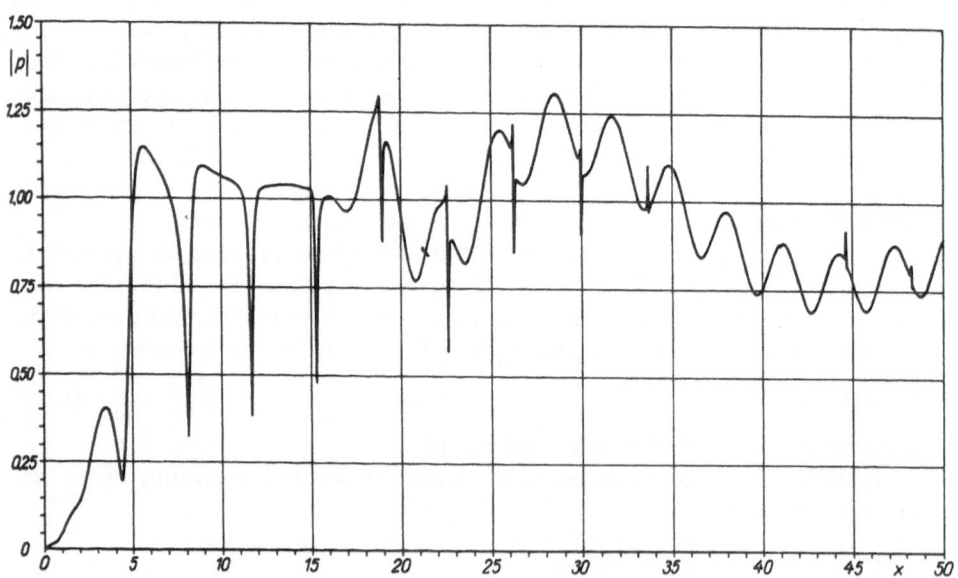

Fig. 13.4. The form function for backscattering by an Armco iron shell (with $h = 1/24$) when the series is truncated at $N = 16$

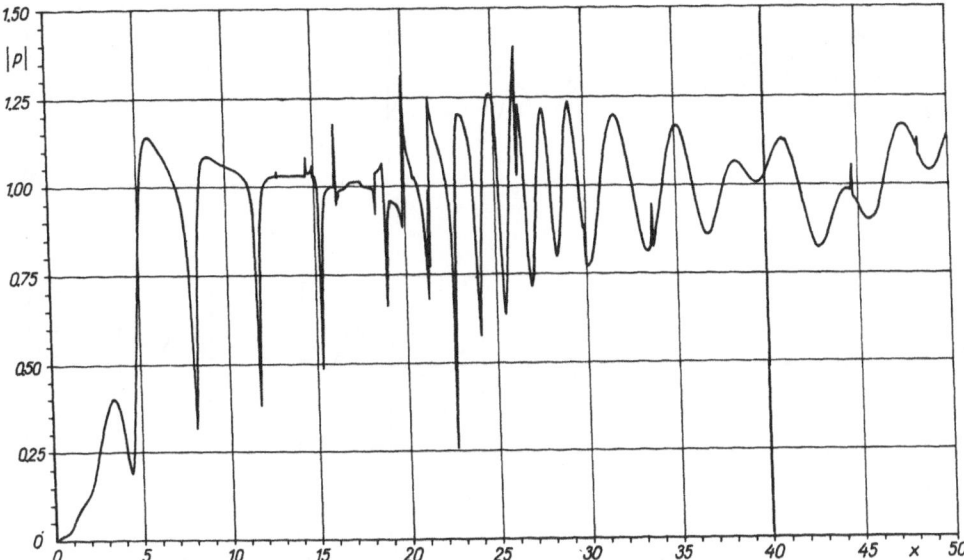

Fig. 13.5. The form function for backscattering by an Armco iron shell (with $h = 1/24$) when the series is truncated at $N = 32$

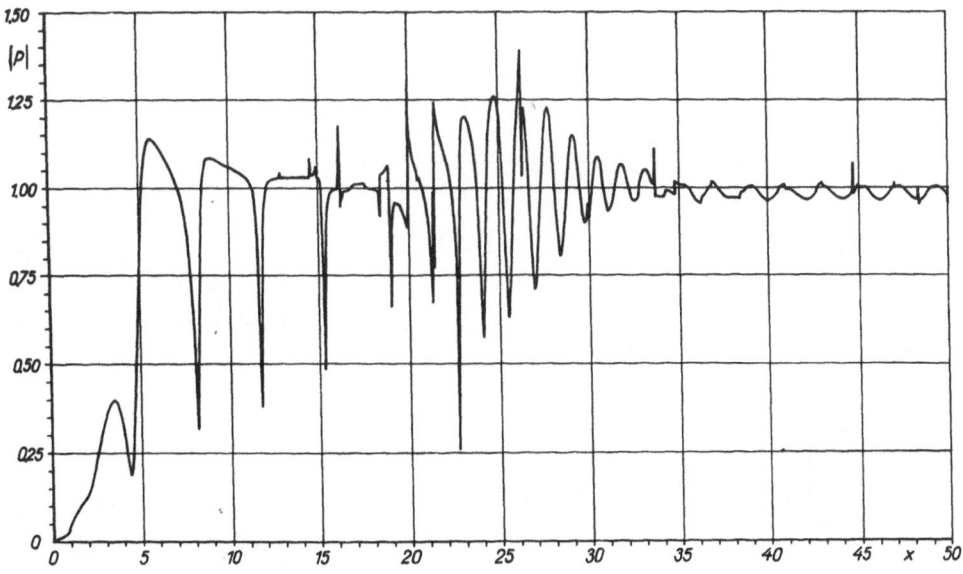

Fig. 13.6. The form function for backscattering by an Armco iron shell (with $h = 1/24$) when the series is truncated at $N = 64$

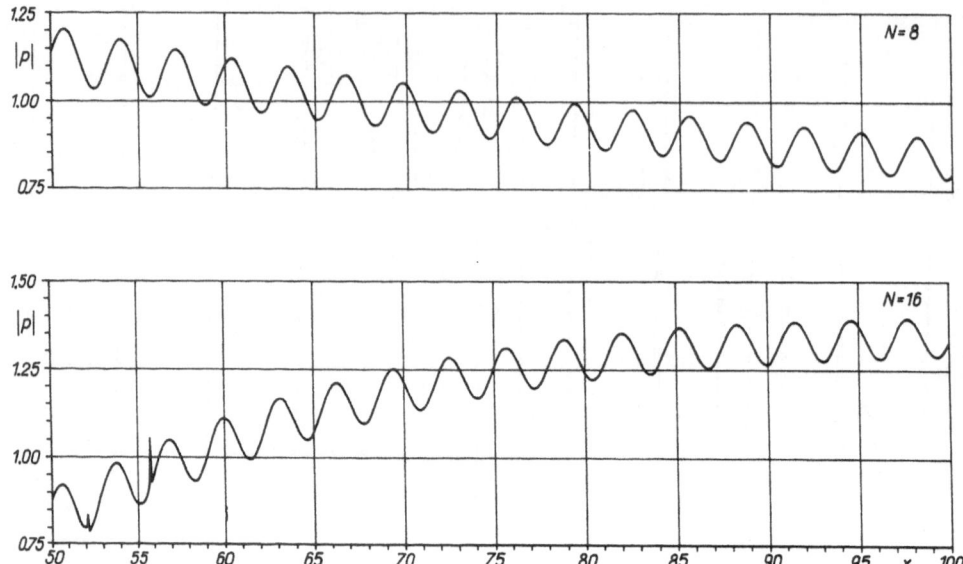

Fig. 13.7. The form function for backscattering by an Armco iron shell (with $h = 1/24$) when the series is truncated at $N = 8$ and $N = 16$, respectively

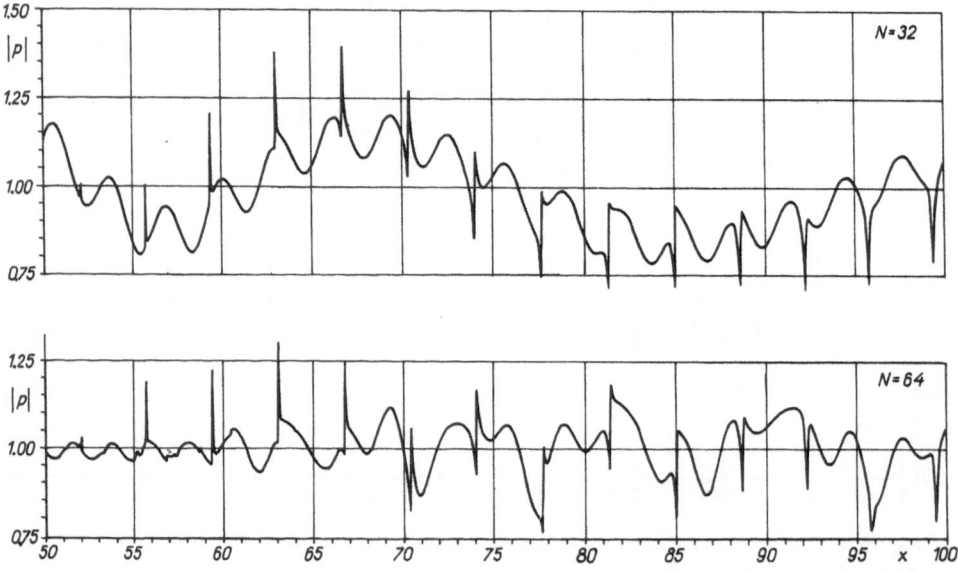

Fig. 13.8. The form function for backscattering by an Armco iron shell (with $h = 1/24$) when the series is truncated at $N = 32$ and $N = 64$, respectively

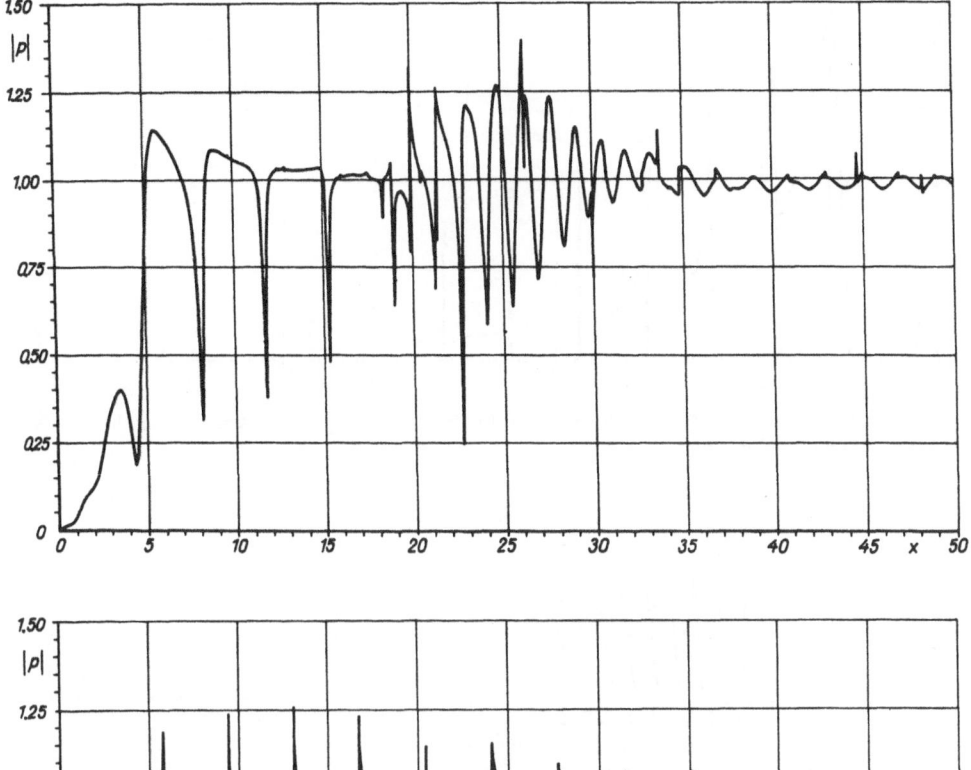

Fig. 13.9. The proper form function for backscattering by an Armco iron shell (with $h = 1/24$) immersed in water

Usually, the form function is computed in a restricted x domain (from x_a to x_b). Therefore the series truncation, fundamentally, introduces a filtering of the form function. The filtering becomes most clearly apparent in the plots of isolated modal resonances.

In Figs. 13.10 and 13.11, the plots of isolated modal resonances are presented for the S_0 and the A wave, respectively.

One can see from Figs. 13.10 and 13.11 that the magnitudes of isolated resonances slowly decrease with increasing n and therefore, formally, there is no reason to truncate the series at $n = 8, 16, 32, \ldots$

When the relative thickness of the shell is $h = 1/24$, only two peripheral waves, namely S_0 and A, contribute to the form function in the domain

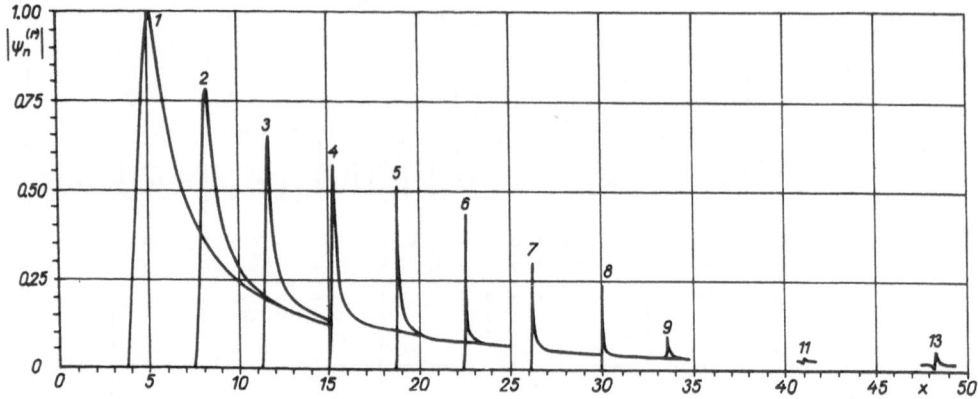

Fig. 13.10. Modal resonances of the S_0 wave for backscattering by an Armco iron shell (with $h = 1/24$) immersed in water

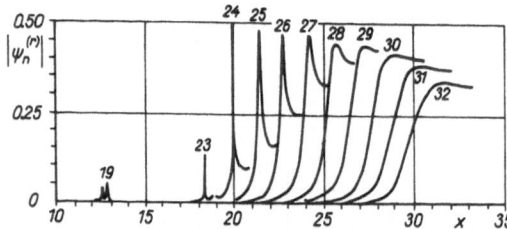

Fig. 13.11. Modal resonances of the A wave for backscattering by an Armco iron shell (with $h = 1/24$) immersed in water

$0 \leqslant x \leqslant 100$. For the case of a thicker shell (say with $h = 1/10$), more peripheral waves will be generated in this x domain (see Chaps. 8 and 10).

As an example, let us consider the problem of backscattering of a plane acoustic wave by a circular–cylindrical shell with $h = 1/10$. The physical parameters of the problem, as before, are given by (8.1). Only the domain $0 \leqslant x \leqslant 150$ will be analyzed. The plots of isolated modal resonances are given in Figs. 8.3–11. If the series (13.7) is truncated at $n = 10$ then all the resonances with n bigger than 10 will be absent. This means that the S_0 wave will be described up to $x < 38$, the A wave up to $x < 8$, the A_1 wave will be described in the domain $66 \leqslant x \leqslant 78$, and the S_2 wave in the domain $137 \leqslant x \leqslant 152$. The resonances of the S_1 wave cannot be considered as having been taken into account, because ten resonances situated in the range $124 \leqslant x \leqslant 132$ are with increasing n shifted on the x axis from the right to the left; the succeeding ones, with indices $n = 16$–35, are with increasing n shifted from the left to the right and are situated in the same region. Magnitudes of the thirty five first isolated resonances of the S_1 wave change slowly and therefore, if the contributions of resonances with $n > 10$ are neglected, the contributions of resonances with $n \leqslant 10$ must also be discarded. Thus, when the series is truncated at $n = 10$ no resonance of the S_1 wave is described.

In fact, when in the series (13.7) we only keep the terms with $n \leqslant 10$, the contribution of all the resonances situated above $x > 8$ should be omitted. The magnitudes of isolated resonances of waves of different types are commensurable in the order, and if the A wave resonances situated at $x > 8$ are rejected (by the restriction $n \leqslant 10$), the contributions of other resonances situated at $x > 8$ must also be discarded. Thus we come to the conclusion that the restriction $n \leqslant n_0$ brings about a filtering of the form function (in the example considered, from the waves A, S_0, S_1, S_2 and A_1 only the two waves S_0 and A are retained), and a restriction on the frequency band. In the example considered the initial band $0 \leqslant x \leqslant 150$ is reduced nineteen times and is now only $0 \leqslant x \leqslant 8$.

As is well known, at small x (say at $x \lesssim 10$) the dispersion curve of the phase velocity of the S_0 (A) wave is situated higher (lower) than c_t. Therefore, for fixed n, the resonance of the S_0 wave is situated to the right in comparison with the resonance of the A wave. If the upper value of the summation index in the series (13.6) is fixed, then the region in which one can trust the obtained form function will be determined by the resonance frequency x_{n_0} of the A wave.

The truncations of the series simplify the computational model. In the example considered the problem at first was formulated with the equations of motion corresponding to linear elasticity theory. In the domain $0 \leqslant x \leqslant 150$, five types of peripheral waves (three symmetric and two antisymmetric) can be generated in the shell. The restriction $n \leqslant 10$ preserves only the waves S_0 and A. Thus, with the assumption $n \leqslant 10$ the original problem could from the very beginning be posed in a simpler form using the two-dimensional thin shell theory (say, of Timoshenko-type) in which these two waves are sufficiently well described.

The restriction $n \leqslant n_0$ not only makes the problem poorer in the sense of the peripheral waves generated in it, but it also strongly reduces the frequency band in which the solution can be used. The situation is especially unfortunate when the transient problem is considered. The band in which one can trust the computed form function is rather narrow. To obtain the correct results in the transient problem one should put restrictions on the loading (the incident wave). Its frequency, when the pulse is a harmonic wave, should be situated in the domain where one can trust the computed form function, and its spectrum must be very narrow. These conditions could not be fulfilled very often, and therefore one should renounce the formulation of the transient problem.

13.6 Ordinary Precision

The coefficients b_n in the series (13.7) are rather complicated (see Chaps. 2 and 3) and have a quotient form. Its numerator and denominator are determinants (of the third rank when the scatterer is solid and the fifth rank when the scatterer is a shell) with elements containing Bessel functions and their derivatives of different arguments. The inverse denominator, through its zeroes, furnishes poles, and the numerator has zeroes which are situated near the poles. Therefore the computation of elements of determinants and their values should be carried out very precisely. Besides that, a sufficiently exact asymptotics of the Bessel functions should be used when the index of the function is larger than the argument; the calculation on a computer with 16 digits must be carried out with double precision, because the ordinary length of the word is not sufficient. Our computational experience has shown that when ordinary precision is used, then with increasing frequency the possibility of false extrema arises. They form a fine structure. The interval of these false extrema allows us to distinguish them from the genuine ones. The proper extrema of the form function reflect the resonances (and antiresonances) of the shock and slip waves caused by the peripheral and creeping waves. Their interval is connected with the typical linear dimension – the wavelength – depending on the ordinal number and type of the resonance. The false extrema do not relate to any meaningful loading. They could be distinguished by successive reduction of the computation step. The number of false extrema in this case will increase and their magnitude (or their deviation from the carrying line) will be practically unchanged. The plot of the form function looks like white noise in this case. The false resonances could be eliminated only by applying double precision (without any change of the algorithm and computation program). Thus here, as in other cases, one should attempt to foresee in a general way what should be obtained as a result of computations, and what could, in principle, not be obtained. In this context the above presented qualitative procedures of analyzing the types of waves generated in the elastic scatterer become extremely useful.

13.7 Errors in Interpretation

13.7.1 Indirect Methods of Finding the Resonance Frequencies

Sometimes, model problems are used to find the resonance frequencies generated by the incident wave in the elastic body. Naturally, being model problems, the simple ones are chosen. The problem of finding eigenfrequencies of a "dry" elastic body is one of them. In the practical case of a metallic body immersed in water, the relative impedance $\varepsilon = \rho c / \rho_1 c_t$ is rather small, and therefore the influence of the liquid on the resonance frequencies, as a rule, is also small (see, for example, the results from Chap. 12). Thus, with sufficient accuracy, the resonance frequencies of a "dry" body can be used instead of the resonance frequencies of a body immersed in a liquid. (It is clear, however, that from such a model one cannot obtain the resonances of the fluid-borne waves.) Since the positions of isolated resonances coincide with the resonances of peripheral waves, the dispersion curves of the phase velocities can also be obtained. They do not differ from those calculated according to the resonance scattering theory. A further note should be made here. The dispersion curves obtained with the use of the resonance scattering theory correspond to the actually generated resonances, and those obtained from the model problem only to some possible ones. Thus in such a case one part of the dispersion curve corresponds to real resonances and the other part to non-existing resonances.

13.7.2   Inadequate Choice of the Background

As a background component the rigid and soft backgrounds are often used, the intermediate one is rarely employed. The usual recommendation says: at $h \gtrsim 1/10$ the rigid background should be used, at $h \lesssim 1/100$ the soft one, and at $1/100 \lesssim x \lesssim 1/10$ the intermediate one. Certainly, the choice of the background should depend not only on the relative thickness of the shell but also on the physical parameters of the scatterer and the liquid. Since it is usually supposed that the shell is made from a malleable metal and immersed in water, in the recommendation nothing is said about the physical parameters.

Choosing the background, we want to eliminate the interference it causes. We are interested in peripheral waves and want to exclude the specularly reflected and diffracted waves by an appropriate choice of the background. In choosing the background, we require an adequate description of the specularly reflected wave and the diffracted waves on the elastic body when obtained from the body of comparison (that provides the background description). Here it is desired that the remainder (the peripheral waves) can be observed in a pure form.

In fact the situation is more complicated. The procedure is successful for the isolated modal resonances with a large magnitude and Q factor. The less the magnitude and the Q factor, the worse is the result. In frequency regions where several peripheral waves contribute to the form function, at some n values the procedure of the resonance scattering theory permits us to observe only the superposition of two or more resonance curves. If resonance frequencies of the same n number, but stemming from different families, are situated near each other, it is nevertheless possible to distinguish the resonances and include them in a proper family. When they coincide, as happens at the points of intersection of dispersion curves from different families (see Figs. 8.2 and 10.8) then the direct application of the resonance scattering theory does not provide the opportunity to separate the resonances and determine the magnitudes of each resonance curve separately (see, for example, the dashed lines in Fig. 8.7).

Let us consider a situation where a superposition of two waves N_1 and N_2 exists. We suppose that the first wave possesses a small magnitude and large Q factor, but that for the second wave both of these are small. The isolated modal resonances of the N_1 wave can be observed in superposition with those belonging to the N_2 wave. This is the case when the "tail" of the isolated resonance of the N_2 wave extends to the resonance frequency of the N_1 wave. Here the resonance curve of the N_2 wave plays the role of the background for the resonance of the N_1 wave. The magnitude of such a background is often bigger than the resonance magnitude of the N_1 wave. In this case the resonance magnitude of the N_1 wave should be counted from the level of this background. Here the error in the obtained resonance magnitude may be rather big.

An erroneous choice of the background can be easily recognized for a wave with a large magnitude and Q factor. With a properly chosen background (say, for example, the rigid one) the resonance curve has a standard form. If instead of the proper (rigid) background the acoustically soft background is chosen, then the configuration of the resonance curve will be in the form of a "dip". It is typical that even in the case of an erroneously chosen background, the resonance frequency and the Q factor will be correctly determined.

13.7.3   Families of Resonances with Equal Cutoff Frequencies

The analysis of the form function is rather easy when in some frequency domain the main contribution is provided by the resonances of one peripheral wave. The typical situation is more complicated. This is the case when, beginning from some cutoff frequency, the incident wave generates in the scatterer two or more peripheral waves. In the two-dimensional scattering problem (e.g., that for spherical and cylindrical shells) this occurs at zero frequency. Beginning from $x = 0$, the water-borne wave A, two zero-order Lamb-type waves S_0 and A_0, and the Franz-type creeping wave can be generated. Depending on the relative thickness of the shell h, the domain in which the resonances of the water-borne wave A can be clearly observed is situated either at low frequencies (for thick-walled shells) or at high frequencies (for thin-walled shells). The low-frequency domain is being investigated in great detail and here, as a rule, nothing unexpected will happen. Partially, this can be explained by the essential difference in the properties of the isolated modal resonances of the waves A, S_0 and A_0 (see Chaps. 8–10). And even in this domain, at small h values, it is sometimes difficult to indicate unambiguously which wave resonances we are dealing with. This is especially the case for the situation when, with n increasing, the resonance curves of one family are degenerating and those belonging to the other family are rising. The typical example of such a situation is the case of the isolated modal resonances of the A and A_0 waves (see Fig. 8.4). The resonances of the A wave degenerate at $n \gtrsim 10$. In turn, at $n \gtrsim 13$ the A_0 wave resonances become more and more apparent. It is difficult to recognize immediately that they belong to different families.

The situation is not so easy when the three-dimensional scattering problem is considered. Obliquely incident plane waves generate shear waves in a thin cylindrical shell. The T_0 wave is generated beginning from $x = 0$. Therefore, in this case, five families of resonances can be simultaneously observed at small x values. They correspond to the A, A_1, S_0, T_0 and Franz waves. At small angles of incidence (at $\alpha \to 0$) the cutoff frequency of the T_1 wave coincides with that of the A_1 wave. Potentially, the isolated modal resonances of the T_1 wave can be observed beginning from the cutoff frequency. However, at small n values the loading (the incident plane wave) can generate the shell resonances with rather small magnitudes. Thus, for example, in Fig. 12.2, the resonances of the T_1 wave can be actually observed only at $n \geqslant 17$. Using the dispersion curves of the phase velocities (or the Regge poles), one can easily separate the resonances belonging to different families.

13.7.4   Superposition of Resonances with Low Q Factor

From the analysis of the acoustic field scattered by cylindrical and spherical shells, it is seen that with n increasing, for every Lamb-type peripheral wave the isolated modal resonances with successive n values become more and more similar to each other (see Figs. 10.3–5). Their magnitude and Q factor change slowly with frequency, the Q factor being rather small, and the resonance curves are overlapping. As a rule, the successive (in n) resonances are in anti-phase and the result of their superposition (on the modulus) is substantially smaller than a single resonance. The contribution of such a superposition is almost not distinguishable in the form function. In this case the extrema of the form function do not coincide with the positions of resonance frequencies. Therefore, without any knowledge of what change the overlapping resonances cause to the family of isolated resonances, one can just omit their contribution beginning from some frequency. At the resonance frequency, the contribution of the hidden resonance can be commensurable with that corresponding to the specularly reflected wave, as it is the case for scattering by a spherical shell (see Sect. 10.4).

We shall now illustrate this situation. Let us refer to Fig. 10.1. On the form function curve, the typical resonance can be observed as a maximum, $(18, 0)$, or a minimum, $(28, 0)$, or as an oscillation, $(22, 0)$. At the resonance frequency the ordinate of the extremum differs strongly from the level of unity corresponding to the contribution of the specularly reflected wave. With the Q factor diminishing (this can be observed as the increase of the angle of opening of the form function curve, in the vicinity of the resonance frequency) on the $p(x)$ curve, the difference between the resonance and antiresonance gradually diminishes. At $x \gtrsim 150$ it is difficult to say which point of the form function curve (maximum or minimum) corresponds to the resonance frequency. At $180 \lesssim x \lesssim 210$ the form function curve looks like a damped sinusoid, lifted to the level $p(x) = 1$. As it is shown in Sect. 10.4, in this restricted frequency band the resonance frequencies of the S_0 wave coincide neither with the maxima, nor with the minima of the form function curve. The above-described character of the form function curve is typical and can be observed in rather broad bands as, for example, at $250 \lesssim x \lesssim 350$, where the result of a superposition of hidden resonances of the A_1 wave becomes apparent.

14   Estimation of the Shell Thickness
from the Form Function

We present here the results of investigating the influence of the relative thickness of the shell h, the density ρ and sound velocity c of the liquid surrounding the shell on the form function for small x values ($x \lesssim 10$). The analysis shows that the form function has stable and informative points, which are proposed to be used for the determination of the parameters of the shell via the acoustic field scattered by it. The approach outlined below still has the character of a test, and the obtained relative thickness should be considered not as an exact value, but as an estimate.

14.1   Form Function Structure in the Domain
of the First Resonances of the S_0 Wave

Let a circular–cylindrical shell of infinite extent be immersed in an unbounded ideal compressible liquid. On this shell let there be an incident plane acoustic pressure wave whose direction of propagation is normal to the longitudinal axis of the shell, and is scattered by the shell. The problem considered is a two-dimensional one.

It is supposed that the form function has been measured in the far field for backscattering. The problem of determining the radius of the shell a from this measurement is a standard one, and will not be considered here. It is thus assumed that the outer radius of the shell is known. The thickness of the shell is to be found from the measured form function.

The experience gained when solving scattering problems by thin-walled cylindrical shells has shown that for scattering by metal shells in water at small x ($x \lesssim 10$), the form function is composed of contributions from: (i) specular reflection, (ii) refraction in the thin cylindrical layer, (iii) diffraction (in the narrow sense) and (iv) re-radiation.

For backscattering, the specular reflection and refraction can be described by simple formulae taken from a model problem, that of reflection of a plane acoustic wave from a plane liquid layer at normal incidence (see Sect. 9.2). Here it is supposed that the sound velocity c and density of the liquid layer ρ are equal to c_1 and ρ_1, respectively. The outer divergence of the specularly reflected wave should be taken into account by an additional geometrical "curvature" factor. In the small-frequency range the reflection and refraction processes do not generate extrema in the form function.

In the case of a thin-walled shell, the diffracted (creeping) wave generated by the incident plane wave is, to a large extent, similar to the diffracted wave on an acoustically soft cylinder. The magnitude of the latter wave is negligibly small in comparison with the magnitude of the specularly reflected wave. Therefore, in the total form function the influence of the diffracted wave can practically not be observed.

Two Lamb-type waves, namely the S_0 and A_0 waves, and the Stoneley-type wave A, do not possess any cutoff frequency and can be generated beginning from $x = 0$. In fact, at small x the resonances of the shock wave radiated by the A wave can be observed in the total form function only for rather thick shells (see, e.g., Fig. 8.1, where $h = 1/10$). With the shell becoming thinner, the resonances of the A wave become apparent only beginning with $n > n_0$, and their resonances are situated at rather high frequencies (see, e.g., Table 12.2 at $\alpha = 0°$ for the shell with $h = 1/32$, and Table 9.2 for the shell with $h = 1/64$). Therefore, at $h \ll 1$ and $x \lesssim 10$ the resonances corresponding to the S_0 wave should be clearly observed in the form function. They can be used for estimating the thickness of the shell.

As was shown before (see, e.g., Fig. 5.9), with h diminishing, the dispersion curve of the S_0 wave generated on the shell tends towards that corresponding to the S_0 wave in the limiting problem of wave propagation in a "dry" plane layer. Except in the domain of small frequencies (at $k_t d \lesssim 0.2$), for shells with $h \lesssim 1/30$ the indicated dispersion curves practically coincide. Thus, the limit curve can be used for the determination of the phase velocity. This limit curve, as is well known, is found from the dispersion equation $E = 0$ (see (8.2)). At small z, coth p_l and coth p_t in this equation can be replaced by their asymptotic representation and the following solution can be obtained

$$y = c^{ph}/c_t = \text{const} = 2\sqrt{1 - \gamma_0^2} \,, \tag{14.1}$$

where $\gamma_0 = c_t/c_l$ is connected only with Poisson's ratio v (see (8.4)).

The relation (14.1) can be recast into the form

$$c^{ph} = c_* = c_t \sqrt{\frac{2}{1 - v}} = c_l \frac{\sqrt{1 - 2v}}{1 - v} \,. \tag{14.2}$$

This is the well-known velocity of the longitudinal wave in a thin plate. The value given by (14.2) can be considered as a physical constant determining the material. For example, instead of c_l, c_t and ρ_1, the material can be described either by c_l, c_* and ρ_1 or by c_*, c_t and ρ_1. The c_* values for different materials are given in [14.1, Table 9.1]. There this velocity is termed ψc_1. According to (14.2),

$$\psi = \frac{\sqrt{1 - 2v}}{1 - v} \,. \tag{14.3}$$

Two misprints are found in the mentioned table. They concern the velocity c_* for zinc and cast iron. The correct values are 3960 m/s and 3990 m/s, respectively.

As follows from this table, the velocity c_* usually determines unambiguously the material. True, even in the limits of this table two different materials – duraluminum and Armco iron – correspond to a single value $c_* = 5440$ m/s.

As is well known, the normalized phase velocity

$$v^{ph}(x) = c^{ph}(x)/c \tag{14.4}$$

is defined by the formula

$$v^{ph}(x) = x_n/n \quad (n = 1, 2, 3, \dots), \tag{14.5}$$

where x_n is the resonance position on the x axis, n is the ordinal number of the resonance, and c is the sound velocity in the liquid. Here, for the sake of simplicity, only one index is used; the wave-family index l is dropped. It is supposed that we are dealing with the S_0 wave resonances. They are clearly distinguished on the form function curve as sharp and deep dips.

As an example, in Table 14.1 the positions of the first three resonances of the S_0 wave obtained directly from the form function are given. The computations are carried out for the case of a copper shell immersed in water at the following values of material parameters

copper: $\rho_1 = 8.9 \times 10^3$ kg/m³, $\quad c_1 = 5060$ m/s, $\quad c_t = 2260$ m/s ,

water: $\quad \rho = 1 \times 10^3$ kg/m³, $\quad c = 1493$ m/s . $\tag{14.6}$

The relative thickness of the shell $h = 1 - b/a$ was changed within broad limits

$$h = 2^{-q} \quad (q = 5, 6, \dots, 10) . \tag{14.7}$$

Table 14.1. Dependence of the position x_* and the value $|p(x_*)|$ of the first maximum of the form function, the positions of the first three resonances of the S_0 wave x_n ($n = 1, 2, 3$) and the phase velocity of this wave, $c^{ph}(x_n)$, on the relative thickness of the shell h in backscattering from a copper shell in water

h	1/32	1/64	1/128	1/256	1/512	1/1024
x_*	1.88	1.21	0.70	0.43	0.27	0.12
$|p(x_*)|$	0.55	1.05	1.38	1.70	2.13	3.29
x_1	2.97	2.73	2.70	2.70	2.70	2.70
$c^{ph}(x_1)$ [m/s]	4430	4080	4030	4030	4030	4030
x_2	5.94	5.70	5.51	5.47	5.43	5.43
$c^{ph}(x_2)$ [m/s]	4430	4260	4110	4080	4050	4030
x_3	8.67	8.47	8.32	8.20	8.16	8.13
$c^{ph}(x_3)$ [m/s]	4310	4220	4140	4080	4060	4050

As can be seen form the table, the thinner the shell, the less the correct phase velocity of the S_0 wave differs from the limiting value $c_* = 4040$ m/s.

By direct computations it can be shown that the phase velocity of the S_0 wave generated on the cylindrical shell depends only weakly on the properties of the liquid in which the shell is immersed. This is confirmed by the tables of the phase velocities, obtained from the first resonances of the S_0 wave in the total form function. The resonances of the S_0 wave are so clearly distinguishable that they can be identified without any separation of modal resonances. As an example, below we present the results of computations for backscattering by an aluminum shell with $h = 1/64$ and with the following material parameters

$$\rho_1 = 2.7 \times 10^3 \text{ kg/m}^3, \qquad c_1 = 6420 \text{ m/s}, \qquad c_t = 3040 \text{ m/s} . \tag{14.8}$$

Table 14.2. Dependence on the sound velocity of the liquid $c_m = 0.1 \, m \, c_t$ of the position x_* and the value $|p(x_*)|$ of the first maximum of the form function, the positions x_n, the values $|p(x_n)|$, and the phase velocities $c^{\text{ph}}(x_n)$ of the first three resonances of the S_0 wave in backscattering from an aluminum shell with $h = 1/64$ in a liquid with $\rho = 1 \times 10^3$ kg/m^3

m	$c_m = \dfrac{m}{10} c_t$ [m/s]	x_* $|p(x_*)|$	x_1 $|p(x_1)|$ $c^{\text{ph}}(x_1)$ [m/s]	x_2 $|p(x_2)|$ $c^{\text{ph}}(x_2)$ [m/s]	x_3 $|p(x_3)|$ $c^{\text{ph}}(x_3)$ [m/s]
4	1216	1.06 1.26	4.38 0.15 5320	8.95 0.36 5440	13.52 0.45 5480
5	1520	0.74 1.39	3.52 0.05 5340	7.15 0.27 5430	10.78 0.38 5460
6	1824	0.55 1.53	2.93 0.08 5340	5.98 0.16 5450	8.98 0.32 5460
7	2128	0.43 1.67	2.54 0.24 5410	5.12 0.12 5440	7.70 0.31 5460
8	2432	0.35 1.82	2.23 0.32 5420	4.49 0.01 5460	6.76 0.28 5480
9	2736	0.31 1.96	1.95 0.44 5340	3.98 0.21 5450	6.02 0.33 5490
10	3040	0.27 2.11	1.76 0.55 5340	3.59 0.17 5460	5.39 0.41 5460
11	3344	0.23 2.24	1.60 0.63 5360	3.28 0.14 5490	4.92 0.39 5490

In the first series of computations (see Table 14.2) the velocity of the hypothetical liquid was changed at a fixed density

$$c_m = 0.1 \, m \, c_t \quad (m = 4, 5, \ldots, 11) \,, \tag{14.9}$$

$$\rho = \text{const} = 1.0 \times 10^3 \, \text{kg/m}^3 \,,$$

in the second series of computations (see Table 14.3) the liquid density was changed at a fixed sound velocity

$$\rho_m = 0.1 \, m \, \rho_1 \quad (m = 3, 4, \ldots, 11), \qquad c = 1493 \, \text{m/s} \,, \tag{14.10}$$

Table 14.3. Dependence on the density of the liquid $\rho_m = 0.1 \, m \, \rho_1$ of the position x_* and the value $|p(x_*)|$ of the first maximum of the form function, the positions x_n, the values $|p(x_n)|$, and the phase velocities $c^{\text{ph}}(x_n)$ of the first three resonances of the S_0 wave in backscattering from an aluminum shell with $h = 1/64$ in a liquid with sound velocity $c = 1493$ m/s

m	$\rho_m = \dfrac{m}{10} \rho_1$ [kg/m³]	x_* $|p(x_*)|$	x_1 $|p(x_1)|$ $c^{\text{ph}}(x_1)$ [m/s]	x_2 $p|(x_2)|$ $c^{\text{ph}}(x_2)$ [m/s]	x_3 $p|(x_3)|$ $c^{\text{ph}}(x_3)$ [m/s]
3	810	0.90 1.31	3.55 0.06 5300	7.31 0.28 5450	11.02 0.39 5500
4	1080	0.70 1.41	3.55 0.07 5310	7.31 0.27 5450	10.98 0.38 5460
5	1350	0.59 1.49	3.60 0.10 5370	7.27 0.25 5420	10.94 0.38 5440
6	1620	0.51 1.57	3.59 0.08 5370	7.27 0.24 5420	10.94 0.36 5440
7	1890	0.47 1.64	3.59 0.05 5370	7.27 0.23 5420	10.94 0.37 5440
8	2160	0.43 1.71	3.59 0.03 5370	7.27 0.23 5420	10.94 0.39 5440
9	2430	0.39 1.77	3.59 0.01 5370	7.27 0.23 5420	10.94 0.39 5420
10	2700	0.35 1.83	3.59 0.03 5370	7.27 0.23 5420	10.94 0.37 5420
11	2970	0.35 1.88	3.59 0.06 5370	7.27 0.23 5420	10.90 0.37 5420

Table 14.4. Dependence on the impedance of the liquid $(\rho c)_m = 0.1\, m\, (\rho_1 c_l)$ of the position x_* and the value $|p(x_*)|$ of the first maximum of the form function, the positions x_n, the values $|p(x_n)|$, and the phase velocities $c^{ph}(x_n)$ of the first three resonances of the S_0 wave in backscattering from an aluminum shell with $h = 1/64$

m	$(\rho c)_m = \dfrac{m}{10}(\rho_1 c_l)$ [kg/m²s]	$\rho_m = \sqrt{\dfrac{m}{10}}\,\rho_1$ [kg/m³]	$c_m = \sqrt{\dfrac{m}{10}}\,c_l$ [m/s]	x_* $\|p(x_*)\|$	x_1 $\|p(x_1)\|$ $c^{ph}(x_1)$ [m/s]	x_2 $\|p(x_2)\|$ $c^{ph}(x_2)$ [m/s]	x_3 $\|p(x_3)\|$ $c^{ph}(x_3)$ [m/s]
4	3283×10^3	1708	1923	0.35 1.86	2.77 0.24 5330	5.66 0.20 5450	8.48 0.35 5430
5	4104×10^3	1909	2150	0.27 2.07	2.50 0.23 5380	5.04 0.28 5420	7.58 0.39 5430
6	4924×10^3	2091	2355	0.23 2.27	2.30 0.42 5430	4.61 0.17 5430	0.95 0.44 5460
7	5746×10^3	2259	2543	0.20 2.44	2.11 0.44 5360	4.30 0.49 5460	6.41 0.47 5430
8	6566×10^3	2719	2719	0.20 2.56	1.99 0.43 5420	3.98 0.54 5420	6.02 0.41 5450
9	7387×10^3	2561	2884	0.16 2.76	1.88 0.46 5410	3.79 0.53 5460	5.66 0.33 5450
10	8208×10^3	2700	3040	0.16 2.85	1.76 0.72 5340	3.59 0.53 5460	5.39 0.60 5460
11	9029×10^3	2832	3188	0.12 2.87	1.68 0.74 5360	3.44 0.71 5480	5.12 0.36 5440

and in the third series of computations (see Table 14.4) the acoustic impedance was changed

$$(\rho c)_m = 0.1 \, m \, (\rho_1 c_t) \quad (m = 4, 5, \ldots, 11) \,. \tag{14.11}$$

In the third series it was supposed that

$$\rho_m = \sqrt{0.1 \, m} \, \rho_1, \qquad c_m = \sqrt{0.1 \, m} \, c_t \,. \tag{14.12}$$

In each of the series the varying values of the velocity, density and impedance of the liquid were proportional to c_t, ρ_1 and $\rho_1 c_t$, respectively, i.e. they changed within very broad limits. Nevertheless, for each of the variants considered, it turns out that the actual velocity of the S_0 wave, although depending on the wave radius x, as a whole differs little from the limit value $c_* = 5360$ m/s. From all the $c^{ph}(x_n)$ values presented in Tables 14.2–4, the smallest is 5300 m/s, and the largest is 5500 m/s.

As the computations have shown, the velocity c_* can be used as an indicator of the shell material. In the table [14.1, Table 9.1] instead of c_* the value $c^{ph}(x_n)$ can be used, where n, as before, defines the ordinal number of the S_0 wave resonance. This number should be chosen from the range where the phase velocity of the S_0 wave changes slowly, i.e. when $c^{ph}(x_n) \cong c^{ph}(x_{n+1})$.

In the analysis of the form functions for backscattering by circular–cylindrical shells, it was noted that the position on the x axis and the value of the first maximum on the form function curve are very sensitive to the relative thickness

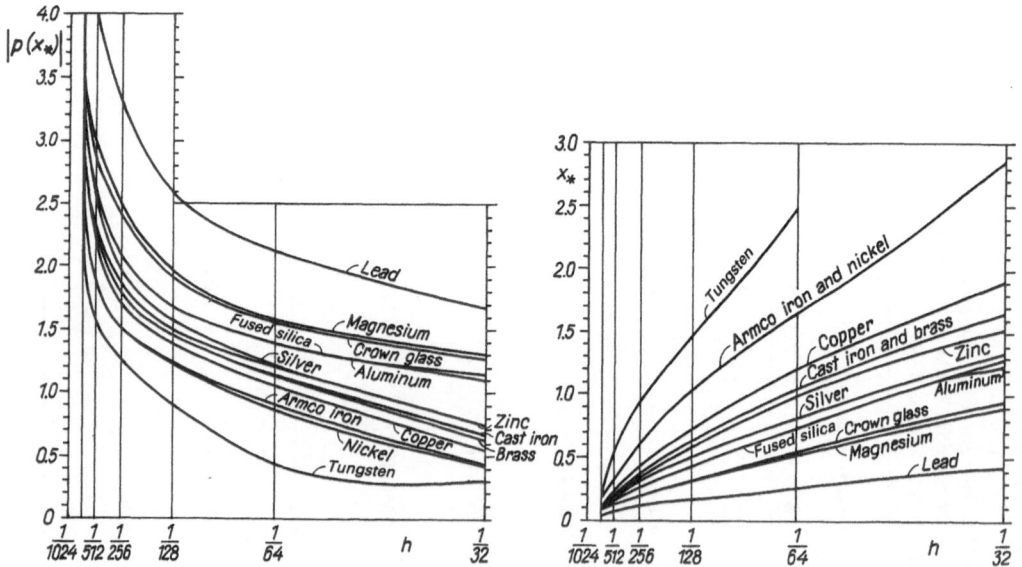

Fig. 14.1. Dependence of the value $|p(x_*)|$ (left) and the position x_* (right) of the first maximum of the form function backscattered by the shell, on its relative thickness h for the case of empty shells of different materials immersed in water [14.2, Fig. 1]

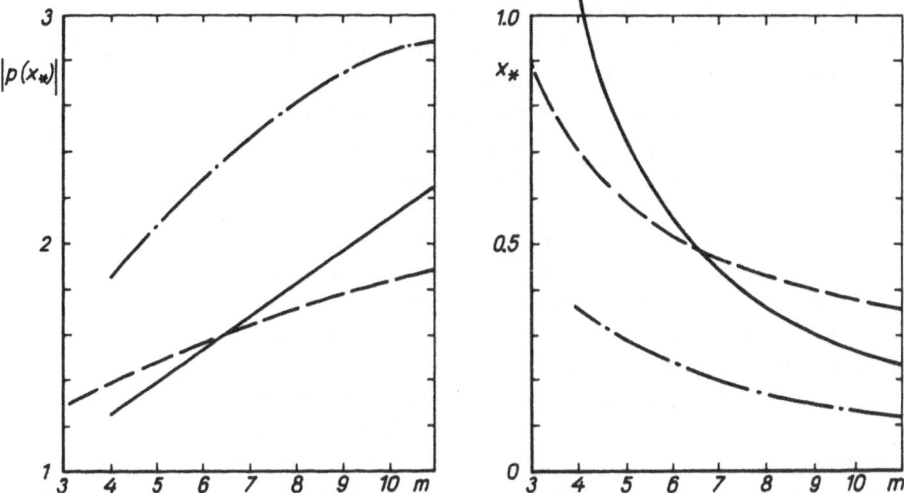

Fig. 14.2. Dependence of the value $|p(x_*)|$ (left) and the position x_* (right) of the first maximum of the form function backscattered by an aluminum shell with $h = 1/64$ on the parameters of the liquid: sound velocity $c_m = 0.1\,m\,c_t$ (solid line), density $\rho_m = 0.1\,m\,\rho_1$ (dashed line), and impedance $(\rho c)_m = 0.1\,m(\rho_1 c_t)$ (dot-and-dash line)

of the shell and its material. It should be mentioned that the first maximum of the form function is neither a resonance nor an antiresonance of the S_0 wave and therefore cannot correspond to its spectral line on the acoustic spectrogram. In Fig. 14.1, for shells of different materials immersed in water, the dependence of the value $|p(x_*)|$ (left) and its position on the x axis x_* (right) on the relative thickness h is shown.

The value $|p(x_*)|$ and the position x_* of the first maximum in the form function curve are sensitive to the parameters of the liquid surrounding the shell. This can be clearly observed from the plots presented in Fig. 14.2. They were obtained for the case of scattering by an aluminum shell with $h = 1/64$.

14.2   Estimation of the Shell Thickness

Suppose that in some x band, beginning with $x = 0$, the form function is obtained from measurements for an empty cylindrical shell. Using this form function, then from the first resonances of the S_0 wave one can obtain c_* [see (14.2)] and then determine the material of the shell from a table, computed beforehand, of the dependence of c_* on the elastic parameters of shell material.

In the left part of Fig. 14.1 the horizontal corresponding to the measured value $|p(x_*)|$ is drawn, its intersection point with the curve corresponding to the chosen material is found, and from this point a perpendicular to the h axis is traced to find the shell thickness. An analog tracing is performed in the right-hand part of Fig. 14.1. The arithmetic mean of both deduced values of h can be taken as the shell thickness.

Using the thus-known elastic parameters of the shell and its thickness as the input data, and computing the form function, one can compare this form function with the measured one, and thus estimate the accuracy of the proposed procedure.

Thus the problem of shell thickness determination is reduced to a simple selection. If some curve of material corresponding to the velocity c_* is missing from Fig. 14.1, then with a sufficiently dense net of curves one can use the curve of a material which has the velocity c_* nearest to the measured one. The list of shell materials, if needed, can be extended.

A similar approach to the shell thickness determination problem can also be used in the case of spherical shells. Evidently, the proposed procedure can be generalized on the problem of sound scattering by multi-layered shells. In the latter case, the number of combinations of shell materials and thicknesses of the layers can be rather large.

14.3   Commentary

Firstly it should be mentioned that the proposed procedure still has the character of a test. Besides, it should be considered as just one of many possible procedures. The result obtained for the relative thickness should be considered not as an exact value, but as an estimate.

One aim of this chapter is to emphasize the fact that in the form function there are active (rich in information) regions (and points), which are closely connected with the inner structure of the elastic scatterer. The resonances of peripheral and creeping waves and the cutoff frequencies can play such a role.

It is clear that the procedure is very sensitive to the choice of the material curve in Fig. 14.1. All measurements are carried out with some (limited) precision and for those materials which have close c_* values it is difficult to choose an appropriate curve in Fig. 14.1. In practice, the matter will be simplified if, when choosing the material for c_*, one draws on additional considerations. Here we bear in mind that for technological and constructional reasons, the elastic scatterers in water are made of a restricted number of materials. Thus some of materials can be eliminated from the examination beforehand.

It is understood that in order to distinguish between the materials with close c_* values, another procedure should be used. Particularly, while for different materials the c_* values can be equal, or close to each other, their dispersion curves of the S_0 wave phase velocity can be different. Therefore in such cases one has to abandon the determination of the material from the parameter c_*, and use for this purpose the acoustic spectrogram. For the case of thin-walled shells it is sufficient to know the position of the resonance lines of the S_0 wave $(n, 0)$ for the first several values n ($n = 1, 2, \ldots \sim 10$). In this case the dispersion curve of the S_0 wave phase velocity is used as an indicator of the material. It is clear that in such a curve there is more information contained on the scatterer than in just its limit point c_*. In fact, the measurements should be carried out at a larger number of points and the precision should be high (the position of each resonance line should be determined with an accuracy of three significant digits).

It is significant that in the proposed approach both spectral and amplitude features of the form functions are used. This, from our point of view, somewhat extends the possibilities of the method.

Certainly the solution depends very critically on the angle of incidence. At oblique incidence, extrema corresponding to the waves propagating along helical paths will be present in the form function (see Chap. 12). This should be taken into account in the procedure.

In some cases it is sufficient to obtain not the value of h itself, but an estimate of it, say in form $h \ll 1$. In this situation the approach outlined above can be useful.

In this proposed approach, the long-wavelength region of the form function is used. Here, in this rather limited domain, the form function can be easily measured and computed.

The material presented in this chapter was first published in part in [14.2].

References

Chapter 1

1.1 T.R. Meeker, A.H. Meitzler: Wave-guide propagation in elongated cylinders and plates, in *Physical Acoustics: Principles and Methods*, Vol. 1, ed. by W.P. Mason (Academic, New York 1964) pp. 140–203

1.2 N.D. Veksler: *Information Analysis in Hydroelasticity* (Valgus, Tallinn 1982) (in Russian)

1.3 R.M. White: Elastic wave scattering at a cylindrical discontinuity in a solid. J. Acoust. Soc. Am. **30**, 771–785 (1958)

1.4 L. Flax, V.K. Varadan, V.V. Varadan: Scattering of an obliquely incident wave by an infinite cylinder. J. Acoust. Soc. Am. **68**, 1832–1835 (1980)

1.5 G. Maze, J.L. Izbicki, J. Ripoche: Acoustic scattering from cylindrical shells: Guided waves and resonances of the liquid column. Ultrasonics **24**, 354–362 (1986)

1.6 A. Nagl, H. Überall, P.P. Delsanto, J.D. Alemar, E. Rosario: Refraction effects in the generation of helical surface waves on a cylindrical obstacle. Wave Motion **5**, 235–247 (1983)

1.7 H. Hönl, A. Maue, K. Westphal: *Theorie der Beugung* (Springer, Berlin 1961)

Chapter 2

2.1 L. Flax, L.R. Dragonette, H. Überall: Theory of elastic resonance excitation by sound scattering. J. Acoust. Soc. Am. **63**, 723–731 (1978)

2.2 L. Flax, W.G. Neubauer: Acoustic reflection from layered elastic absorptive cylinders. J. Acoust. Soc. Am. **61**, 307–312 (1977)

2.3 J.J. Faran Jr.: Sound scattering by solid cylinders and spheres. J. Acoust. Soc. Am. **23**, 405–418 (1951)

2.4 N.D. Veksler: *Information Analysis in Hydroelasticity* (Valgus, Tallinn 1982) (in Russian)

2.5 G. Maze, J. Ripoche, A. Derem, J.L. Rousselot: Diffusion d'une onde ultrasonore par des tubes remplis d'air immergés dans l'eau. Acustica **55**, 69–85 (1984)

2.6 R.H. Vogt, W.G. Neubauer: Relationship between acoustic reflection and vibrational modes of elastic spheres. J. Acoust. Soc. Am. **60**, 15–22 (1976)

2.7 H. Überall, L.R. Dragonette, P. Flax: Relation between creeping waves and normal modes of vibration of a curved body. J. Acoust. Soc. Am. **61**, 711–715 (1977)

2.8 M.C. Junger: Sound scattering by thin elastic shells. J. Acoust. Soc. Am. **24**, 366–373 (1952)

2.9 M.C. Junger, D. Feit: *Sound, Structures and Their Interaction* (MIT, Cambridge, MA 1972)

2.10 H. Überall, H. Huang: Acoustical response of submerged elastic structures obtained through integral transforms, in *Physical Acoustics*, Vol. 12 ed. by W.P. Mason, R.N. Thurston (Academic, New York 1976) pp. 217–275

2.11 J.D. Murphy, J. George, A. Nagl, H. Überall: Isolation of the resonant component in acoustic scattering from fluid-loaded elastic spherical shells. J. Acoust. Soc. Am. **65**, 368–373 (1979)

2.12 P. Uginčius, H. Überall: Creeping-wave analysis in acoustic scattering by elastic cylindrical shells. J. Acoust. Soc. Am. **43**, 1025–1035 (1968)

2.13 R.D. Doolittle, H. Überall: Sound scattering by elastic cylindrical shells. J. Acoust. Soc. Am. **39**, 272–275 (1966)

2.14 J.D. Murphy, E.D. Breitenbach, H. Überall: Resonance scattering of acoustic waves from cylindrical shells. J. Acoust. Soc. Am. **64**, 677–683 (1978)

2.15 D. Brill, G. Gaunaurd: Acoustic resonance scattering by penetrable cylinder. J. Acoust. Soc. Am. **73**, 1448–1455 (1983)

2.16 G.C. Gaunaurd, D. Brill: Acoustic spectrogram and complex-frequency poles of a resonantly excited elastic tube. J. Acoust. Soc. Am. **75**, 1680–1693 (1984)

2.17 E.D. Breitenbach, H. Überall, K.B. Yoo: Resonant scattering from elastic cylindrical shells. J. Acoust. Soc. Am. **74**, 1267–1273 (1983)

Chapter 3

3.1 J.J. Faran Jr.: Sound scattering by solid cylinders and spheres. J. Acoust. Soc. Am. **23**, 405–418 (1951)

3.2 R. Hickling: Analysis of echoes from a solid elastic sphere in water. J. Acoust. Soc. Am. **34**, 1582–1592 (1962)

3.3 R. Hickling: Analysis of echoes from a hollow metallic sphere in water. J. Acoust. Soc. Am. **36**, 1124–1137 (1964)

3.4 K.J. Diercks, R. Hickling: Echoes from hollow aluminum spheres in water. J. Acoust. Soc. Am. **41**, 380–393 (1967)

3.5 R. Hickling, R.W. Means: Scattering of frequency-modulated pulses by spherical elastic shells in water. J. Acoust. Soc. Am. **44**, 1246–1252 (1968)

3.6 L. Flax, L.R. Dragonette, H. Überall: Theory of elastic resonance excitation by sound scattering. J. Acoust. Soc. Am. **63**, 723–731 (1978)

3.7 N.D. Veksler, U.K. Nigul, R.A. Pukk: On the algorithm of computation by a Fourier series of echo-signals from elastic spherical objects in unbounded acoustic media. Trans. Acad. Sci. USSR, Mekh. Tverd. Tela, 71–83 (1970)

3.8 G.C. Gaunaurd, H. Überall: RST analysis of monostatic and bistatic acoustic echoes from an elastic sphere. J. Acoust. Soc. Am. **73**, 1–12 (1983)

3.9 A.E.H. Love: *A Treatise on the Mathematical Theory of Elasticity* (Dover, New York 1944)

3.10 W.G. Neubauer, R.H. Vogt, L.R. Dragonette: Acoustic reflection from elastic spheres. I. Steady-state signals. J. Acoust. Soc. Am. **55**, 1123–1129 (1974)

3.11 L.R. Dragonette, S.K. Numrich, L.J. Frank: Calibration technique for acoustic scattering measurements. J. Acoust. Soc. Am. **69**, 1186–1189 (1981)

3.12 R.H. Vogt, W.G. Neubauer: Relationship between acoustic reflection and vibrational modes of elastic spheres. J. Acoust. Soc. Am. **60**, 15–22 (1976)

3.13 J.D. Murphy, J. George, A. Nagl, H. Überall: Isolation of the resonant component in acoustic scattering from fluid-loaded elastic spherical shells. J. Acoust. Soc. Am. **65**, 368–373 (1979)

3.14 R.R. Goodman, R. Stern: Reflection and transmission of sound by elastic spherical shells. J. Acoust. Soc. Am. **34**, 338–344 (1962)

3.15 U.K. Nigul, Ja.A. Metsaveer, N.D. Veksler, M.E. Kutser: *Echo Signals from Elastic Bodies* (Academy of Sciences of the Estonian SSR, Tallinn 1974) (in Russian)

3.16 M.F. Werby: The isolation of resonances and the ideal acoustical background for submerged elastic shells. Acoust. Lett. **15**, 65–69 (1991); The acoustical background for a submerged elastic shell. J. Acoust. Soc. Am. **90**, 3279–3282 (1991)

3.17 H. Überall, G.C. Gaunaurd, J.D. Murphy: Acoustic surface wave pulses and the ringing of resonances. J. Acoust. Soc. Am. **72**, 1014–1017 (1982)

Chapter 4

4.1 K.L. Williams, P.L. Marston: Backscattering from an elastic sphere: Sommerfeld–Watson transformation and experimental confirmation. J. Acoust. Soc. Am. **78**, 1093–1102 (1985)

4.2   K.L. Williams, P.L. Marston: Synthesis of backscattering from an elastic sphere using Sommerfeld–Watson transformation and giving a Fabry–Perot analysis of resonances. J. Acoust. Soc. Am. **79**, 1702–1708 (1986)

4.3   J.J. Faran Jr.: Sound scattering by solid cylinders and spheres. J. Acoust. Soc. Am. **23**, 405–418 (1951)

4.4   D. Brill, H. Überall: Acoustic waves transmitted through solid elastic cylinders. J. Acoust. Soc. Am. **50**, 921–939 (1971)

4.5   J.D. Murphy, J. George, A. Nagl, H. Überall: Isolation of the resonant component in acoustic scattering from fluid-loaded elastic spherical shells. J. Acoust. Soc. Am. **65**, 368–373 (1979)

4.6   N.D. Veksler: *Resonance Scattering in Hydroacoustics* (Valgus, Tallinn 1984) (in Russian)

4.7   N.D. Veksler: *Information Analysis in Hydroelasticity* (Valgus, Tallinn 1982) (in Russian)

4.8   G. Gaunaurd, H. Überall: Relation between creeping wave acoustic transients and the complex-frequency poles of the singularity expansion method. J. Acoust. Soc. Am. **78**, 234–243 (1985)

Chapter 5

5.1   G. Maze, B. Taconet, J. Ripoche: Influence des ondes de "Galerie à echo" sur la diffusion d'une onde ultrasonore plane par un cylindre. Phys. Lett. **84A**, 309–312 (1981)

5.2   G. Maze, J. Ripoche: Méthode d'isolement et d'identification des résonances (M.I.I.R.) de cylindres et de tubes soumis à une onde acoustique plane dans l'eau. Rev. Phys. Appl. **18**, 319–326 (1983)

5.3   G. Maze, J. Ripoche: Visualization of acoustic scattering by elastic cylinders at low *ka*. J. Acoust. Soc. Am. **73**, 41–43 (1983)

5.4   A. Derem, J.L. Rousselot, G. Maze, J. Ripoche, A. Faure: Diffusion d'une onde acoustique plane par des cylindres solides immergés: étude expérimentale et théorie des résonances. Acustica **50**, 39–50 (1982)

5.5   G. Maze, J. Ripoche, A. Derem, J.L. Rousselot: Diffusion d'une onde ultrasonore par des tubes remplis d'air immergés dans l'eau. Acustica **55**, 69–85 (1984)

5.6   G. Maze, J.L. Izbicki, J. Ripoche: Resonances of plates and cylinders: Guided waves. J. Acoust. Soc. Am. **77**, 1352–1357 (1985)

5.7   J. Ripoche, G. Maze, J.L. Izbicki: New research in nondestructive testing: Resonance acoustic spectroscopy, in *Ultrasonics International 85*, Conf. Proc. (Butterworth Scientific, Guildford, UK 1985) pp. 364–369

5.8   R. Burvingt, J.L. Rousselot, A. Derem, G. Maze, J. Ripoche: Résponse résonnante de cylindres élastiques. Rev. CETHEDEC – Ondes Signal **78**, 73–93 (1984)

5.9   J.L. Izbicki, G. Maze, J. Ripoche: Diffusion acoustique par des tubes immergés. Origine d'une famille de résonances. Phys. Lett. A **115**, 393–397 (1986)

5.10   G. Maze, J.L. Izbicki, J. Ripoche: Identification de résonances liées à l'interface eau–tube élastique par diffusion acoustique. Rev. Acoust. **78**, 15–18 (1986)

5.11   N.D. Veksler: *Information Analysis in Hydroelasticity* (Valgus, Tallinn 1982) (in Russian)

5.12   G. Gaunaurd, H. Überall: Relation between creeping wave acoustic transients and the complex-frequency poles of the singularity expansion method. J. Acoust. Soc. Am. **78**, 234–243 (1985)

5.13   J.D. Murphy, E.D. Breitenbach, H. Überall: Resonance scattering of acoustic waves from cylindrical shells. J. Acoust. Soc. Am. **64**, 677–683 (1978)

5.14   E.D. Breitenbach, H. Überall, K.B. Yoo: Resonant scattering from elastic cylindrical shells. J. Acoust. Soc. Am. **74**, 1267–1273 (1983)

Chapter 6

6.1   G.R. Barnard, C.M. McKinney: Scattering of acoustic energy by solid and air-filled cylinders in water. J. Acoust. Soc. Am. **33**, 226–238 (1961)

6.2   L.D. Hampton, C.M. McKinney: Experimental study of the scattering of acoustic energy from solid metal spheres in water. J. Acoust. Soc. Am. **33**, 664–673 (1961)

6.3   R. Hickling: Analysis of echoes from a solid elastic sphere in water. J. Acoust. Soc. Am. **34**, 1582–1592 (1962)

6.4   R. Hickling: Analysis of echoes from a hollow metallic sphere in water. J. Acoust. Soc. Am. **36**, 1124–1137 (1964)

6.5   R. Hickling, R.W. Means: Scattering of frequency-modulated pulses by spherical elastic shells in water. J. Acoust. Soc. Am. **44**, 1246–1252 (1968)

6.6   L.R. Dragonette, S.K. Numrich, L.J. Frank: Calibration technique for acoustic scattering measurements. J. Acoust. Soc. Am. **69**, 1186–1189 (1981)

6.7   M. de Billy: Determination of the resonance spectrum of elastic bodies via the use of short pulses and Fourier transform theory. J. Acoust. Soc. Am. **79**, 219–221 (1986)

6.8   L. Flax, L.R. Dragonette, H. Überall: Theory of elastic resonance excitation by sound scattering. J. Acoust. Soc. Am. **63**, 723–731 (1978)

6.9   A. Derem, J.L. Rousselot, G. Maze, J. Ripoche, A. Faure: Diffusion d'une onde acoustique plane par des cylindres solides immergés: étude expérimentale et théorie des résonances. Acustica **50**, 39–50 (1982)

6.10 G. Maze, J. Ripoche: Méthode d'isolement et d'identification des résonances (M.I.I.R.) de cylindres et de tubes soumis à une onde acoustique plane dans l'eau. Rev. Phys. Appl. **18**, 319–326 (1983)

6.11 P. Pareige, J.L. Rembert, G. Maze, J. Ripoche: Méthode impulsionnelle numerisée (MIN) pour l'isolement et l'identification des resonances de tubes immergés. Phys. Lett. A **135**, 143–146 (1989)

6.12 G. Quentin, A. Cand: Pulsed resonance identification method. Electron. Lett. **25**, 353–354 (1989)

6.13 G. Quentin, I. Molinero, M. de Billy: Etude du spectre acoustique de quelques échantillons élastiques de forme cylindrique. II Etude expérimentale. Acustica **65**, 275–283 (1988)

6.14 J.L. Rousselot: Etude du spectre acoustique de quelques échantillons élastiques de forme cylindrique. Acustica **65**, 267–274 (1988)

6.15 M. Talmant, G. Quentin: Backscattering of short ultrasonic pulses from thin cylindrical shells. J. Appl. Phys. **63**, 1857–1863 (1988)

6.16 M. Talmant, G. Quentin, J.L. Rousselot, J.V. Subrahmanyam, H. Überall: Acoustic resonances of thin cylindrical shells and the resonance scattering theory. J. Acoust. Soc. Am. **84**, 681–688 (1988)

6.17 D.E. Busson: Diffraction of a plane acoustic wave by a layered elastic sphere. Proc. Inst. Acoust. **7**, 160–168 (1985)

6.18 N.D. Veksler: *Information Analysis in Hydroelasticity* (Valgus, Tallinn 1982) (in Russian)

Chapter 7

7.1   G. Maze, J.L. Izbicki, J. Ripoche: Resonances of plates and cylinders: Guided waves. J. Acoust. Soc. Am. **77**, 1352–1357 (1985)

7.2   G.C. Gaunaurd, D. Brill: Acoustic spectrogram and complex-frequency poles of a resonantly excited elastic tube. J. Acoust. Soc. Am. **75**, 1680–1693 (1984)

7.3   J. Ripoche, G. Maze, J.L. Izbicki: New research in nondestructive testing: Resonance acoustic spectroscopy, in *Ultrasonics International 85*, Conf. Proc. (Butterworth Scientific, Guildford, UK 1985) pp. 364–369

7.4   G.C. Gaunaurd, H. Überall: RST analysis of monostatic and bistatic acoustic echoes from an elastic sphere. J. Acoust. Soc. Am. **73**, 1–12 (1983)

7.5   D.E. Busson: Diffraction of a plane acoustic wave by a layered elastic sphere. Proc. Inst. Acoust. **7**, 160–168 (1985)

7.6   N.D. Veksler: Transverse whispering gallery waves in scattering by elastic cylinders. Ultrasonics **28**, 67–76 (1990)

7.7   N.D. Veksler, V.M. Korsunskii: Analysis of peripheral waves in scattering of an acoustic wave by an elastic sphere, in *Interaction of Acoustic Waves with Elastic Bodies*, Symp. Proc. (Academy of Sciences of the Estonian SSR, Tallinn 1989) pp. 41–44 (in Russian)

Chapter 8

8.1   R.D. Doolittle, H. Überall: Sound scattering by elastic cylindrical shells. J. Acoust. Soc. Am. **39**, 272–275 (1966)
8.2   H. Hönl, A. Maue, K. Westphal: *Theorie der Beugung* (Springer, Berlin 1961)
8.3   H. Überall, G.C. Gaunaurd, J.D. Murphy: Acoustic surface wave pulses and the ringing of resonances. J. Acoust. Soc. Am. **72**, 1014–1017 (1982)
8.4   G.C. Gaunaurd, D. Brill: Acoustic spectrogram and complex-frequency poles of a resonantly excited elastic tube. J. Acoust. Soc. Am. **75**, 1680–1693 (1984)
8.5   G. Gaunaurd, H. Überall: Relation between creeping wave acoustic transients and the complex-frequency poles of the singularity expansion method. J. Acoust. Soc. Am. **78**, 234–243 (1985)
8.6   G. Maze, J. Ripoche, A. Derem, J.L. Rousselot: Diffusion d'une onde ultrasonore par des tubes remplis d'air immergés dans l'eau. Acustica **55**, 69–85 (1984)
8.7   E.D. Breitenbach, H. Überall, K.B. Yoo: Resonant scattering from elastic cylindrical shells. J. Acoust. Soc. Am. **74**, 1267–1273 (1983)
8.8   M. Talmant, G. Quentin, J.L. Rousselot, J.V. Subrahmanyam, H. Überall: Acoustic resonances of thin cylindrical shells and the resonance scattering theory. J. Acoust. Soc. Am. **84**, 681–688 (1988)
8.9   I.A. Viktorov: Ultrasonic Lamb-type waves. Survey. Akust. Zh. **11**, 1–18 (1965) (in Russian)
8.10 I.A. Viktorov : *Acoustic Surface Waves in Solid Bodies* (Nauka, Moscow 1981) (in Russian)
8.11 V.T. Grinchenko, V.V. Meleshko: *Harmonic Vibrations and Waves in Elastic Bodies* (Naukova Dumka, Kiev 1981) (in Russian)
8.12 P.L. Marston: GTD for backscattering from elastic spheres and cylinders in water and the coupling of surface elastic waves with the acoustic field. J. Acoust. Soc. Am. **83**, 25–37 (1988)
8.13 N.D. Veksler, J.D. Kaplunov, V.M. Korsunskii: Asymptotic formulas for resonance frequencies in the scattering of an acoustic wave from a cylindrical shell at normal incidence. Akust. Zh. **36**, 399–404 (1990)

Chapter 9

9.1   M.C. Junger: Sound scattering by thin elastic shells. J. Acoust. Soc. Am. **24**, 366–373 (1952)
9.2   L.M. Ljamshev: *Sound Reflection by Thin Plates and Shells in a Liquid* (Acad. Sci. USSR, Moscow 1955) (in Russian)
9.3   L.M. Ljamshev: Non-specular sound reflection by thin cylindrical shells. Akust. Zh. **2**, 188–193 (1956)
9.4   N.D. Veksler: *Information Analysis in Hydroelasticity* (Valgus, Tallinn 1982) (in Russian)
9.5   W.G. Neubauer, P. Uginčius, H. Überall: Theory of creeping waves in acoustics and their experimental demonstration. Z. Naturforsch. **24a**, 691–700 (1969)
9.6   E.I. Grigoljuk: Shell-fluid interaction problems, in Proc. of the VII All-Union Conference on Shells and Plates (Nauka, Moscow 1970) pp. 755–778 (in Russian)
9.7   E.N. Mnyev, A.K. Pertsev: *Hydroelasticity of Shells* (Sudostroenie, Leningrad 1970) (in Russian)
9.8   M.C. Junger, D. Feit: *Sound, Structures and Their Interaction* (MIT, Cambridge, MA 1972)
9.9   E.L. Shenderov: *Wave Problems in Hydroacoustics* (Sudostroenie, Leningrad 1972) (in Russian)
9.10 H. Überall: Surface waves in acoustics, in *Physical Acoustics*, Vol. 10, ed. by W.P. Mason, R.N. Thurston (Academic, New York 1973) pp. 1–60 .

9.11   E.I. Grigoljuk, A.G. Gorshkov: *Transient Hydroelasticity of Shells* (Sudostroenie, Leningrad 1974) (in Russian)

9.12   V.D. Kubenko: *Transient Interaction of Elements of Structures with Media* (Naukova Dumka, Kiev 1979) (in Russian)

9.13   A.S. Vol'mir: *Shells in a Flow of Liquid and Gas. Hydroelasticity Problems* (Nauka, Moscow 1979) (in Russian)

9.14   A.G. Gorshkov: Interaction of shock waves with deformable obstacles, in *Results of Science and Engineering*, Ser. Mechanics of Deformable Solid Bodies, Vol. 13 (VINITI, Moscow 1980) pp. 105–186 (in Russian)

9.15   C.W. Horton, Sr.: A review of reverberation, scattering and echo structure. J. Acoust. Soc. Am. **51**, 1049–1061 (1972)

9.16   W.G. Neubauer: Observation of acoustic radiation from plane and curved surfaces, in *Physical Acoustics*, Vol. 10, ed. by W.P. Mason, R.N. Thurston (Academic, New York 1973) pp. 61–126

9.17   H. Überall, H. Huang: Acoustical response of submerged elastic structures obtained through integral transforms, in *Physical Acoustics*, Vol. 12, ed. by W.P. Mason, R.N. Thurston (Academic, New York 1976), pp. 217–275

9.18   L. Flax, G. Gaunaurd, H. Überall: The theory of resonance scatterning, in *Physical Acoustics*, Vol. 15, ed. by W.P. Mason, R.N. Thurston (Academic, New York 1976) pp. 191–275

9.19   S.K. Numrich, L.R. Dragonette, L. Flax: Classification of submerged targets by acoustic means, in *Elastic Wave Scattering and Propagation*, ed. by V.V. Varadan, V.K. Varadan (Ann Arbor Science Wobum, MA 1982) pp. 149–175

9.20   G.C. Gaunaurd, D. Brill: Acoustic spectrogram and complex-frequency poles of a resonantly excited elastic tube. J. Acoust. Soc. Am. **75**, 1680–1693 (1984)

9.21   A.N. Guz', V.D. Kubenko, A.E. Babaev: *Hydroelasticity of Shell Systems* (Vischa Sckola, Kiev 1984) (in Russian)

9.22   Ja.S. Podstrigach, A.P. Poddubnjak: *Scattering of Acoustic Beams by Elastic Bodies of Spherical and Cylindrical Shape* (Naukova Dumka, Kiev 1986) (in Russian)

9.23   V.T. Grinchenko, I.V. Vovk: *Wave Problems of Scattering by Elastic Shells* (Naukova Dumka, Kiev 1986) (in Russian)

9.24   V.B. Poruchikov: *Methods of Dynamical Elasticity Theory* (Nauka, Moscow 1986) (in Russian)

9.25   G.C. Gaunaurd: Elastic and acoustic resonance wave scattering. Appl. Mech. Rev. **42**, 143–192 (1989)

9.26   E.L. Shenderov: *Sound Radiation and Scattering* (Sudostroenie, Leningrad 1989) (in Russian)

9.27   N.D. Veksler: *Acoustic Spectroscopy* (Valgus, Tallinn 1989) (in Russian)

9.28   K.L. Williams, P.L. Marston: Backscattering from an elastic sphere: Sommerfeld–Watson transformation and experimental confirmation. J. Acoust. Soc. Am. **78**, 1093–1102 (1985)

9.29   K.L. Williams, P.L. Marston: Synthesis of backscattering from an elastic sphere using Sommerfeld–Watson transformation and giving a Fabry–Perot analysis of resonances. J. Acoust. Soc. Am. **79**, 1702–1708 (1986)

9.30   P.L. Marston: GTD for backscattering from elastic spheres and cylinders in water and the coupling of surface elastic waves with the acoutic field. J. Acoust. Soc. Am. **83**, 25–37 (1988)

9.31   K.L. Williams, P.L. Marston: Axially focused (glory) scattering due to surface wave generated on spheres: Model and experimental confirmation using tungsten carbide spheres. J. Acoust. Soc. Am. **78**, 722–728 (1985)

9.32   M. Talmant, G. Quentin, J.L. Rousselot, J.V. Subrahmanyam, H. Überall: Acoustic resonances of thin cylindrical shells and the resonance scattering theory. J. Acoust. Soc. Am. **84**, 681–688 (1988)

9.33   M. Talmant, H. Überall, R.D. Miller, M.F. Werby, J.W. Dickey: Lamb waves and fluid-borne waves on water-loaded, air-filled thin spherical shells. J. Acoust. Soc. Am. **86**, 278–289 (1989)

9.34   G. Maze, J. Ripoche: Méthode d'isolement et d'identification des résonances (M.I.I.R.) de cylindres et de tubes soumis à une onde acoustique plane dans l'eau. Rev. Phys. Appl. **18**, 319–326 (1983)

9.35   G. Maze, J. Ripoche: Visualization of acoustic scattering by elastic cylinders at low ka. J. Acoust. Soc. Am. **73**, 41–43 (1983)

9.36   J.D. Murphy, E.D. Breitenbach, H. Überall: Resonance scattering of acoustic waves from cylindrical shells. J. Acoust. Soc. Am. **64**, 677–683 (1978)

9.37   E.D. Breitenbach, H. Überall, K.B. Yoo: Resonant scattering from elastic cylindrical shells. J. Acoust. Soc. Am. **74**, 1267–1273 (1983)

9.38   R. Hickling, R.W. Means: Scattering of frequency-modulated pulses by spherical elastic shells in water. J. Acoust. Soc. Am. **44**, 1246–1252 (1968)

9.39   L. Flax: High ka scattering of elastic cylinders and spheres. J. Acoust. Soc. Am. **62**, 1502–1503 (1977)

9.40   G. Maze, B. Taconet, J. Ripoche: Etude experimentale des resonances de tubes cylindriques immergés dans l'eau. Rev. CETHEDEC-Ondes Signal. **72**, 103–119 (1982)

9.41   M. Kuz'michev, Ju., V.I. Makarov: Cylindrical shell excitation by ultrasound. Akust. Zh. **4**, 282–283 (1958)

9.42   E.P. Smirnov, E.I. Heifets, E.L. Shenderov: Quantitative analysis of acoustic fields by their schlieren visualization. Akust. Zh. **19**, 240–250 (1973)

9.43   K.J. Diercks, T.G. Goldsberry, C.W. Horton: Circumferential waves in thin-walled air-filled cylinders in water. J. Acoust. Soc. Am. **35**, 59–64 (1963)

9.44   W.G. Neubauer, L.R. Dragonette: Observation of waves radiated from circular cylinders caused by an incident pulse. J. Acoust. Soc. Am. **48**, 1135–1149 (1970)

9.45   C.W. Horton, M.V. Mechler: Circumferential waves in a thin-walled, air-filled cylinder in a water medium. J. Acoust. Soc. Am. **51**, 295–309 (1972)

9.46   E.I. Heifets, E.L. Shenderov: Caustics formation during sound interaction with shells. Akust. Zh. **18**, 456–463 (1972) (in Russian)

9.47   L.R. Dragonette: The influence of the Rayleigh surface wave on the backscattering by submerged aluminum cylinders. J. Acoust. Soc. Am. **65**, 1570–1572 (1979)

9.48   A. Derem, J.L. Rousselot, G. Maze, J. Ripoche, A. Faure: Diffusion d'une onde acoustique plane par des cylindres solides immergés: étude expérimentale et théorie des résonances. Acustica **50**, 39–50 (1982)

9.49   G. Maze, B. Taconet, J. Ripoche: Influence des ondes de "Galerie à echo" sur la diffusion d'une onde ultrasonore plane par un cylindre. Phys. Lett. A **84**, 309–312 (1981)

9.50   J.L. Izbicki, G. Maze, J. Ripoche: Diffusion acoustique par des tubes immergés. Origine d'une famille de résonances. Phys. Lett. A **115**, 393–397 (1986)

9.51   J.L. Izbicki, G. Maze, J. Ripoche: Influence of the free modes of vibration on the acoustic scattering of a circular cylindrical shell. J. Acoust. Soc. Am. **80**, 1215–1219 (1986)

9.52   J.L. Izbicki, G. Maze, J. Ripoche: Diffusion acoustique par des tubes immergés dans l'eau: nouvelles résonances observées en basse fréquence. Acustica **61**, 137–139 (1986)

9.53   C.Y. Tsui, G.N. Reid, G. Gaunaurd: Resonance scattering by elastic cylinders and their experimental verification. J. Acoust. Soc. Am. **80**, 382–390 (1986)

9.54   A. Gérard, J.L. Rousselot, J.L. Izbicki, G. Maze, J. Ripoche: Résonances d'ondes d'interface de coques cylindriques minces immergées: détermination et interprétation. Rev. Phys. Appl. **23**, 289–299 (1988)

9.55   G. Maze, J. Ripoche, A. Derem, J.L. Rousselot: Diffusion d'une onde ultrasonore par des tubes remplis d'air immergés dans l'eau. Acustica **55**, 69–85 (1984)

9.56   V.F. Humphrey, S.M. Knapp, C. Beckett: Laboratory studies of acoustic scattering. Proc. Inst. Acoust. **12**, 91–98 (1990)

9.57   S.M. Knapp, V.F. Humphrey: Schlieren visualisation of low frequency ultrasonic fields, in *Ultrasonics International 89*, Conf. Proc. (Butterworth, Guildford, UK 1989) pp. 1089–1094

9.58   C. Beckett, V.F. Humphrey: Experimental form function determination for normally and obliquely incident wavefields on a solid cylinder, in *Ultrasonics International 89*, Conf. Proc. (Butterworths, Guildford, UK 1989) pp. 382–387

9.59   P. Pareige, J.L. Rembert, G. Maze, J. Ripoche: Méthode impulsionnelle numerisée (MIN) pour l'isolement et l'identification des resonances de tubes immergés. Phys. Lett. A **135**, 143–146 (1989)

9.60 G. Quentin, A. Cand: Pulsed resonance identification method. Electron. Lett. **25**, 353–354 (1989)

9.61 G. Quentin, I. Molinero, M. de Billy: Etude du spectre acoustique de quelques échantillons élastiques de forme cylindrique. II Etude expérimentale. Acustica **65**, 275–283 (1988)

9.62 M. Talmant, G. Quentin: Backscattering of short ultrasonic pulses from thin cylindrical shells. J. Appl. Phys. **63**, 1857–1863 (1988)

9.63 N.D. Veksler, V.M Korsunskii: On the acoustic pressure formula for backscattering by thin elastic circular cylindrical shells. Acoust. Lett. **6**, 70–78 (1982)

9.64 N.D. Veksler, V.M. Korsunskii: The analysis and synthesis of the acoustic pressure scattered by a circular cylindrical shell. Preprint. Acad. Sci. Estonian SSR (1986)

9.65 N.D. Veksler, V.M. Korsunskii: Analysis and synthesis of backscattering from a circular cylindrical shell. J. Acoust. Soc. Am. **87**, 943–962 (1990)

Chapter 10

10.1 R.R. Goodman, R. Stern: Reflection and transmission of sound by elastic spherical shells. J. Acoust. Soc. Am. **34**, 338–344 (1962)

10.2 N.D. Veksler: Phase velocities of the Lamb-type and Stoneley-type waves in the problem of acoustic wave scattering on a hollow elastic sphere. Akust. Zh. **37**, 42–45 (1991) (in Russian)

10.3 N.D. Veksler: Scattering of an acoustic wave by an elastic sphere. Proc. Estonian Acad. Sci., Phys. Math. **39**, 133–138 (1990) (in Russian)

10.4 M. Talmant, H. Überall, R.D. Miller, M.F. Werby, J.W. Dickey: Lamb waves and fluid-borne waves on water-loaded, air-filled thin spherical shells. J. Acoust. Soc. Am. **86**, 278–289 (1989)

10.5 N.D. Veksler: Analysis of the frequency dependence of plane pressure wave scattering by an elastic tube. Acoust. Lett. **12**, 21–27 (1988)

10.6 N.D. Veksler: The analysis of peripheral waves in the problem of plane acoustic pressure wave scattering by a circular cylindrical shell. Acustica **69**, 63–72 (1989)

10.7 J.D. Kaplunov: On the integration of equations of a dynamic boundary-layer. Trans. Acad. Sci. USSR, Mekh. Tverd. Tela. 148–160 (1990)

10.8 N.D. Veksler, J.D. Kaplunov, V.M. Korsunskii: Asymptotic formulas for resonance frequencies in the scattering of an acoustic wave from a cylindrical shell at normal incidence. Akust. Zh. **36**, 399–404 (1990)

10.9 N.D. Veksler: Ultrasonic scattering from cylindrical and spherical shells. Proc. Inst. Acoust. **12**, 799–807 (1990)

Chapter 11

11.1 L. Flax, V.K. Varadan, V.V. Varadan: Scattering of an obliquely incident wave by an infinite cylinder. J. Acoust. Soc. Am. **68**, 1832–1835 (1980)

11.2 A. Nagl, H. Überall, P.P. Delsanto, J.D. Alemar, E. Rosario: Refraction effects in the generation of helical surface waves on a cylindrical obstacle. Wave Motion **5**, 235–247 (1983)

11.3 G. Maze, J.L. Izbicki, J. Ripoche: Resonances of plates and cylinders: Guided waves. J. Acoust. Soc. Am. **77**, 1352–1357 (1985)

11.4 G. Maze, J.L. Izbicki, J. Ripoche: Acoustic scattering from cylindrical shells: Guided waves and resonances of the liquid column. Ultrasonics **24**, 354–362 (1986)

11.5 J.L. Izbicki, G. Maze, J. Ripoche: Influence of the free modes of vibration on the acoustic scattering of a circular cylindrical shell. J. Acoust. Soc. Am. **80**, 1215–1219 (1986)

11.6 E.D. Breitenbach, H. Überall, K.B. Yoo: Resonant scattering from elastic cylindrical shells. J. Acoust. Soc. Am. **74**, 1267–1273 (1983)

11.7 V.A. Borovikov, N.D. Veksler: Scattering of sound waves by smooth convex elastic cylindrical shells. Wave Motion **7**, 143–152 (1985)

11.8   K.L. Williams, P.L. Marston: Synthesis of backscattering from an elastic sphere using Sommerfeld–Watson transformation and giving a Fabry–Perot analysis of resonances. J. Acoust. Soc. Am. **79**, 1702–1708 (1986)

11.9   P.L. Marston: GTD for backscattering from elastic spheres and cylinders in water and the coupling of surface elastic waves with the acoustic field. J. Acoust. Soc. Am. **83**, 25–37 (1988)

11.10   N.D. Veksler: Scattering of a plane acoustic wave obliquely incident on a solid elastic cylinder. Acustica **71**, 111–120 (1990)

Chapter 12

12.1   L. Flax, V.K. Varadan, V.V. Varadan: Scattering of an obliquely incident wave by an infinite cylinder. J. Acoust. Soc. Am. **68**, 1832–1835 (1980)

12.2   N.D. Veksler: The analysis of peripheral waves in the problem of plane acoustic pressure wave scattering by a circular cylindrical shell. Acustica **69**, 63–72 (1989)

12.3   L. Flax, L.R. Dragonette, H. Überall: Theory of elastic resonance excitation by sound scattering. J. Acoust. Soc. Am. **63**, 723–731 (1978)

12.4   A. Nagl, H. Überall, P.P. Delsanto, J.D. Alemar, E. Rosario: Refraction effects in the generation of helical surface waves on a cylindrical obstacle. Wave Motion **5**, 235–247 (1983)

12.5   A.L. Gol'denveizer, J.D. Kaplunov: Dynamic boundary layer in problems of shell vibrations. Trans. Acad. Sci. USSR. Mekh. Tverd, Tela. 152–162 (1988) (in Russian)

12.6   V.L. Berdichevskii: High-frequency long-wave vibrations of plates. Dokl. Akad. Nauk SSSR **236**, 1319–1326 (1977) (in Russian)

12.7   J.D. Kaplunov: On the integration of equations of a dynamic boundary-layer. Trans. Acad. Sci. USSR, Mekh. Tverd. Tela. 148–160 (1990)

12.8   N.D. Veksler, J.D. Kaplunov: Peripheral waves in cylindrical shells immersed in water. Acustica **72**, 131–139 (1990)

Chapter 13

13.1   A.I. Lurie: *Statics of thin-walled elastic shells* (State Publ. Techn.-Theor. Lit., Moscow 1947)

13.2   V.V. Novozhilov: *Theory of Thin Shells* (State Union Publ. Shipbuilding Industry, Leningrad 1962) (in Russian)

13.3   A.L. Gol'denveizer: *Theory of Elastic Thin Shells* (Nauka, Moscow 1976) (in Russian)

13.4   L. Ainola, U. Nigul: Wave processes of deformation in elastic plates and shells. Proc. Acad. Sci. Estonian SSR. Ser. Fis.-Mat. Techn. Sci. **14**, 3–63 (1965) (in Russian)

13.5   E.H. Kennard: The new approach to shell theory: circular cylinders. J. Appl. Mech. **20**, 33–40 (1953)

13.6   V.L. Berdichevskii: *Variational Principles in Mechanics of Continuous Media* (Nauka, Moscow 1983) (in Russian)

13.7   N.D. Veksler: *Information Analysis in Hydroelasticity* (Valgus, Tallinn 1982) (in Russian)

13.8   B.R. Levy, J.B. Keller: Diffraction by a smooth object. Comm. Pure Appl. Math. **12**, 159–209 (1959)

13.9   W. Franz: Über die Greenschen Funktionen des Zylinders und der Kugel. Z. Naturforsch. **9a**, 705–716 (1954)

13.10   N.D. Veksler: *Resonance Scattering in Hydroacoustics* (Valgus, Tallinn 1984) (in Russian)

13.11   D. Brill, H. Überall: Transmitted waves in the diffraction of sound from liquid cylinders. J. Acoust. Soc. Am. **47**, 1467–1469 (1970)

13.12   D. Brill., H. Überall: Acoustic waves transmitted through solid elastic cylinders. J. Acoust. Soc. Am. **50**, 921–939 (1971)

13.13   G.V. Frisk, J.W. Dickey, H. Überall: Surface wave modes on elastic cylinders. J. Acoust. Soc. Am. **58**, 996–1008 (1975)

Chapter 14

14.1   N.D. Veksler: *Information Analysis in Hydroelasticity* (Valgus, Tallinn 1982) (in Russian)
14.2   N.D. Veksler: On shell thickness determination using echo-signals. Acoust. Lett. **9**, 139–142 (1985)

Subject Index